Operator Theory and Its Applications

In Memory of V. B. Lidskii (1924–2008)

American Mathematical Society

TRANSLATIONS

Series 2 • Volume 231

Advances in the Mathematical Sciences—65

(*Formerly Advances in Soviet Mathematics*)

Operator Theory and Its Applications

In Memory of V. B. Lidskii (1924–2008)

Michael Levitin
Dmitri Vassiliev
Editors

American Mathematical Society
Providence, Rhode Island

2010 *Mathematics Subject Classification.* Primary 01A70, 35P05, 35P15, 35Q40, 35Q60, 35Q74, 47A10, 47B35, 47E05, 47G30.

Library of Congress Card Number 91-640741
ISBN-13: 978-0-8218-5272-9
ISBN-10: 0-8218-5272-8
ISSN 0065-9290

∞ The paper used in this book is acid-free and falls within the guidelines
established to ensure permanence and durability.
Visit the AMS home page at http://www.ams.org/

10 9 8 7 6 5 4 3 2 1 15 14 13 12 11 10

Victor Borisovich Lidskii (1924–2008)

Contents

Amer. Math. Soc. Transl.
Volume **231**, 2010

Victor Borisovich Lidskii (1924–2008)

Michael Levitin and Dmitri Vassiliev

This volume is dedicated to Victor Borisovich Lidskii who died on July 29, 2008. It is a collection of papers in subject areas related to Lidskii's work, some of which are written by people who knew him well. The editors of this volume, Michael Levitin and Dmitri Vassiliev, are former students of Lidskii, both at undergraduate and Ph.D. levels.

Lidskii was born in Odessa in the Soviet Union on May 4, 1924[1]. As with most men of his generation, his life was severely affected by the Second World War. Lidskii finished secondary school on June 20, 1941, two days before Germany invaded the Soviet Union. After his parents' divorce he remained with his father and stepmother, so at that time Lidskii was living in the city of Bobruisk (now the Minsk region, Belarus). His mother lived separately and died in the Minsk ghetto in 1942.

Lidskii escaped the advancing German troops and found himself, on his own, in the city of Saratov. Shortly afterwards he was drafted into the Red Army and sent to a military intelligence school. The main reason for his being assigned to military intelligence was the fact that he knew German. Lidskii studied German at school and he knew some Yiddish as well (which was helpful in learning German). After graduating as an officer, Lidskii spent the last two years of the war on the front line serving in field reconnaissance and going behind enemy lines. Here the German he mastered at intelligence school proved handy in allowing him, on one occasion, to be taken for a German by Nazi soldiers when he lead his group back from behind enemy lines. During these two years he was decorated on four occasions; see [**1**] for details. Lidskii's war experience ended on VE Day in Prague and he was demobilized in 1946.

As a decorated officer and member of the Communist Party (he joined the Party at the front line), Lidskii was offered a job in the NKVD—the secret police. Instead, he chose to study mathematics and enrolled as an undergraduate at the Faculty of Mechanics and Mathematics of Moscow State University. As with many war veterans, studying mathematics was not easy for him, but in the end he proved to be good enough to be recommended for postgraduate study under the supervision of I. M. Gelfand. However, by that time (1951) problems had already started for

---

2010 *Mathematics Subject Classification.* Primary 01A70.

[1]The correct date of his birth was established only in the mid 1970s. Lidskii's original birth certificate was lost, and until he got an authorized copy, he considered May 5 to be his birthday and put May 5 on all forms and documents.

students of Jewish origin and he was refused a full-time place at the Graduate School of Moscow State University. Lidskii managed to secure a teaching post at an engineering college and did his Ph.D. work at Moscow State University as an external part-time student. Despite the difficulties, Lidskii successfully completed his Ph.D. thesis *Questions of spectral theory for systems of second order differential equations* in three years (1954).

In 1954 Lidskii started teaching at the Moscow Institute of Physics and Technology, universally known as FizTech (we write more about FizTech below), where he worked for the rest of his life. In 1959 he was awarded a D.Sc. by the Steklov Mathematical Institute for his thesis *Conditions for the completeness of the system of root subspaces of non-self-adjoint operators with discrete spectra* and shortly afterwards (1961) he became a full professor at FizTech. From 1966 until 1988 he also held a part-time research position at the Institute for Problems in Mechanics of the USSR Academy of Sciences.

V. B. Lidskii made significant contributions to various branches of analysis. In this short note we will not attempt to review all his work. For a survey see [**2**] with additional bibliographies in [**3, 4**]. We will only highlight two research topics.

The first—with which Lidskii's name is most closely associated—is the spectral theory of nonsymmetric matrices and nonselfadjoint operators in a Hilbert space. The famous result of 1959 which is widely referred to as Lidskii's Theorem states that the matrix trace of a trace class operator in a separable Hilbert space equals its spectral trace (sum of its eigenvalues). Despite its apparent simplicity, it is a deep and difficult achievement; see, for example, B. Simon's book [**5**]. Later Lidskii devised a summation method for series over root vectors of an operator (with discrete spectrum or compact) which is not necessarily close to selfadjoint and established sufficient conditions for summability. In some sense Lidskii's works on nonselfadjoint operators together with the works of I. Gohberg and M. G. Krein were far ahead of their time. The recent resurgence of interest in nonselfadjoint problems and the deeper understanding of the underlying difficulties (see some of the papers in this volume) is strongly influenced by Lidskii's scientific heritage.

The other, rather unexpected, subject, to which Lidskii devoted the second half of his career, is the *rigorous* analysis of differential operators arising in elasticity and hydroelasticity. His favorite research object was the operator of shell theory. This operator describes the dynamics of a thin elastic structure such as, say, a hull of a ship or a fuselage of an airplane. In mathematical terms, dealing with the operator of shell theory means looking at a selfadjoint Agmon–Douglis–Nirenberg elliptic system of partial differential equations of mixed order acting on a two-dimensional manifold with or without boundary. This system depends singularly on a small asymptotic parameter (relative shell thickness) and can be reduced to a higher-order analogue of the scalar stationary Schrödinger equation with a pseudodifferential potential which feels in a complicated way the geometry of the manifold. In this highly nontrivial subject area Lidskii and his collaborators obtained a collection of delicate rigorous results which at that time required the use of and contributed to the development of cutting-edge methods of analysis. Regrettably, these achievements remain largely unknown: the statements of the problems are too complicated to attract pure mathematicians and the techniques used are too advanced for the applied mathematics community. Nevertheless, some of the results buried in Lidskii's works on the subject later became classical through the work of other authors.

For example, Lidskii was probably the first to prove that the essential spectrum of a selfadjoint Agmon–Douglis–Nirenberg elliptic operator A (say, on a manifold without boundary) coincides with the set of values of the spectral parameter λ for which the principal symbol of $A - \lambda I$ is not invertible [**6**]. The fact that Lidskii's works on elasticity and hydroelasticity are not widely known in the West is not surprising given that in Soviet times he was allowed to travel abroad only once, that a trip to Romania.

Lidskii's conversion to shell theory was spontaneous and happened at Gelfand's seminar at Moscow State University. One day in the mid 1960s the speaker was A. L. Gol′denveizer, a prominent specialist in shell theory. After the talk Gelfand stood up and said, "You, people, forever keep studying the Laplace operator. Is anyone prepared to handle something more challenging, like shell theory?" And Lidskii was the (only) participant of Gelfand's seminar who accepted this challenge.

One of the defining aspects of Lidskii's career which is not widely known was his teaching at FizTech. Here we need to explain the unique status of this university within the Soviet higher education system. FizTech was created in 1946 (for the first few years it was a department of Moscow State University before becoming an independent institution in 1951) as a result of an initiative of a group of prominent Soviet scientists which included the future Nobel laureate P. L. Kapitsa. The purpose was to provide elite training of researchers, mostly in physics, who would later contribute to the Soviet nuclear and ballistic missile programs. The university was based just outside Moscow, in the town of Dolgoprudny, on the site of the former airship factory of Umberto Nobile. This location served two purposes: it was a first attempt to create a Western-style university campus where all the students and many of the staff lived, but it also stopped foreigners from visiting as most needed special permission to leave Moscow. Despite the secrecy and close association with the military industrial complex, academic freedom at FizTech was unparalleled. Students at FizTech received additional extensive training working directly in leading research establishments of the USSR Academy of Sciences or industry as part of their undergraduate degree programs. This gave academics a lot of freedom in choosing and building their own courses. An interesting feature of the FizTech "system" was the entrance test which involved not only very strict examinations but also an informal interview with world-leading scientists whose opinion ultimately determined who was admitted. Unfortunately, as with many other Soviet achievements, this became distorted in later years and strong anti-Semitic tendencies developed.

The unique features of FizTech attracted not only the best students from the whole of the USSR but also the best lecturers. The names of physicists, mathematicians, engineers, chemists, and later also biologists who taught there make an illustrious list. It suffices to say that Nobel laureates V. Ginzburg, P. L. Kapitsa, L. D. Landau, A. M. Prokhorov, and N. N. Semyonov lectured at FizTech. But even amongst these Lidskii stands out as a true FizTech legend. His unique passionate teaching style, deep involvement with the subject and students, and strong personality made him one of the most popular and loved lecturers, even though he was not by any means a soft touch as an examiner.

To this day, many years after his peak in the 1960s and 1970s, anecdotes about Lidskii's teaching and personality still circulate on the Web. We cite [**7**] two such stories here.

Once in a lecture Lidskii gave an example of an everywhere discontinuous function and sketched its graph. Later on, in the same lecture, he mentioned that in physics all functions are continuous and differentiable. He was immediately asked how this agrees with the quantization of energy. Lidskii replied that quantization occurs precisely because the wave function and its derivative are assumed to be continuous. At this point he noticed the graph still displayed on the board. "This is not a function," he said and stopped and waved his hand in the air trying to find a proper word, "but some sort of pathology!" He got so angry with this "pathology" that he smashed the board with his fist so that chalk dust flew everywhere.

On another occasion, in an oral examination for his course, Lidskii became so upset with a student's answers that in order to calm down he dashed towards the grand piano standing in the corner of the exam hall, played angrily some classical piece, and only then rejoined the unfortunate examinee.

It must be said that overall classical music played an important role in Lidskii's life. It is probably no accident that his youngest son, Mikhail Lidsky, from his second marriage to literary translator Inna Bernshtein (born 1929), became a well-known concert pianist. Lidskii's first marriage was to Militsa Neuhaus (1929–2008), a fellow student at Moscow State University and daughter of the famous pianist Heinrich Neuhaus.

Lidskii also coauthored the legendary exercise book Problems in Elementary Mathematics (authors V. Lidsky, L. Ovsyannikov, A. Tulaikov, and M. Shabunin, Mir Publishers, Moscow, 1973) used by Soviet high school students preparing for university entrance examinations. Everyone who went on to study mathematics in the Soviet Union at university level in the 1960s and 1970s associates Lidskii's name with this exercise book.

The other place where Lidskii worked, part-time, maintaining his full-time job at FizTech, was the Institute for Problems in Mechanics (IPMech). IPMech was founded in 1965 by A. Yu. Ishlinsky, a distinguished applied mathematician who gained prominence by developing gyroscopes for the Soviet space program. Most of the staff of IPMech had very applied research interests and many were experimentalists. However, being a well-educated and sophisticated person, Ishlinsky recognized that his institute needed a small group of "proper" mathematicians. The group of mostly part-timers which came into being as a result included, apart from Lidskii, O. A. Oleinik and M. I. Vishik. IPMech was a purely research establishment so staff were not obliged to teach and, moreover, the institute had no undergraduates of its own. Nevertheless, like many other institutes of the USSR Academy of Sciences, IPMech had close links with Fiztech: full-timers from IPMech had the opportunity of teaching part-time at FizTech and full-timers from FizTech had the opportunity of doing research part-time at IPMech. Some lecture courses for FizTech students were taught at IPMech and many student projects and theses were supervised there. Thus, Lidskii's part-time research job at IPMech was a natural extension of his full-time teaching job at FizTech.

For several decades Lidskii served as Deputy Editor of *Functional Analysis and Its Applications*, an influential Soviet journal founded by I. M. Gelfand. His efforts contributed to this journal becoming one of the best know in the West and helped to maintain its reputation in the difficult times following the collapse of the Soviet Union.

Lidskii had around 20 Ph.D. students, most of whom became professional mathematicians. We were amongst his very last students and remember fondly not only his academic guidance but his openness and friendship. Coming to his family apartment for discussions was always a pleasure. Lunch or dinner was immediately served, questions regarding family members and events of our lives kindly discussed, and immediate help offered if needed. In many ways Victor Borisovich Lidskii influenced our careers and shaped our lives.

References

[1] http://ru.wikipedia.org/wiki/Лидский,_Виктор_Борисович.

[2] M. S. Agranovich, V. I. Arnol′d, I. M. Gel′fand, A. L. Gol′denveizer, A. G. Kostyuchenko, L. D. Kudryavtsev, D. Vassiliev, and M. I. Vishik, *Victor Borisovich Lidskii (on his 75th birthday)*, Russian Mathematical Surveys, 1999, vol. **54**, no. 5, pp. 1077–1085.

[3] Mathematics in the USSR over the forty years 1917–1957, vol. **2**, pp. 409–410 (Fizmatgiz, Moscow, 1959).

[4] Mathematics in the USSR, 1958–1967 (Nauka, Moscow, 1969).

[5] B. Simon, Trace Ideals and Their Applications (American Mathematical Society, Providence, RI, 2005).

[6] A. G. Aslanyan and V. B. Lidskii, Distribution of eigenfrequencies of thin elastic shells (Nauka, Moscow, 1974) (in Russian).

[7] O. Dashevskiy, Legends and myths of the Moscow Institute of Phsyics and Technology (in Russian),
`http://ffke1975.narod.ru/s/s6/beholder/beholder-1.htm`,
`http://ffke1975.narod.ru/s/s6/beholder/beholder-2.htm`,
`http://ffke1975.narod.ru/s/s6/beholder/beholder-3.htm`,
`http://ffke1975.narod.ru/s/s6/beholder/beholder-4.htm`,
`http://ffke1975.narod.ru/s/s6/beholder/beholder-5.htm`.

Department of Mathematics and Statistics, University of Reading, Whiteknights, PO Box 220, Reading RG6 6AX, United Kingdom
E-mail address: `M.Levitin@reading.ac.uk`

Department of Mathematics, University College London, Gower Street, London WC1E 6BT, United Kingdom
E-mail address: `D.Vassiliev@ucl.ac.uk`

Amer. Math. Soc. Transl.
Volume **231**, 2010

In Memory of Victor Borisovich Lidskii (1924–2008)

Israel Gohberg

Israel Gohberg passed away on October 12, 2009, shortly after submitting this article.

Victor Borisovich was a talented mathematician and an outstanding expert in pure and applied mathematics. His achievements in operator theory, in the theory of Hamiltonian systems, and in the theory of differential operators of elasticity and hydroelasticity were outstanding. His influence was fundamental and will be felt for many years to come. He is remembered as an excellent lecturer and teacher. Though he was living in Moscow and I in Kishinev, we nevertheless met often. We were very friendly, and for a long time we had common research interests. In the second half of his research career our mathematical interests parted and also I emigrated to Israel. Victor Borisovich was four years older than I was, and wiser, and in many cases I was receptive to his advice. This short note contains reminiscences about our work and friendship, which I always enjoyed and appreciated.

The first time I heard about the mathematician Victor Borisovich Lidskii was from my teacher M. G. Krein, when he presented the very nice results of I. M. Gelfand and V. B. Lidskii about zones of stability for Hamiltonian systems. These results were deep extensions of the earlier papers of Krein. Much later, when writing the book [**GLR**] with my colleagues, a book which is essentially based on these achievements, I came to understand the importance and to appreciate the beauty of this piece of mathematics.

A while later Lidskii sent to Krein the draft of his (doctoral) habilitation thesis dedicated to linear nonselfadjoint operators. During these years together with Krein we were busy writing the book on nonselfadjoint operators. We found the thesis very interesting and we immediately included some of its results in our book [**GK**], for instance theorems on completeness and the famous Lidskii trace theorem. The latter theorem states that for any trace class operator in a separable Hilbert space the trace is equal to the sum of its eigenvalues (multiplicities taken into account). Formally, this can be written in the form

$$(1)\qquad (Ax_1, x_1) + (Ax_2, x_2) + \cdots + (Ax_n, x_n) + \cdots = \lambda_1 + \lambda_2 + \cdots + \lambda_n + \cdots ,$$

2010 *Mathematics Subject Classification.* Primary 01A70.

where $x_1, x_2, \ldots, x_n, \ldots$ is any orthonormal basis of the Hilbert space and λ_n, $n = 1, 2, \ldots$, are the eigenvalues of the operator A with the multiplicities taken into account. This is an infinite-dimensional generalization of the very well-known theorem for matrices from linear algebra. This generalization is not simple at all. Krein and I were thinking about this generalization. Initially we thought that the equality (1) holds always for any compact operator and we were trying to prove it. Later we understood that this is doubtful and to separate the two traces we even introduced for a while two traces, the "spectral trace", the right-hand side of (1), and the "matricial trace", the left-hand side of (1). Later it became clear that for operators outside the trace class the left- and right-hand sides of (1) in general are not equal, even when both sides converge. The sum on the left-hand side depends on the basis $x_1, x_2, \ldots, x_n, \ldots$ and the sum on the right-hand side depends on the ordering of the eigenvalues of A. Only for trace class operators is the left-hand side of (1) absolutely convergent for any basis and then the sum on the right-hand side is also absolutely convergent and the two sums coincide for any orthonormal basis. Lidskii proved this theorem via complex analysis. The theory of trace and determinant of operators in an infinite-dimensional case was extended and developed. I would mention here the book [**GGK**] and its bibliography.

The next encounter with Lidskii is connected with the Soviet Mathematical Congress in Leningrad in 1961. At this congress a joint plenary presentation by M. V. Keldysh and V. B. Lidskii on the theory of nonselfadjoint operators was planned. Evaluation of the resolvent, completeness theorems, and applications were the main topics in this talk. Keldysh did not come to the congress, so the talk had to be presented by Lidskii. He arrived in Leningrad in time, but there was a problem—Lidskii had a cold and he was losing his voice. When only two to three days remained till the talk, one could hardly hear him at all. He then asked me to give the talk in his place. The last two days before the talk we spent long hours during the beautiful White Nights preparing my presentation of their talk. Lidskii gave me all the details and I rehearsed in front of him to show him how I would replace him. Before the talk he announced to the audience that he had a problem with his voice and was not convinced that he would be able to give the entire talk. Nevertheless he was going to begin speaking and would continue as long as possible. The moment he felt that he could not lecture any more, I would replace him and continue the presentation. He also announced that I was prepared for this mission. Fortunately his voice improved while speaking and he gave a beautiful talk, without my help. Later Lidskii told me that on the morning of the day of the talk somebody introduced him to a physician who was a specialist in helping singers and artists on such occasions. This physician treated Lidskii and helped him considerably, which for me was a great relief. I was released from the great responsibility of having to stand in for the president of the USSR Academy of Sciences. I was also happy to have had an excellent opportunity to discuss in depth interesting mathematics with a good friend and expert.

By the way, I also remember the Leningrad Congress as the place where a fifteen-minute talk was given which, according to the author, contained the proof of the Riemann hypothesis on the zeros of the zeta function. The speaker was Nikolai Ivanovich Gavrilov, professor of Odessa University. It was included in the program as the result of a fight between the president of the Congress, Professor Yu. V. Linnik, and the Odessa Communist Party Regional Committee. Professor

Linnik did not want to include this talk in the program, explaining that it would be confusing since the author wrongly claimed that it contained the solution of the famous Riemann problem. The committee strongly insisted on including this important talk by the member of the Communist Party, N. I. Gavrilov. Of course, Professor Linnik lost the fight with the Communist Party officials and a fifteen-minute talk was scheduled. Foreign guests were asked not to attend this talk, but it attracted a very large audience and the largest auditorium was, as expected, used for it. First the author presented in the very short time allocated the main points of the solution of the problem. Then came the questions and later the comments. Gavrilov gave answers, but each answer was never a direct one and often did not refer at all to the questions and comments. This in turn drew a strong reaction from the audience. Tempers during the session became heated and the audience behaved like a crowd at a football game. Finally the chairman had to stop the meeting. Unfortunately, the Riemann hypothesis has not been proven to this day.

The International Congress of Mathematicians was scheduled to take place in Moscow in 1966. One of the sections of this Congress was devoted to functional analysis. Professor I. M. Gelfand, who was responsible for this section, invited Lidskii to organize it. At that time I had already received a habilitation degree (Doctor of Sciences); by the way, I was happy to learn that the external report on my thesis was signed by M. V. Keldysh and V. B. Lidskii. Lidskii invited A. S. Dynin and myself to help him organize this section. So our leader was Lidskii and he was also responsible for communication with Gelfand. The work of the three of us consisted of reading the abstracts of the submitted short talks, discussing them, and sending them to referees if necessary. We also had to make a decision about every talk, and Lidskii went with our decisions to Gelfand for final approval. We worked hard in Moscow for a month. I enjoyed this period. I would like to recall an incident that happened during our work. As always, Lidskii went with a sheaf of papers to Gelfand. One of the talks was about numerical ranges for some operators. The paper was of a very specialized nature and we had doubts that it would be of general interest, so we rejected it. Gelfand controlled us strictly. Probably he liked the paper that we rejected and he asked Lidskii to prove the main result of it, but he could not do it. Then Gelfand gave him 24 hours to provide a proof. This would substantiate our rejection; otherwise the talk would be accepted. The result is that the talk was accepted and we subsequently became more careful in our work.

I would also like to mention another unusual situation which happened during the Congress. A chairman, vice chairman, and secretary were appointed to run each session of the sections. In one of the sessions the three leaders were N. Levinson, I. Gohberg, and I. Feldman. We all turned up a little early to prepare the session. Unfortunately, we could not even start. The reason was purely linguistic—we could not communicate. The English of Gohberg and Feldman was very poor, and the Russian of Levinson was totally nonexistent. In these complicated circumstances a solution was found. I asked Professor Levinson, "Do you speak Yiddish?" He replied, "Avody", which means "of course" or "certainly" in Yiddish. So the talks were announced by Norman Levinson in English and the members of the presidium communicated between themselves in Yiddish. (I knew that my colleague Feldman spoke Yiddish.)

In the end I would like to tell a short story about how we lost a fight with the "Higher Attestation Commission". In the Soviet Union all degrees (kandidat nauk

= Ph.D. and doctor nauk = highest degree, habilitation degree) had to be approved by the so-called Higher Attestation Commission. The procedure was secret and went as follows. After a successful viva the thesis had to be submitted to this Commission. For the highest degree the Commission always sent the thesis to an additional secret expert referee (sometimes called the black referee). After receiving back the report from the black referee, the Commission had to decide whether to approve or to reject the thesis. Each subject had its own panel. The head of the mathematics panel at the time was Professor Yu. V. Linnik, an outstanding mathematician and an open and sincere scientist. He invited me to serve as a black referee for a habilitation thesis. I accepted this assignment. Together with my friend and colleague A. S. Markus we spent long hours reading and discussing this thesis. In the end we concluded that it was a very weak thesis and in the best case could be considered on the borderline of approval. After deeper examination we also found errors and arrived at the opinion that it would only be fair to recommend the rejection of this thesis. I wrote a negative referee report but did not post it. I understood that as it was my first experience in this capacity I could not be sure that it would be the right decision. The situation was even more complicated because there were some political strings attached.

This was a difficult situation so I decided to ask Lidskii for advice. I planned that before going ahead with my report I would discuss the problem with him. On my next trip to Moscow I realized this plan. I gave Lidskii all the details concerning my decision. He listened very carefully and after that gave many reasons for not agreeing with my decision. Some of the reasons were of a political nature, to which I had not paid enough attention. He suggested I should write, if possible, a lukewarm (bittersweet) but positive report. I followed his advice. It became clear to me that Lidskii was in fact concerned with our well-being. Later, I learned that the thesis was approved. But this is not the end of the story.

Sometime later, Markus went through a viva at Voronezh University with a very good habilitation thesis and a very strong team of referees (opponents). Lidskii wrote the external report. The viva went very well and we celebrated this success in Kishinev with Krein and with Heinz Langer. During this period Krein was elected as a foreign member of the US National Academy of Sciences. My first toast was to myself that I had become a student of an American academician and teacher of a habilitation doctor. We know now that the celebration was premature. The thesis was sent to the above-mentioned Higher Attestation Commission for the next secret step, as a result of which the thesis was rejected. Here is the story behind the rejection. The atmosphere in the Commission was now very different. It was strongly influenced by the anti-Semitic policy of the Soviet government. Professors A. G. Kurosh and G. E. Shilov resigned from the Committee. A group of leading Soviet mathematicians was manipulating the choice of the black referees in such a way that Jewish habilitation theses were, as a rule, rejected. Lidskii made strenuous efforts to prevent an unfair decision in this case, but to no avail. The head of the mathematics panel of the Higher Attestation Commission was now Professor (later academician) V. A. Il′in. We know now some details of how he organized the procedure of the secret step, but I do not want to speculate about this. The person who was approved in the habilitation degree earlier due to my report which I mentioned in the previous paragraph, together with Il′in, were instrumental in this procedure. Fortunately Markus emigrated to Israel and the rejection of his habilitation degree by the Higher Attestation Commission meant nothing to him.

Victor Borisovich was highly decorated during World War II when he was fighting at the front line. He had a long and fruitful academic career and his scientific achievements were appreciated by government authorities and institutions in which he worked. In the latter years of his career he received a number of prestigious awards including the State Prize of the Russian Federation.

We will always remember Victor Borisovich as a kind person, an excellent mathematician, and a very good friend and colleague.

References

[GGK] I. Gohberg, S. Goldberg, N. Krupnik, *Traces and Determinants of Linear Operators.* Operator Theory: Advances and Applications, Vol. 116, Birkhauser Verlag, Basel, 270 pages, 2000.

[GK] I. Gohberg, M. Krein, *Introduction to the theory of linear nonselfadjoint operators.* American Mathematical Society, Providence, 378 pages (translated from Russian), 1969.

[GLR] I. Gohberg, P. Lancaster, L. Rodman, *Matrices and Indefinite Scalar Products.* Operator Theory: Advances and Applications, Vol. 8, Birkhauser Verlag, Basel, 374 pages, 1983.

School of Mathematical Sciences, Tel Aviv University, Ramat Aviv 69978, Israel

Amer. Math. Soc. Transl.
Volume **231**, 2010

List of Publications by V. B. Lidskii

This list has been compiled by the editors of this volume on the basis of the references given in the first article of this volume. It is somewhat longer than the list produced by MathSciNet as it includes conference proceedings, papers from lesser known Russian journals, as well as Lidskii's publications in "mechanical" (rather than strictly mathematical) literature.

Monographs

[B1] V. B. Lidskii, *Differential forms, surface orientation and Stokes formula*, Moscow Institute of Physics and Technology, Moscow, 1967 (in Russian).

[B2] A. G. Aslanyan and V. B. Lidskii, *The distribution of eigenfrequencies of thin elastic shells*, Nauka, Moscow, 1974 (in Russian).

[B3] V. B. Lidskii, *Separation of regular branches of multivalued functions*, Moscow Institute of Physics and Technology, Moscow, 1977 (in Russian).

[B4] A. L. Goldenveizer, V. B. Lidskii, and P. V. Tovstik, *Free vibrations of thin elastic shells*, Nauka, Moscow, 1979 (in Russian).

Theses

[T5] V. B. Lidskii, *Questions of spectral theory for systems of second order differential equations*, Ph.D. Thesis, Moscow State University, 1954 (in Russian).

[T6] V. B. Lidskii, *Conditions for the completeness of the system of root subspaces of non-self-adjoint operators with discrete spectra*, D.Sc. Thesis, Mathematical Institute of the USSR Academy of Sciences, Moscow, 1959 (in Russian).

Papers

[P7] V. B. Lidskii, *On eigenvalues of the sum and product of symmetric matrices*, Doklady AN SSSR **75** (1950), 769–772.

[P8] V. B. Lidskii, *The proper values of the sum and product of symmetric matrices*, Uspekhi Mat. Nauk **6** (1951), 169–170, English transl.: C. D. Benster. (transl.), N.B.S. Report 2248, U.S. Department of Commerce, National Bureau of Standards, Washington, DC (1953).

[P9] V. B. Lidskii and M. G. Neigauz, *On the boundedness of solutions of systems of linear differential equations with periodic coefficients*, Doklady AN SSSR **77** (1951), 189–192.

[P10] V. B. Lidskii, *On the number of the square integrable solutions of the system of differential equations* $y''_{tt} + P(t)y = 0$, Doklady AN SSSR **95** (1954), 217–220.

[P11] I. M. Gelfand and V. B. Lidskii, *On the structure of the regions of stability of linear canonical systems of differential equations with periodic coefficients*, Uspekhi Mat. Nauk **10** (1955), 3–40, English translation in Amer. Math. Soc. Transl. (2) **8** (1958), 143–181.

[P12] V. B. Lidskii, *Oscillatory theorems for canonical systems of differential equations*, Doklady AN SSSR **102** (1955), 877-880.

[P13] V. B. Lidskii and M. G. Neigauz, *On the stability criteria for systems of differential equations with periodic coefficients*, Prikl. Mat. Mekh. **19** (1955), 625–627.

[P14] V. B. Lidskii, *On the completeness of the system of eigenfunctions and adjoint functions of a non-self-adjoint differential operator*, Doklady AN SSSR **110** (1956), 172–175.

2010 *Mathematics Subject Classification.* Primary 01-00.

[P15] V. B. Lidskii, *A theorem on the spectrum of a perturbed differential operator*, Doklady AN SSSR **112** (1957), 994–997.

[P16] V. B. Lidskii, *Conditions for compactness of the resolvent of a non-self-adjoint differential operator*, Doklady AN SSSR **113** (1957), 28–31.

[P17] V. B. Lidskii, *On the completeness of the system of eigenvectors and adjoint vectors of a compact operator*, Doklady AN SSSR **115** (1957), 234–236.

[P18] V. B. Lidskii, *Theorems on the completeness of the system of eigenvectors and adjoint vectors of an operator with discrete spectrum*, Doklady AN SSSR **119** (1958), 1088–1091.

[P19] V. B. Lidskii, *Theorems on the completeness of the system of eigenvectors and adjoint vectors of non-self-adjoint operators with discrete spectra*, Uspekhi Mat. Nauk **13** (1958), 229–230.

[P20] V. B. Lidskii, *Conditions for the completeness of the system of root subspaces of operators with discrete spectra*, Trudy Mosk. Mat. Obshch. **8** (1959), 83–120.

[P21] V. B. Lidskii, *Non-self-adjoint operators with a trace*, Doklady AN SSSR **125** (1959), 485–487.

[P22] V. B. Lidskii, *On the summation of series in principal vectors of non-self-adjoint operators*, Doklady AN SSSR **132** (1960), 275–278, English translation in Sov. Math. Dokl. **1**, 540–543.

[P23] V. B. Lidskii, *Non-self-adjoint Sturm-Liouville operators*, Trudy Mosk. Mat. Obshch. **9** (1960), 45–79.

[P24] V. B. Lidskii, *Eigenfunction expansions associated with equations with periodic coefficients*, in Russian translation of E. C. Titchmarsh's *Eigenfunction expansions associated with second-order differential equations*, vol. 2, 1961, 529–532.

[P25] V. B. Lidskii, *On the summability of series in principal vectors of non-self-adjoint operators*, Trudy Mosk. Mat. Obshch. **11** (1962), 3–35.

[P26] V. B. Lidskii, *On Fourier expansions in principal vectors of an elliptic operator*, Mat. Sbornik **57** (1962), 137–150.

[P27] V. B. Lidskii and M. G. Neigauz, *On the sweep method in the case of a second-order self-adjoint system*, Zh. Vychisl. Mat. i Mat. Fiz. **2** (1962), 161–165, English translation in Comp. Math. Math. Phys. **2** (1963), 171–174.

[P28] M. V. Keldysh and V. B. Lidskii, *Spectral theory of non-self-adjoint operators*, in *Proceedings of the USSR Mathematical Congress*, vol. 1, 1963, 101–120.

[P29] V. B. Lidskii and P. A. Frolov, *On the topological structure of stability domains of a self-adjoint system of differential equations with periodic coefficients*, Doklady AN SSSR **161** (1965), 764–766.

[P30] V. B. Lidskii, *On the perturbation theory for non-self-adjoint operators*, Zh. Vychisl. Mat. i Mat. Fiz. **69** (1966), 52–60, English translation in Comp. Math. Math. Phys., **6** (1966), 73–85.

[P31] V. B. Lidskii, *Inequalities for eigenvalues and singular numbers of matrices*, Appendix in F. R. Gantmakher's *The theory of matrices*, Gostekhizdat, Moscow, 1966, 535–559.

[P33] V. B. Lidskii and P. A. Frolov, *Stability domains of complex self-adjoint systems of differential equations*, Doklady AN SSSR **171** (1966), 41–43, English translation in Sov. Math. Dokl..

[P34] V. B. Lidskii and P. A. Frolov, *The structure of stability domains of a self-adjoint system of differential equations with periodic coefficients*, Mat. Sbornik **71** (1966), 48–64, English transl. in Math. USSR-Sb..

[P35] V. B. Lidskii and P. A. Frolov, *Stability domains of self-adjoint systems of differential equations with periodic coefficients*, in *Abstracts of reports at the International Congress of Mathematicians*, Section 6, Moscow, 1966.

[P36] V. B. Lidskii and B. V. Fedosov, *On the vanishing of solutions of a second-order differential equation with operator-valued coefficients*, Mat. Zametki **2** (1967), 307–314, English transl. in Math. Notes **2** (1967), 670–674.

[P37] V. B. Lidskii and V. A. Sadovnichii, *Regular sums of the roots of a class of entire functions*, Funkc. Anal. Prilozhen. **1** (1967), 52–59, English transl. in Func. Anal. Appl. **1** (1967) 133–139.

[P38] V. B. Lidskii and V. A. Sadovnichii, *Regularized sums of the roots of a class of entire functions*, Doklady AN SSSR **176** (1967), 259–262.

[P39] V. B. Lidskii and V. A. Sadovnichii, *Asymptotic formulae for the roots of a class of entire functions*, Mat. Sbornik **75** (1968), 558–566, English transl. in Math. USSR-Sb. **4** (1968), 519–527.

[P40] V. B. Lidskii and V. A. Sadovnichii, *Trace formulae for the Orr-Sommerfeld equation*, Izv. AN SSSR, Ser. Mat. **32** (1968), 633–648, English transl. in Math. USSR-Izv. **2**, 585–600.

[P41] A. G. Aslanyan and V. B. Lidskii, *On the spectrum of an elliptic equation*, Mat. Zametki **7** (1970), 495–502, English transl. in Math. Notes **7** (1970), 300–304.

[P42] V. B. Lidskii and P. A. Frolov, *Connected components of normally solvable elliptic systems on the plane*, Doklady AN SSSR **192** (1970), 728–731, English translation in Sov. Math. Dokl. **11** (1970), 712–716.

[P43] V. B. Lidskii and N. V. Kharkova, *The spectrum of membrane equations in the case of axisymmetrical vibrations of a shell of revolution*, Doklady AN SSSR **194** (1970), 786–789, English translation in Sov. Phys. Dokl. **15** (1970–71), 982–984.

[P44] A. G. Aslanyan and V. B. Lidskii, *Oscillatory theorem in the theory of vibrations of thin elastic shells of revolution*, Doklady AN SSSR **196** (1971), 1040–1042, English translation in Sov. Phys. Dokl. **16** (1971), 95–97.

[P45] A. G. Aslanyan and V. B. Lidskii, *Spectral properties of the system describing free vibrations of shells of revolution*, Doklady AN SSSR **201** (1971), 300–303, English translation in Sov. Phys. Dokl. **16** (1971), 946.

[P46] A. G. Aslanyan and V. B. Lidskii, *Asymptotic formulae for the lower series in the theory of shells of revolution*, Izv. AN Arm. SSR, Ser. Mat. **6** (1971), 113–130.

[P47] A. O. Kravitskii and V. B. Lidskii, *Asymptotic series for the solution of the Cauchy problem*, Sibirsk. Mat. Zh. **12** (1971), 748–759, English transl. in Siberian Math. J. **12** (1971), 537–540.

[P48] A. G. Aslanyan and V. B. Lidskii, *The spectrum of the system describing vibrations of a shell of revolution*, Prikl. Mat. Mekh. **35** (1971), 701–717, English translation in J. Appl. Math. Mech. **35** (1971), 648–664.

[P49] A. G. Aslanyan and V. B. Lidskii, *Asymptotics of the eigenfrequency distribution function of a shell of revolution*, Izv. AN Arm. SSR, Ser. Mat. **7** (1972), 236–249.

[P50] A. G. Aslanyan and V. B. Lidskii, *Eigenfrequency distribution of thin shells*, Izv. AN SSSR, Ser. Mech. Tverdogo Tela (1972), 155.

[P51] G. R. Gulgazaryan, V. B. Lidskii, and G. I. Eskin, *The spectrum of the membrane system in the case of a thin shell of an arbitrary shape*, Sibirsk. Mat. Zh. **14** (1973), 978–986, English transl. in Siberian Math. J. **14** (1973), 681–687.

[P52] A. G. Aslanyan, Z. N. Kuzina, V. B. Lidskii, and V. N. Tulovskii, *Distribution of the natural frequencies of a thin elastic shell of arbitrary outline*, Prikl. Mat. Mekh. **37** (1973), 604–617, English translation in J. Appl. Math. Mech. **37** (1973), 571–584.

[P53] A. G. Aslanyan and V. B. Lidskii, *The spectrum of eigenfrequencies of a shell of revolution*, in *Proceedings of the 8th Conference on the Theory of Plates and Shells*, Nauka, Moscow, 1973, 371–375 (in Russian).

[P54] A. L. Goldenveizer and V. B. Lidskii, *Some general properties of free vibrations of a thin elastic shell*, Izv. AN SSSR, Ser. Mech. Tverdogo Tela **9** (1974), 120–124.

[P55] A. G. Aslanyan, V. B. Lidskii, S. A. Lukovenko, and N. V. Kharkova, *The asymptotics of the eigenfrequency distribution function of a shell of revolution with a fixed number of waves along a parallel*, Preprint no. 38, Institute for Problems in Mechanics, Moscow, 1974 (in Russian).

[P56] A. G. Aslanyan and V. B. Lidskii, *Formula for the number of eigenfrequencies of a shell of revolution*, Doklady AN SSSR **222** (1975), 790–792, English translation in Sov. Phys. Dokl. **20** (1976), 461–462.

[P57] A. G. Aslanyan, V. B. Lidskii, S. A. Lukovenko, and N. V. Kharkova, *The asymptotics of the eigenfrequency distribution function of a shell of revolution with a fixed number of waves along a parallel*, in *Proceedings of the 9th Conference on the Theory of Plates and Shells*, Sudizdat, Leningrad, 1975, 157–160 (in Russian).

[P58] V. B. Lidskii, *On an estimate for the resolvent of an elliptic differential operator*, Funkc. Anal. Prilozhen. **10** (1976), 89–90, English transl. in Func. Anal. Appl. **10** (1976), 324–325.

[P59] V. B. Lidskii, *Mathematical problem in the theory of oscillations of thin shells*, in *Unsolved problems of mechanics and applied mathematics*, Moscow State University, Moscow, 1977, 94–99 (in Russian).

[P60] A. G. Aslanyan, V. B. Lidskii, and G. G. Tarposhjan, *The spectra of free vibrations of thin elastic shells*, in *Proceedings of the 11th Conference on the Theory of Plates and Shells*, Nauka, Moscow, 1977, 22 (in Russian).

[P61] A. G. Aslanyan and V. B. Lidskii, *Formula for the number of eigenfrequencies of axisymmetrical vibrations of shells of revolution*, Diff. Uravn. **13** (1977), 1355–1365, English translation in Diff. Eqs. **13** (1977), 939–946.

[P62] A. G. Aslanyan, A. L. Goldenveizer, and V. B. Lidskii, *Free vibrations of thin elastic shells*, in *Proceedings of the I. G. Petrovskii Conference on Partial Differential Equations*, Moscow State University, Moscow, 1978, 33–35 (in Russian).

[P63] A. L. Goldenveizer and V. B. Lidskii, *Some mathematical problems in the theory of vibrations of thin elastic shells*, Uspekhi Mat. Nauk **33** (1978), 225–226.

[P65] A. G. Aslanyan, D. Vassiliev, and V. B. Lidskii, *Vibrations of elastic shells interacting with fluid*, Uspekhi Mat. Nauk **35** (1980), 255-256.

[P66] A. G. Aslanyan, D. Vassiliev, and V. B. Lidskii, *Frequencies of free vibrations of shells interacting with fluid*, in *Proceedings of the 12th Conference on the Theory of Plates and Shells*, Yerevan State University, Yerevan, 1980, Vol. 2, 86–92 (in Russian).

[P67] A. G. Aslanyan, D. Vassiliev, and V. B. Lidskii, *The distribution of eigenfrequencies of thin elastic shells interacting with fluid*, Uspekhi Mat. Nauk **36** (1981), 216.

[P68] A. G. Aslanyan, D. Vassiliev, and V. B. Lidskii, *Frequencies of free vibrations of a shell interacting with fluid*, Funkc. Anal. Prilozhen. **15** (1981), 1–9, English transl. in Func. Anal. Appl. **15** (1981), 157–164.

[P69] A. G. Aslanyan, D. Vassiliev, and V. B. Lidskii, *Free vibrations of a shell interacting with fluid*, in *Proceedings of the II Symposium on Physics of Acousto-hydrodynamical Phenomena and Optoacoustics*, Nauka, Moscow, 1982, 264–266 (in Russian).

[P70] V. B. Lidskii, *Shell theory*, in *Mathematical Encyclopedia*, vol. 3, 1982, 1127–1129 (in Russian).

[P71] G. R. Gulgazaryan and V. B. Lidskii, *Density of free vibration frequencies of a thin anisotropic shell composed of anisotropic layers*, Izv. AN SSSR, Ser. Mech. Tverdogo Tela (1982), 171–174.

[P72] D. Vassiliev and V. B. Lidskii, *Vibrations of a thin shells contacting fluid*, Uspekhi Mat. Nauk **37** (1982), 224–225.

[P73] V. B. Lidskii, *Spectra in shell theory*, in *International Conference on Partial Differential Equations*, Nauka, Novosibirsk, 1983, pp. 59–60 (in Russian).

[P74] V. B. Lidskii and P. V. Tovstik, *Spectra in shell theory*, Usp. Mekhaniki **7** (1984), 25–54.

[P75] D. Vassiliev and V. B. Lidskii, *Forced vibrations of a thin shell immersed in a fluid*, Uspekhi Mat. Nauk **39** (1984), 131–132.

[P76] D. Vassiliev and V. B. Lidskii, *Asymptotics of solutions of the radiation problem for a shell immersed in an inviscid compressible liquid*, in *Mathematical methods in mechanics of deformable rigid body*, Nauka, Moscow, 1986, pp. 30–36 (in Russian).

[P77] D. Vassiliev and V. B. Lidskii, *Quasiresonances in the problem of forced vibrations of a thin elastic shell interacting with fluid*, Funkc. Anal. Prilozhen. **20** (1986), 17–28, English transl. in Func. Anal. Appl. **20** (1986), 267–276.

[P78] D. Vassiliev and V. B. Lidskii, *Forced vibrations of a thin shell interacting with fluid*, Uspekhi Mat. Nauk **41** (1986), 157–158.

[P79] D. Vassiliev, M. Levitin, and V. B. Lidskii, *Vibrations of a shell filled with a viscous compressible fluid*, in *Proceedings of the 14th Conference on the Theory of Plates and Shells*, Tbilisi State University, Tbilisi, 1987, vol. 1, 279–284 (in Russian).

[P80] D. Vassiliev, M. Levitin, and V. B. Lidskii, *Forced vibrations of a shell filled with a viscous compressible fluid*, Uspekhi Mat. Nauk **43** (1988), 164.

[P81] V. B. Lidskii, Appendix in F. R. Gantmakher's *The theory of matrices*, 4th edition, Nauka, Moscow, 1988, 502–528 (in Russian).

[P82] D. Vassiliev, M. Levitin, and V. B. Lidskii, *Forced vibrations of a shell filled with a viscous compressible fluid*, Doklady AN SSSR **305** (1989), 329–332, English translation in Sov. Phys. Dokl. **34** (1989), 215-217.

[P83] A. L. Goldenveizer, Yu. D. Kaplunov, V. B. Lidskii, and D. Vassiliev, *Vibrations of shells filled with fluid*, in *Engineering Hydroelasticity*, Praha, 1989, 185–190.

[P84] D. Vassiliev, M. Levitin, and V. B. Lidskii, *On forced vibrations of a shell immersed in a viscous compressible fluid*, Funkc. Anal. Prilozhen. **25** (1991), 93–95, English transl. in Func. Anal. Appl. **25** (1991), 309–311.

[P85] A. G. Aslanyan, D. Vassiliev, and V. B. Lidskii, *A formula for the distribution function of the frequencies of a thin shell of revolution in the case of a simple turning point*, Funkc. Anal. Prilozhen. **28** (1994), 74–77. English transl. in Funct. Anal. Appl. **28** (1994), 282–284.

[P86] V. B. Lidskii, D. Vassiliev, and M. Levitin, *The oscillation of a thin elastic shell*, in *The 7th All-Union Conference on Mechanics, 15–21 September 1994*, pp. 93–95.

[P87] V. B. Lidskii, *Kirchoff's formula*, FizTech Journal **2** (1996), 21–34.

[P88] A. A. Aslanyan, A. G. Aslanyan, and V. B. Lidskii, *Asymptotic formulas for frequencies of axisymmetric oscillations of a shell of revolution*, Zh. Vychisl. Mat. Mat. Fiz. **38** (1998), 298–309, English transl. in Comput. Math. Math. Phys. **38** (1998), 288–299.

Amer. Math. Soc. Transl.
Volume **231**, 2010

Eigenvalue Inequalities for Mixed Steklov Problems

Rodrigo Bañuelos, Tadeusz Kulczycki, Iosif Polterovich, and Bartłomiej Siudeja

ABSTRACT. We extend some classical inequalities between the Dirichlet and Neumann eigenvalues of the Laplacian to the context of mixed Steklov–Dirichlet and Steklov–Neumann eigenvalue problems. The latter one is also known as the sloshing problem and has been actively studied for more than a century due to its importance in hydrodynamics. The main results of the paper are used to obtain certain geometric information about nodal sets of sloshing eigenfunctions. The key ideas of the proofs include domain monotonicity for eigenvalues of mixed Steklov problems, as well as an adaptation of the techniques developed originally to compare the Dirichlet and Neumann eigenvalues.

1. Introduction and main results

1.1. Mixed Steklov problems. Let W be a bounded domain in $\mathbb{R}^d$ satisfying the following assumptions:

(I) *W is Lipschitz and $W \subset \{(x, y) : x \in \mathbb{R}^{d-1},\ y < 0\}$.*

(II) *Its boundary ∂W consists of two sets F and B with $B = \partial W \setminus F$ and $F = F' \times \{0\} \subset \mathbb{R}^{d-1} \times \{0\}$, where F' is a bounded Lipschitz domain in $\mathbb{R}^{d-1}$ (see Figure 1).*

Consider the following two eigenvalue problems on the domain W:

(1) The mixed Steklov–Neumann problem:

$$\begin{cases} \Delta v(x,y) = 0, & (x,y) \in W, \\ \dfrac{\partial v}{\partial y}(x,0) = \mu\, v(x,0), & (x,0) \in F, \\ \dfrac{\partial v}{\partial \nu}(x,y) = 0, & (x,y) \in B. \end{cases} \tag{1.1.1}$$

2010 *Mathematics Subject Classification.* Primary 35P15; Secondary 35Q35.
The first author was supported in part by NSF Grant #0603701-DMS.
The second author was supported in part by MNiSW grant #N N201 373136.
The third author was supported in part by NSERC, FQRNT, and the Canada Research Chairs program.

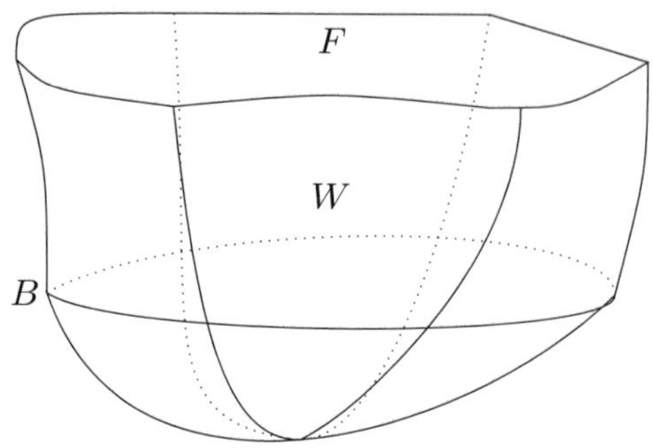

Figure 1: Domain W with boundary $\partial W = F \cup B$.

(2) The mixed Steklov–Dirichlet problem:

$$\begin{cases} \Delta u(x,y) = 0, & (x,y) \in W, \\ \dfrac{\partial u}{\partial y}(x,0) = \lambda\, u(x,0), & (x,0) \in F, \\ u(x,y) = 0, & (x,y) \in B. \end{cases} \tag{1.1.2}$$

We note that because of the Lipschitz boundary, the $(d-1)$-dimensional Lebesgue measure (surface area measure) σ is well defined on ∂W and that the outward unit normal vector field ν is well defined at almost all points of B with respect to σ. We understand that the equality $\partial v/\partial \nu = 0$ in (1.1.1) is satisfied for all points $(x,y) \in B$ for which ν is defined. At the same time, as indicated in Remark 1.1.5 below, for the weak formulation of the Steklov–Neumann problem one does not need to assume that B is Lipschitz.

It is well known (see, for example, [**3**, p. 69], [**26**, Theorem 1, p. 270], [**15**, p. 34, p. 250]) that under our assumptions the Steklov–Neumann eigenvalue problem (1.1.1) has discrete spectrum $\{\mu_n\}_{n=1}^{\infty}$,

$$0 = \mu_1 < \mu_2 \leq \mu_3 \leq \ldots \to \infty.$$

Note that the first nonzero Steklov–Neumann eigenvalue is denoted by μ_2 (some other papers use a different convention). The eigenvalues μ_n admit the following variational characterization:

$$\mu_n = \inf_{V_n \subset H^1(W)} \sup_{0 \neq v \in V_n} \frac{\int_W |\nabla v(x,y)|^2\, dx\, dy}{\int_{F'} v^2(x,0)\, dx}, \tag{1.1.3}$$

where the infimum is taken over all n-dimensional subspaces V_n of the Sobolev space $H^1(W)$. The corresponding eigenfunctions we denote by v_n, $n = 1, 2, \ldots$.

Similarly, it is known (see [**1**]) that the Steklov–Dirichlet eigenvalue problem (1.1.2) has discrete spectrum $\{\lambda_n\}_{n=1}^{\infty}$,

$$0 < \lambda_1 \leq \lambda_2 \leq \lambda_3 \leq \ldots \to \infty$$

and the eigenvalues admit the following variational characterization:

$$\lambda_n = \inf_{U_n \subset H_0^1(W,B)} \sup_{0 \neq u \in U_n} \frac{\int_W |\nabla u(x,y)|^2\, dx\, dy}{\int_{F'} u^2(x,0)\, dx}, \tag{1.1.4}$$

where the infimum is taken over all n-dimensional subspaces U_n of the space $H^1_0(W,B) = \{u \in H^1(W) : u \equiv 0 \text{ on } B\}$. The corresponding eigenfunctions we denote by u_n, $n = 1, 2, \dots$.

REMARK 1.1.5. One may weaken the assumption (I) that W is a Lipschitz domain and still guarantee that the problems (1.1.2) and (1.1.1) have discrete spectrum. In fact, in the Steklov–Dirichlet case, no regularity of the boundary is required. In the Steklov–Neumann case it suffices to assume that F' is Lipschitz and that there exists a Lipschitz domain $V \subset W$ such that $\partial V \cap \{y = 0\} = F'$. In both cases the result follows immediately from domain monotonicity for mixed Steklov eigenvalues (see Sections 3.1 and 3.2) and [**5**, Theorem 4.5.2].

1.2. Main results. Let $W \subset \mathbb{R}^d$ be a domain satisfying the assumptions (I) and (II). Following [**20**, Section 3.2.1], we say that W satisfies the *(standard) John's condition* if $W \subset F' \times (-\infty, 0)$.

DEFINITION 1.2.1. We say that W satisfies the *weak John's condition* if

$$\int_W e^{ay}\,dx\,dy \le \frac{|F'|}{a}, \quad \text{for any } a > 0, \tag{1.2.2}$$

where $|F'|$ is the $(d-1)$-dimensional Lebesgue measure of F'.

It is easy to check that the standard John's condition implies the weak John's condition (the converse is not true, as seen from the example constructed in Section 2.2). Indeed, if $W \subset F' \times (-\infty, 0)$, then for any $a > 0$,

$$\int_W e^{ay}\,dx\,dy \le \int_{F'} dx \int_{-\infty}^0 e^{ay}\,dy = \frac{|F'|}{a}.$$

Let us formulate the main results of the paper.

THEOREM 1.2.3. *Consider the eigenvalue problems* (1.1.1) *and* (1.1.2) *on a domain $W \subset \mathbb{R}^d$ satisfying the weak John's condition* (1.2.2). *Then for any $n \in \mathbb{N}$ we have*

$$\mu_{n+1} < \lambda_n \ \text{ if } \ d \ge 3,$$
$$\mu_{n+1} \le \lambda_n \ \text{ if } \ d = 2.$$

The proof of Theorem 1.2.3 is presented in Section 4.2 and is based on an adaptation of the argument due to Filonov [**8**]. It can be also obtained using the results of Safarov [**30**]; see Remark 4.2.10. Note that the weak John's condition is essential for Theorem 1.2.3 to hold; see Sections 2.3 and 2.4.

THEOREM 1.2.4. *Consider the eigenvalue problems* (1.1.1) *and* (1.1.2) *on a domain $W \subset \mathbb{R}^d$, $d \ge 2$, satisfying the standard John's condition. Then for any $n \in \mathbb{N}$ we have*

$$\mu_{n+1} < \lambda_n. \tag{1.2.5}$$

Moreover, if $F' \subset \mathbb{R}^{d-1}$ is a convex set, then

$$\mu_{n+d-1} < \lambda_n. \tag{1.2.6}$$

The proof of Theorem 1.2.4 presented in Section 4.1 is quite short: it uses domain monotonicity for eigenvalues of mixed Steklov problems (see Sections 3.1 and 3.2), the properties of mixed Steklov eigenvalues for cylindrical domains (see Section 2.1), and the classical inequalities between the Dirichlet and Neumann eigenvalues [**11, 8, 23**].

REMARK 1.2.7. Theorems 1.2.3 and 1.2.4 are also valid for unbounded domains W satisfying the weak and the standard John's conditions, respectively, provided the problems (1.1.1) and(1.1.2) have discrete spectra. In order to guarantee that the spectrum is discrete, one has to impose an additional constraint: the solutions must have a gradient decaying sufficiently fast at infinity (see [**13**]). A classical example of a sloshing problem on an unbounded domain is the "ice fishing problem" described in Section 2.4.

1.3. Discussion. The eigenvalue problem (1.1.1) has important applications to hydrodynamics and is also known as the *sloshing problem* (see, for example, [**14**] and references therein).

For $d = 3$, it models free fluid oscillations in a container W with bottom B and a free surface of a steady fluid F (see Figure 1). This problem was first studied by Euler [**7**] as early as 1761 and has since been the topic of a great number of papers. We refer to [**9**] for a historical review of this subject. Earlier results on the sloshing problem are described by Lamb [**22**] in his book *Hydrodynamics*. For more recent developments, the reader may consult the books [**15**] and [**20**], as well as the papers [**16**], [**17**], and [**21**]. The sloshing problem is the main motivation to study (1.1.1) and, in particular, it justifies our assumptions on the domains W, F, and B.

For $d = 2$, the eigenvalue problem (1.1.1) describes oscillations of a 2-dimensional free fluid in a channel with uniform cross-section W. Here B is the uniform cross-section of the bottom of the channel and F is the uniform cross-section of the free surface of the steady fluid. Free fluid oscillations are assumed here to be 2-dimensional and identical for all the cross-sections of the channel. In [**18**], some properties of the first nontrivial eigenfunction for the 2-dimensional sloshing problem were established. To obtain these properties, the inequality $\mu_2 \le \lambda_1$ was proved there for the case $d = 2$. It was conjectured in [**18**, Conjecture 4.3] that the inequality $\mu_{n+1} \le \lambda_n$ should hold for $d = 2$ (note that in [**18**] a different notation $\nu_n = \mu_{n+1}$ was used). Theorem 1.2.3 for $d = 2$ gives a positive answer to this conjecture.

Eigenvalue problems (1.1.1) and (1.1.2) are used as well to model some other physical processes. For instance, they describe the stationary heat distribution in W under the conditions that the heat flux through F is proportional to the temperature (see [**3**]), and part B of the boundary is either perfectly insulated (in (1.1.1)) or kept under zero temperature (in (1.1.2)).

The boundary value problems (1.1.1) and (1.1.2) also have interesting probabilistic interpretations in terms of jump processes on F which arise as traces of Brownian motion in W. Roughly speaking, μ_n and $v_n|_F$ are the eigenvalues and eigenfunctions of the generator of the jump process which is the trace on F of the Brownian motion in W with reflection on ∂W. Similarly, λ_n and $u_n|_F$ are the eigenvalues and eigenfunctions of the generator of the jump process which is the trace on F of the Brownian motion in W with killing on B and reflection on F. The connection between the mixed Steklov problem (1.1.2) and the eigenvalues and eigenfunctions of the generator of the d-dimensional Cauchy process (which is the trace of the $(d+1)$-dimensional Brownian motion) in some domains is described in detail in [**2**].

Finally, it is worth pointing out here that Steklov type eigenvalue problems have attracted considerable attention in recent years. For some of this literature, see [**3**], [**6**], [**1**], [**16**], [**17**], [**21**], [**12**], [**10**].

1.4. Nodal sets of sloshing eigenfunctions. Let

$$\mathcal{N}_f = \{x | \, f(x) = 0\}$$

denote the *nodal set* of a function f. The following lemma is a simple consequence of domain monotonicity for eigenvalues of mixed Steklov problems. Recall that $B = \partial W \setminus F$ and set $B_0 = \partial W \setminus \overline{F}$.

LEMMA 1.4.1. *Let ϕ be an eigenfunction of the sloshing problem* (1.1.1) *on a domain W satisfying the assumptions* (I) *and* (II). *Suppose that ϕ corresponds to an eigenvalue $\mu \leq \lambda_1$, where λ_1 is the first eigenvalue of the Steklov–Dirichlet problem* (1.1.2). *Let $C \subset \mathcal{N}_\phi$ be a connected component of the nodal set of ϕ. Then $C \cap B_0 \neq \emptyset$. Moreover, if $d = 2$, then $C \cap \partial F' = \emptyset$.*

The proof of Lemma 1.4.1 is analogous to the proof of the fact that the second Neumann eigenfunction cannot have a closed nodal line [**28**, p. 546]. We present the details in Section 4.3.

Lemma 1.4.1 together with Theorems 1.2.3 and 1.2.4 immediately imply the following result.

COROLLARY 1.4.2. *Let $W \subset \mathbb{R}^d$ be as above and let ϕ_n be an eigenfunction of the sloshing problem* (1.1.1) *on W corresponding to an eigenvalue μ_n.*

(i) *If W satisfies the weak John's condition, then $N_{\phi_2} \cap B_0 \neq \emptyset$. If, moreover, $d = 2$, then $N_{\phi_2} \cap \partial F' = \emptyset$.*

(ii) *If W satisfies the standard John's condition and $F' \subset \mathbb{R}^{d-1}$ is a convex set, then $C \cap B_0 \neq \emptyset$ for any connected component of the set $C \subset \mathcal{N}_{\phi_k}$, $k = 2, \ldots, d$.*

Geometric properties of nodal sets of sloshing eigenfunctions in two dimensions have been previously studied in [**19, 17**]. In fact, it was claimed in [**19**] that any nodal line of any sloshing eigenfunction must intersect the bottom of the container (i.e., the set B in our notation). However, in [**17**] a counterexample to this statement was constructed. At the same time, it was shown in [**17**, Theorem 3.1(ii)] that this is indeed true for the nodal set $\mathcal{N}_{\phi_2}$. The first statement of Corollary 1.4.2 (i) can be viewed as a higher-dimensional generalization of this result for domains satisfying the weak John's condition. Note that in any dimension, the set $\mathcal{N}_{\phi_2}$ consists of a single connected component. This follows from the analogue of Courant's nodal domain theorem for sloshing problems (we note that while this theorem is stated in [**19, 17**] for planar domains only, the argument extends to higher dimensions in a straightforward way).

It was also shown in [**17**, Corollary 3.4] that if a planar domain W satisfies the standard John's condition, then the nodal line for the first nontrivial eigenfunction does not contain the endpoints of the free boundary F. The second part of Corollary 1.4.2 (i) extends this result to domains satisfying the *weak* John's condition.

Let us also remark that the proof of Corollary 1.4.2 is based on more elementary ideas (such as domain monotonicity of mixed Steklov eigenvalues) than the argument in [**17**].

1.5. Plan of the paper. In Sections 2.1–2.4 we discuss some examples illustrating Theorems 1.2.3 and 1.2.4 and shedding more light on the geometric assumptions on the sets W, F, and B. In Sections 3.1 and 3.2 we review the results regarding domain monotonicity of eigenvalues for the mixed Steklov problems.

While this property is well known, it seems that it has not been stated in the literature in full strength. In particular, we show that domain monotonicity is *strict*, which requires some extra work. Finally, in Sections 4.1–4.3 the proofs of the main results are presented.

2. Examples

2.1. Cylindrical domains. Let $F' \subset \mathbb{R}^{d-1}$ be a bounded Lipschitz domain. Set $F = F' \times \{0\}$, $W = F' \times (-l, 0)$, $l > 0$, and $B = \partial W \setminus F$. Clearly, the cylindrical domain W satisfies the standard John's condition. By separation of variables it is easy to see that the eigenfunctions and eigenvalues of the problems (1.1.1) and (1.1.2) are given by

$$v_n(x,y) = \tilde{v}_n(x)\cosh(\sqrt{\tilde{\mu}_n}(y+l)), \quad \mu_n = \sqrt{\tilde{\mu}_n}\tanh(\sqrt{\tilde{\mu}_n} l) \tag{2.1.1}$$

and, respectively,

$$u_n(x,y) = \tilde{u}_n(x)\sinh(\sqrt{\tilde{\lambda}_n}(y+l)), \quad \lambda_n = \sqrt{\tilde{\lambda}_n}\coth(\sqrt{\tilde{\lambda}_n} l), \tag{2.1.2}$$

where $\{\tilde{v}_n\}$ and $\{\tilde{\mu}_n\}$ are the eigenfunctions and eigenvalues of the Neumann problem for the Laplacian on F' and $\{\tilde{u}_n\}$ and $\{\tilde{\lambda}_n\}$ are the eigenfunctions and eigenvalues of the Dirichlet problem for the Laplacian on F'.

By the classical inequalities between the Neumann and Dirichlet eigenvalues of the Laplacian, conjectured in [**27**] and proved in [**11, 8**], we have $\tilde{\mu}_{n+1} < \tilde{\lambda}_n$ for all $n = 1, 2\ldots$ in dimensions $d \geq 3$ (for $n = 1$ and $d = 3$ this inequality can also be deduced from [**29**] and [**31**, Section 1.5]). Moreover, if F' is convex, then $\tilde{\mu}_{n+d-1} \leq \tilde{\lambda}_n$ by [**23**]. If $d = 2$, then F' is just an interval of the real line, and an elementary calculation yields $\tilde{\mu}_{n+1} = \tilde{\lambda}_n$, $n = 1, 2, \ldots$. Since $\tanh(\alpha) < 1 < \coth(\beta)$ for all $\alpha, \beta > 0$, we immediately obtain from (2.1.1) and (2.1.2) the assertions of Theorem 1.2.4 for cylindrical domains in any dimension.

As follows from Section 1.3, for $d = 3$ the eigenvalue problem (1.1.1) in this example describes free fluid oscillations in a glass-like container $W = F' \times (-l, 0)$ with the free fluid surface $F = F' \times \{0\}$.

2.2. Sloshing in a vase. In this section we show that there exist domains satisfying the weak John's condition but not the standard John's condition.

For any $r > 0$, $h_2 < h_1 \leq 0$, let $L(r, h_2, h_1) \subset \mathbb{R}^d$ be a cylinder given by

$$L(r, h_2, h_1) = \{(x, y) \in \mathbb{R}^{d-1} \times (-\infty, 0] : |x| < r,\ h_2 \leq y \leq h_1\}.$$

Let $W \subset \mathbb{R}^3$ satisfy assumptions (I) and (II) and

$$F' = \{(x_1, x_2) \in \mathbb{R}^2 : x_1^2 + x_2^2 < 1\}, \tag{2.2.1}$$

$$W \subset L(1, -1, 0) \cup L(0.5, -2, -1) \cup L(1.5, -4, -2). \tag{2.2.2}$$

Let us show that W satisfies the weak John's condition. It suffices to verify that for any $a > 0$,

$$\int_{L(1,-1,0)\cup L(0.5,-2,-1)\cup L(1.5,-4,-2)} e^{ay}\,dx\,dy \leq \frac{|F'|}{a} = \int_{F'}\int_{-\infty}^{0} e^{ay}\,dy\,dx.$$

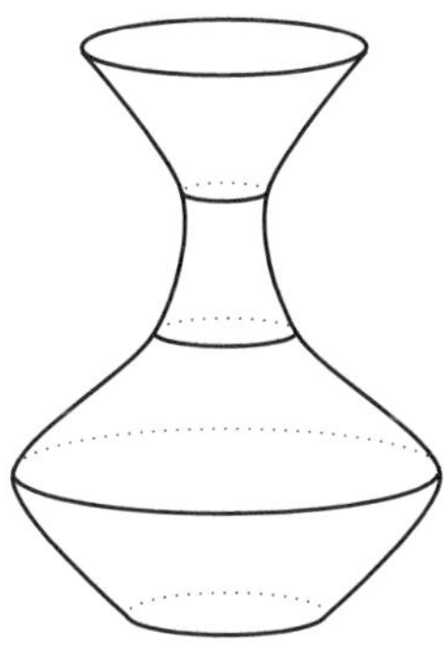

(a) Vase

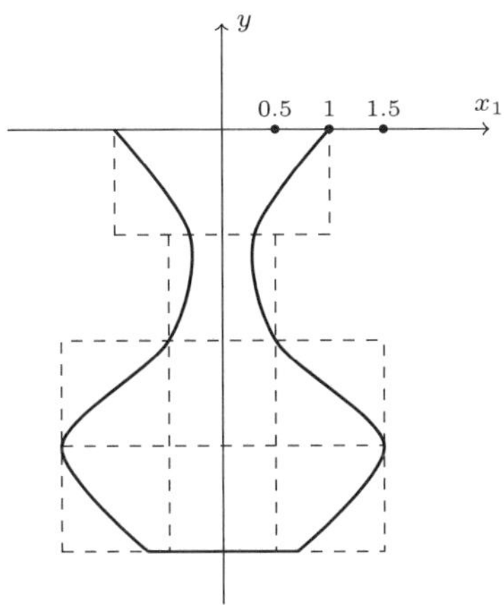

(b) Its projection

Figure 2: Vase-like container satisfying the weak but not the standard John's condition.

This is equivalent to the following inequalities:

$$\pi\left(\int_{-1}^{0} e^{ay}\,dy + (0.5)^2\int_{-2}^{-1} e^{ay}\,dy + (1.5)^2\int_{-4}^{-2} e^{ay}\,dy\right) \le \pi\int_{-\infty}^{0} e^{ay}\,dy,$$

$$\frac{1-e^{-a}}{a} + \frac{1}{4}\frac{e^{-a}-e^{-2a}}{a} + \frac{9}{4}\frac{e^{-2a}-e^{-4a}}{a} \le \frac{1}{a},$$

$$8e^{-2a} \le 3e^{-a} + 9e^{-4a},$$

$$0 \le 3e^{3a} - 8e^{2a} + 9. \tag{2.2.3}$$

Consider the function $f(x) = 3x^3 - 8x^2 + 9$. In order to prove (2.2.3) we need to show that $f(x) > 0$ for any $x \ge 0$. Indeed, by elementary calculus, $\min\{f(x) : x \ge 0\} = f(16/9) > 0$.

In Figure 2(a) we give an example of a domain W satisfying (2.2.1) and (2.2.2) that has the shape of a vase. Its projection on the (x_1, y)-plane is presented in Figure 2(b). The domain W is symmetric with respect to the y-axis and clearly does not satisfy the standard John's condition.

One can modify the previous example and construct domains in any dimension which satisfy the weak but not the standard John's condition. Indeed, let $W \subset \mathbb{R}^d$ be an arbitrary domain, satisfying the assumptions (I) and (II) of Section 1.1, such that

$$F' = \{x \in \mathbb{R}^{d-1} : |x| < r_1\}, \tag{2.2.4}$$

$$W \subset L(r_1, h_1, 0) \cup L(r_2, h_2, h_1) \cup L(r_3, h_3, h_2), \tag{2.2.5}$$

$$0 < r_2 < r_1 < r_3,\ r_2^{d-1} + r_3^{d-1} \le 2r_1^{d-1}, \tag{2.2.6}$$

$$0 \ge h_1 > h_2 > h_3,\ h_1 - h_2 \ge h_2 - h_3. \tag{2.2.7}$$

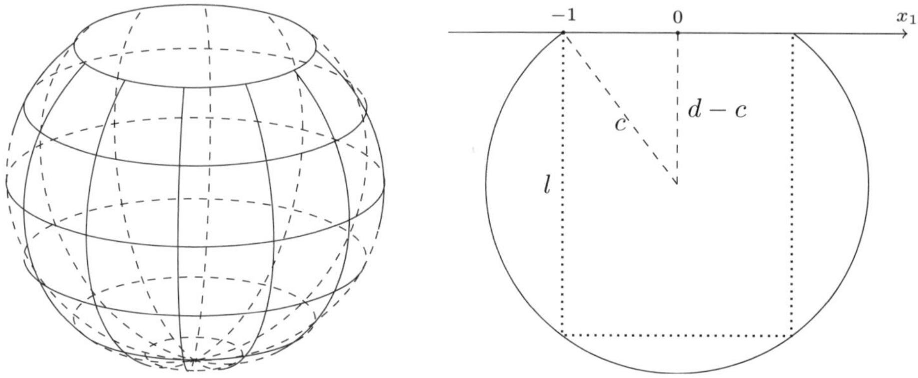

(a) Spherical container (b) Its projection

Figure 3: Spherical container with radius c filled up to level d, with circular aperture of radius 1. Dotted lines show cylindrical container of height l.

Since $r_1 < r_3$, we have $L(r_3, h_3, h_2) \not\subset F' \times (-\infty, 0)$ so W does not have to satisfy the condition $W \subset F' \times (-\infty, 0)$. One can show by an explicit calculation that under these assumptions W satisfies the weak John's condition. We leave the details to the interested reader.

REMARK 2.2.8. Note that the conditions (2.2.1) and (2.2.2) are not a special case of (2.2.4)–(2.2.7) for $d = 3$ and some choice of r_1, r_2, r_3, h_1, h_2, h_3. We use a particular choice of the parameters in (2.2.1) and (2.2.2) in order to construct a domain W that has a vase-like shape which is natural in the context of the sloshing problem.

2.3. Sloshing in a spherical container. In this section we present an example showing that one cannot remove the weak John's condition from the formulation of Theorem 1.2.3. We use spherical containers studied by McIver in [**24**]. Figure 3 shows such a container and its projection with the free surface contained in the x_1-axis.

McIver gives numerical results for various ratios of parameters d/c defined in Figure 3; see [**24**, Table 2] . In particular the second Steklov–Neumann eigenvalue for $d/c = 1.8$ equals 2.376 (after rescaling). This is not enough to get a contradiction by just comparing it with the first Steklov–Dirichlet eigenvalue of the cylinder contained in the spherical container. Indeed, the semi–infinite cylinder has the first Steklov–Dirichlet eigenvalue equal to 2.4048, and the corresponding eigenvalue of a truncated cylinder is even a little bit larger (see (2.1.2)).

We implemented McIver's numerical method to get the second Steklov–Neumann eigenvalue for $d/c = 1.9$. Our algorithm differs slightly from the original one. We used the standard numerical integration function in *Mathematica* to find values of integrals I_m instead of approximating integrands with Chebyshev polynomials (see [**24**, Appendix B]). Numerical results obtained using our method are virtually identical to those found by McIver.

For $d/c = 1.9$, we found that the second Steklov–Neumann eigenvalue is equal to 2.51105. The biggest cylinder contained in such a spherical container has height $l = 4.13$. Using formula (2.1.2), we get that the first Steklov–Dirichlet eigenvalue of this cylinder is equal to 2.4048 (practically the same as for the semi–infinite cylinder). By domain monotonicity (see Section 3.1) it is larger than the value of the first Steklov–Dirichlet eigenvalue of the spherical container, yet it is smaller than its second Steklov–Neumann eigenvalue. This gives a "counterexample" to our main result if the domain does not satisfy the weak John's condition. The latter could be verified directly: a numerical calculation shows that for the spherical container presented on Figure 3 the inequality (1.2.2) fails if $a > 0.08$ (in fact, one can check that if $d/c > 1$, the inequality (1.2.2) does not hold for sufficiently large a).

Note that our estimate of the Steklov–Dirichlet eigenvalue is quite crude; it would be interesting to establish the precise value of a "critical ratio" $\alpha \in (1, 1.9)$, such that Theorem 1.2.3 holds for spherical containers with $d/c < \alpha$ and fails for $d/c \geq \alpha$.

2.4. The "ice fishing problem". In this section we present another "counterexample" to Theorem 1.2.3, this time for an unbounded domain (cf. Remark 1.2.7). For $d = 3$, let $F' = \{x \in \mathbb{R}^2 : |x| < 1\}$ be the unit disk and set $F = F' \times \{0\}$, $W = \{(x, y) : x \in \mathbb{R}^2, y < 0\}$, and $B = \partial W \setminus F$. Even though the domain W is unbounded, it is well known (see [**25**] or [**16**]) that the eigenvalue problem (1.1.1) considered in the function space

$$\mathcal{K} = \left\{ \int_{F'} u^2(x, 0)\, dx < \infty, \int_W |\nabla u(x, y)|^2\, dx\, dy < \infty \right\}$$

has discrete spectrum satisfying

$$0 = \mu_1 < \mu_2 \leq \mu_3 \leq \ldots \to \infty.$$

A similar statement holds for problem (1.1.2).

Clearly, W does not satisfy the weak John's condition, and numerical calculations show that the assertion of Theorem 1.2.3 does not hold in this case. In fact, by [**25**, Table 2], $\mu_2 \approx 2.7547$ and by [**2**, eq. (2.15)], $\lambda_1 \leq 2\pi/3 \approx 2.094$. Thus, $\mu_2 > \lambda_1$.

We remark that the eigenvalue problem (1.1.1) in this example describes the so-called "ice fishing problem" (see [**16**]). That is, it describes free-fluid oscillations in the lower half-space $W = \{(x, y) : x \in \mathbb{R}^2, y < 0\}$ covered above by ice with an ice hole F.

We conclude by noting that in two dimensions, unbounded domains providing "counterexamples" to Theorem 1.2.3 can be constructed using infinite cylindrical domains; see Remark 4.1.1.

3. Domain monotonicity of mixed Steklov eigenvalues

In this section we sum up some facts regarding domain monotonicity of eigenvalues of mixed Steklov problems. These results are well known and in various forms can be found in the literature (see, for example, [**18**, Section 2.2] and references therein); however, since they are essential for the proofs of Theorem 1.2.4 and Lemma 1.4.1, we present them here in detail. A particular emphasis is made on *strict* domain monotonicity of mixed Steklov eigenvalues.

3.1. Steklov–Dirichlet problem. The eigenvalues of the mixed Steklov–Dirichlet problem satisfy strict domain monotonicity in the following sense:

PROPOSITION 3.1.1. *Let (W, F, B) and (W^*, F^*, B^*) be two triples of sets satisfying the assumptions* (I) *and* (II) *of Section* 1.1, *such that* $W \subset W^*$ *and* $F \subset F^*$. *Let* λ_n *and* λ_n^*, $n = 1, 2, \ldots$, *be the eigenvalues of the problem* (1.1.2) *on* W *and* W^*, *respectively. Then* $\lambda_n \geq \lambda_n^*$ *for all* $n \geq 1$. *If, moreover, either* W *is a proper subset of* W^* *or* F *is a proper subset of* F^*, *then* $\lambda_n > \lambda_n^*$ *for all* $n \geq 1$.

PROOF. First, note that the nonstrict domain monotonicity follows immediately from the variational characterization (1.1.4). Indeed, continuing any test-function for the Steklov–Dirichlet problem on W by zero, one gets a test-function on W^* with the same Rayleigh quotient. Therefore, $\lambda_n \geq \lambda_n^*$.

In order to prove strict monotonicity, we follow the argument presented in an abstract form in [**32**, Theorem 2.3]. First, assume by contradiction that W is a proper subset of W^* and $\lambda_n = \lambda_n^*$ for some $n \geq 1$. Let k be such that

$$\lambda_k^* > \lambda_n = \lambda_n^*. \tag{3.1.2}$$

Consider k triples (W_i, F_i, B_i), $i = 1, \ldots, k$, such that $W = W_1 \subset W_2 \subset \cdots \subset W_k = W^*$ and $F = F_1 \subset F_2 \subset \cdots \subset F_k = F^*$. Assume that W_i and F_i, $i = 1, \ldots, k$, are Lipschitz and all the inclusions $W_i \subset W_{i+1}$, $i = 1, 2, \ldots, k-1$, are proper. By nonstrict domain monotonicity we have $\lambda_n = \lambda_n^{(1)} \geq \lambda_n^{(2)} \geq \cdots \geq \lambda_n^{(k)} = \lambda_n^*$, where $\lambda_n^{(i)}$, $i = 1, \ldots, k$, is the nth eigenvalue of the corresponding Steklov–Dirichlet problem on W_i. Therefore, all the inequalities in the previous formula are equalities. Let $u_n^{(i)}$ be an eigenfunction corresponding to the eigenvalue $\lambda_n^{(i)}$; extending it by zero to $W^* \setminus W_i$, we may consider it as a function on W^*. Clearly, the functions $u_n^{(i)}$, $i = 1, \ldots, k$, are admissible for the variational characterization (1.1.4) for λ_k^*. Let us show that they are all linearly independent. Suppose $\sum_{i=1}^k c_i u_n^{(i)} = 0$ on W^* and $c_k \neq 0$. Then $u_n^{(k)}$ is identically zero on $W^* \setminus W_{k-1}$, and hence by the unique continuation property of harmonic functions $u_n^{(k)} \equiv 0$ on W^*, which is impossible. Therefore, $c_k = 0$. Arguing the same way, we show that all the other coefficients $c_i = 0$. Taking the subspace generated by $u_n^{(1)}, \ldots, u_n^{(k)}$ in the variational characterization of λ_k^*, we obtain $\lambda_k^* \leq \lambda_n$ which contradicts (3.1.2). This completes the proof of strict domain monotonicity in the case of the proper inclusion $W \subset W^*$.

If $W = W^*$ and F is a proper subset of F^*, the proof is analogous. In the construction of auxiliary triples (W, F_i, B_i) we must assume that all the inclusions $F = F_1 \subset F_2 \subset \cdots \subset F_k = F^*$ are proper. In order to prove linear independence of test-functions $u_n^{(1)}, \ldots, u_n^{(k)}$ we must show that if for some $i = 1, \ldots, k$ the function $u_n^{(i)}$ vanishes on $F_i \setminus F_{i-1}$, then it should vanish identically. Indeed, in this case the derivatives of $u_n^{(i)}$ are zero on $F_i \setminus F_{i-1}$ in all directions tangential to F_i. Moreover, since $u_n^{(i)}$ is an eigenfunction of the Steklov–Dirichlet problem on (W, F_i, B_i), its normal derivative also vanishes on $F_i \setminus F_{i-1}$. Therefore, $\nabla u_n^{(i)}$ vanishes on $F_i \setminus F_{i-1}$. Hence, a harmonic function u_n^i vanishes together with its gradient on a set of codimension one, which by [**4**, Section 3] implies that it is identically zero. This completes the proof of the Proposition 3.1.1. □

3.2. Steklov–Neumann problem. For eigenvalues of the Steklov–Neumann problem, domain monotonicity holds in a more restrictive sense than in the Steklov–Dirichlet case: namely, the "free boundary" parts of ∂W and ∂W^* (i.e., the sets F and F^*) must coincide.

PROPOSITION 3.2.1. *Let (W, F, B) and (W^*, F, B^*) be two triples of sets satisfying the assumptions* (I) *and* (II) *of Section* 1.1, *such that W is a proper subset of W^*. Let μ_n and μ_n^*, $n = 1, 2, \ldots$, be the eigenvalues of the problem* (1.1.1) *on W and W^*, respectively. Then $\mu_n < \mu_n^*$ for all $n \geq 2$.*

PROOF. Let $v_1^*, \ldots, v_n^*$ be the first n eigenfunctions of the problem (1.1.1) on W^*, $n \geq 2$. Consider the restrictions $v_1, \ldots, v_n$ of these functions on the domain W. Clearly, they are linearly independent: if some linear combination of $v_1, \ldots, v_n$ vanishes on W, it should vanish on the whole of W^* by the unique continuation property of harmonic functions. Take the subspace generated by $v_1, \ldots, v_n$ and plug it in the variational characterization (1.1.3) for μ_n. Suppose, by contradiction, that there exists an element v of this subspace such that

$$\frac{\int_W |\nabla v(x,y)|^2\, dx\, dy}{\int_F v^2(x,0)\, dx} \geq \mu_n^*. \tag{3.2.2}$$

Let v^* be the extension of v to W^*—that is, the corresponding linear combination of $v_1^*, \ldots, v_n^*$. The denominator of its Rayleigh quotient is exactly the same as in (3.2.2), since the boundaries of the domains W and W^* have the same "free surface" F. At the same time, since $v^*|_W = v$ and $W \subset W^*$, we immediately get

$$\int_{W^*} |\nabla v^*(x,y)|^2\, dx\, dy \geq \int_W |\nabla v(x,y)|^2\, dx\, dy.$$

Comparing this with (3.2.2) and using the variational characterization for μ_n^*, one gets that the inequality above has to be an equality. Therefore, ∇v vanishes on $W^* \setminus W$ and hence v^* is constant everywhere on W^* by the unique continuation property. Therefore, $\mu_n = \mu_n^* = 0$ which is impossible for $n \geq 2$ (note that for $n = 1$ this is indeed true). This completes the proof of Proposition 3.2.1. □

REMARK 3.2.3. Proposition 3.1.1 is not surprising. The similar property holds for Dirichlet eigenvalues of the Laplacian and its proof is exactly the same. On the other hand, Proposition 3.2.1 is somewhat unexpected at first glance since the Neumann eigenvalues of the Laplacian do not have the domain monotonicity property. Even more counterintuitive in this case is the fact that monotonicity holds in the "unusual" direction, namely that smaller sets have smaller eigenvalues.

4. Proofs of the main results

4.1. Proof of Theorem 1.2.4. We develop the idea used in [**18**, Theorem 2.6]. Let $W \subset \mathbb{R}^d$ be a domain satisfying the standard John's condition. Then there exists $L > 0$ such that $W \subset F' \times (-L, 0)$. The result then immediately follows from Propositions 3.2.1 and 3.1.1 combined with the results of Section 2.1, where the assertions of Theorem 1.2.4 have been established for cylindrical domains.

REMARK 4.1.1. If W is an unbounded domain satisfying the standard John's condition (with the sloshing problem being understood in the appropriate sense; see Remark 1.2.7), one has to set $L = \infty$, i.e., consider a semi-infinite cylindrical domain. Note that in two dimensions $\lambda_n = \mu_{n+1}$, $n = 1, 2, \ldots$, for the semi–infinite strip. Hence, in order to prove Theorem 1.2.4, it is necessary to use strict domain monotonicity of eigenvalues for either the Steklov–Dirichlet or the Steklov–Neumann problem. Note that Propositions 3.1.1 and 3.2.1 remain true for mixed Steklov eigenvalues of unbounded domains: one can check that the proofs go through without changes.

The same example also shows that even a slight violation of the standard John's condition may force Theorem 1.2.4 to fail. Indeed, by strict domain monotonicity, an arbitrary enlargement of a semi–infinite cylindrical domain in two dimensions away from the line $\{y = 0\}$ yields $\lambda_n < \mu_{n+1}$.

4.2. Proof of Theorem 1.2.3. The argument presented below is an adaptation of the method introduced in [**8**].

Recall that the eigenfunctions $\{v_n\}_{n=1}^{\infty}$ of the Neumann problem (1.1.1) belong to the Sobolev space $H^1(W)$ and that they may be chosen so that $\{v_n(x,0)\}_{n=1}^{\infty}$ is an orthonormal basis in $L^2(F')$.

Moreover, if we define the Neumann counting function by

$$\Lambda_N(\mu) = \#\{\mu_n : \mu_n \le \mu\},$$

we have

$$\Lambda_N(\mu) = \max\left\{\dim(L) : \frac{\int_W |\nabla v(x,y)|^2\, dx\, dy}{\int_{F'} v^2(x,0)\, dx} \le \mu, \;\; v \in L\right\}, \tag{4.2.1}$$

where the maximum is taken over all linear subspaces L of $H^1(W)$. This follows from the variational principle (1.1.3).

The eigenfunctions of the Steklov–Dirichlet problem (1.1.2), $\{u_n\}_{n=1}^{\infty}$, also belong to $H^1(W)$ and, similarly, u_n may be chosen in such a way that $\{u_n(x,0)\}_{n=1}^{\infty}$ is an orthonormal basis in $L^2(F')$.

As in Section 1.1, we use the notation

$$H_0^1(W,B) = \{u \in H^1(W) : u \equiv 0 \quad \text{on} \quad B\}.$$

For any $\mu \in \mathbb{R}$, let $K_N(\mu)$ be the corresponding eigenspace of the problem (1.1.1) if μ is an eigenvalue, and let $K_N(\mu) = \{0\}$ otherwise.

Denote by $\mathcal{H} = \{(x,y) : x \in \mathbb{R}^{d-1}, \ y < 0\}$ the lower half-space of $\mathbb{R}^d$. The following lemma will be used in the sequel.

LEMMA 4.2.2. *Let $W \subset \mathcal{H}$ be a domain satisfying the assumptions* (I) *and* (II) *of Section* 1.1. *Then*

$$H_0^1(W,B) \cap K_N(\mu) = \{0\}$$

for any $\mu > 0$.

PROOF. Let $v \in H_0^1(W,B) \cap K_N(\mu)$. Consider the function $w : \mathcal{H} \to \mathbb{R}$ defined by

$$w(x,y) = \begin{cases} v(x,y), & (x,y) \in W, \\ 0, & (x,y) \in \mathcal{H} \setminus W. \end{cases}$$

Since $w \in H^1(\mathcal{H})$, for any $\psi \in C_0^\infty(\mathcal{H})$ we have by the Green's formula

$$\int_H \nabla w\, \overline{\nabla \psi}\, dx\, dy = \int_W \nabla v\, \overline{\nabla \psi}\, dx\, dy = \int_{\partial W} \frac{\partial v}{\partial \nu}\, \overline{\psi}\, d\sigma = 0, \tag{4.2.3}$$

where σ is the $(d-1)$-dimensional Lebesgue measure on ∂W. It is used here that $v \subset K_N(\mu)$ is harmonic and $\partial v/\partial \nu$ vanishes on $B \subset \partial W$, while ψ vanishes on $\partial W \setminus B \subset \partial \mathcal{H}$. It is well known that a weakly harmonic function is harmonic, and therefore (4.2.3) implies that $\Delta w \equiv 0$ in $\mathcal{H}$. Since $\mathcal{H} \setminus W$ has a nonempty interior, the relation $w \equiv 0$ on $\mathcal{H} \setminus W$ implies $w \equiv 0$ on $\mathcal{H}$. This completes the proof of the lemma. □

Let us now fix an arbitrary $k \in \mathbb{N}$ and set $\mu = \lambda_k$. Take $U = \operatorname{span}\{u_1, \ldots, u_k\}$, where u_n are the eigenfunctions of the mixed Steklov–Dirichlet problem (1.1.2). We have $U \subset H_0^1(W, B) \subset H^1(W)$ and $\dim(U) = k$. For any $u \in U$, we have

$$\begin{aligned}\int_W |\nabla u(x,y)|^2\,dx\,dy &= \int_{F'} \frac{\partial u}{\partial y}(x,0)u(x,0)\,dx \\ &\leq \mu \int_{F'} u^2(x,0)\,dx.\end{aligned} \tag{4.2.4}$$

The rest of the proof is split into two cases: (i) $d \geq 3$ and (ii) $d = 2$.

Case (i), $d \geq 3$. By Lemma 4.2.2 we get that $U \dot{+} K_N(\mu)$ is a direct sum. Given $\mu > 0$, consider the family of exponential functions

$$\left\{e^{i\omega x}e^{\mu y} : |\omega| = \mu,\ \omega \in \mathbb{R}^{d-1}, \quad (x,y) \in \mathbb{R}^{d-1} \times (-\infty, 0]\right\}.$$

It is well known that these functions are linearly independent. Thus there exists $\omega \in \mathbb{R}^{d-1}$, $|\omega| = \mu$ such that $e^{i\omega x}e^{\mu y}$ does not belong to $U \dot{+} K_N(\mu)$. Set

$$G = U \dot{+} K_N(\mu) \dot{+} \left\{c e^{i\omega x}e^{\mu y} : c \in \mathbb{C}\right\} \subset H^1(W).$$

Since W satisfies the weak John's condition (1.2.2), we have

$$\begin{aligned}\int_W |\nabla(c e^{i\omega x}e^{\mu y})|^2\,dx\,dy &= 2\mu^2|c|^2 \int_W e^{2\mu y}\,dx\,dy \\ &\leq \mu|c|^2|F'| = \mu \int_{F'} |c e^{i\omega x}e^{\mu\cdot 0}|^2\,dx.\end{aligned} \tag{4.2.5}$$

Let $u + v + c e^{i\omega x}e^{\mu y}$ be an element of G, where $u \in U$, $v \in K_N(\mu)$. We have

$$\begin{aligned}&\int_W |\nabla(u(x,y) + v(x,y) + c e^{i\omega x}e^{\mu y})|^2\,dx\,dy \\ &\quad= \int_W |\nabla u(x,y)|^2 + |\nabla v(x,y)|^2 |\nabla(c e^{i\omega x}e^{\mu y})|^2\,dx\,dy \\ &\qquad+ 2\operatorname{Re} \int_W \nabla v(x,y)\overline{\nabla(u(x,y) + c e^{i\omega x}e^{\mu y})} \\ &\qquad+ \nabla(c e^{i\omega x}e^{\mu y})\overline{\nabla u(x,y)}\,dx\,dy = \mathcal{I}_1 + 2\operatorname{Re}\mathcal{I}_2,\end{aligned} \tag{4.2.6}$$

where $\mathcal{I}_1$ and $\mathcal{I}_2$ denote, respectively, the first and the second integral in the right-hand side of (4.2.6). By (4.2.4), (4.2.5), and the definition of $K_N(\mu)$, we have

$$\mathcal{I}_1 \leq \mu \int_{F'} u^2(x,0) + v^2(x,0) + |c e^{i\omega x}e^{\mu\cdot 0}|^2\,dx. \tag{4.2.7}$$

Note that the functions u, v, and $e^{i\omega x}e^{\mu y}$ are harmonic in W. Furthermore, $u \equiv 0$ and $\partial v/\partial\nu \equiv 0$ on B. Hence, integrating by parts, we get

$$\begin{aligned}
\mathcal{I}_2 = &\int_{F'} \frac{\partial v}{\partial y}(x,0)\overline{u(x,0) + ce^{i\omega x}} + \frac{\partial}{\partial y}(ce^{i\omega x}e^{\mu y})\,|_{y=0}\,\overline{u(x,0)}\,dx \\
&+ \int_B \frac{\partial v}{\partial \nu}(x,y)\overline{u(x,y) + ce^{i\omega x}e^{\mu y}} + \left(\frac{\partial}{\partial\nu}(ce^{i\omega x}e^{\mu y})\right)\overline{u(x,y)}\,d\sigma(x,y) \\
&- \int_W \Delta v(x,y)\overline{u(x,y) + ce^{i\omega x}e^{\mu y}} + \Delta(ce^{i\omega x}e^{\mu y})\overline{\nabla u(x,y)}\,dx\,dy
\end{aligned}$$

$$= \mu \int_{F'} v(x,0)\overline{u(x,0) + ce^{i\omega x}} + ce^{i\omega x}\overline{u(x,0)}\,dx. \tag{4.2.8}$$

It follows from (4.2.6), (4.2.7), and (4.2.8) that

$$\begin{aligned}
&\int_W |\nabla(u(x,y) + v(x,y) + ce^{i\omega x}e^{\mu y})|^2\,dx\,dy \\
&\qquad \le \mu \int_{F'} |u(x,0) + v(x,0) + ce^{i\omega x}e^{\mu\cdot 0}|^2\,dx.
\end{aligned}$$

Therefore, from (4.2.1) we have

$$\Lambda_N(\mu) \ge \dim G = k + \dim K_N(\mu) + 1.$$

Since $\mu = \lambda_k$, we get

$$\#\{\mu_n : \mu_n < \mu\} = \Lambda_N(\mu) - \dim K_N(\mu) \ge k + 1,$$

which implies $\mu_{k+1} < \lambda_k$.

Case (ii), $d = 2$. Consider the functions $e^{i\mu x}e^{\mu y}$, $(x,y) \in \mathbb{R} \times (-\infty, 0]$. Note that this function does not belong to U because it does not vanish on B. Set

$$G = U \dot{+} \left\{ce^{i\mu x}e^{\mu y} : c \in \mathbb{C}\right\} \subset H^1(W).$$

By the same estimates as in the case $d \ge 3$, we obtain that for any $u + ce^{i\mu x}e^{\mu y}$, where $u \in U$,

$$\int_W |\nabla(u(x,y) + ce^{i\mu x}e^{\mu y})|^2\,dx\,dy \le \mu \int_{F'} |u(x,0) + ce^{i\mu x}e^{\mu\cdot 0}|^2\,dx.$$

Hence, $\Lambda_N(\mu) \ge \dim\{G\} = k + 1$. Since $\mu = \lambda_k$, we get

$$\#\{\mu_n : \mu_n \le \mu\} = \Lambda_N(\mu) \ge k + 1,$$

which implies $\mu_{k+1} \le \lambda_k$. This completes the proof of Theorem 1.2.3.

REMARK 4.2.9. Theorem 1.2.3 can be viewed as a generalization of the main result of [**8**]. Indeed, in order to obtain the classical inequalities between the Neumann and Dirichlet eigenvalues one has to apply Theorem 1.2.3 to a cylindrical domain of depth L and take $L \to \infty$ (see Section 2.1). Note that unlike the proof of Theorem 1.2.4, the proof of Theorem 1.2.3 uses the methods of [**8**], but not the results themselves.

REMARK 4.2.10. As was pointed out to us by Yu. Safarov, Theorem 1.2.3 can be also proved using the results on the comparison of Dirichlet and Neumann eigenvalues obtained in [**30**]. Indeed, in dimensions $d \ge 3$, instead of applying formulas (4.2.6)–(4.2.8), one can plug in the functions $e^{i\omega x}e^{i\mu y}$ with $|\omega| = \mu$ to [**30**, Remark 1.11(4)] and verify that the mixed Steklov eigenvalues fit the abstract setting considered in [**30**]. Similarly, [**30**, Remark 1.11(3)] can be applied in the case $d = 2$.

REMARK 4.2.11. It is likely that for $d = 2$, the assertion of Theorem 1.2.3 could be replaced by a strict inequality. However, this cannot be proved using our argument.

4.3. Proof of Lemma 1.4.1. Let ϕ be an eigenfunction of the Steklov–Neumann problem (1.1.1) with the eigenvalue μ and let C be a connected component of its nodal set. First, note that $C \cap \partial W \neq \emptyset$. Indeed, otherwise C would enclose a bounded domain, and the harmonic function ϕ would vanish on its boundary, implying $\phi \equiv 0$.

Suppose that C has a nonempty intersection only with the part $\overline{F}$ of the boundary. Consider the domain D bounded by C and $\overline{F}$. Since ϕ does not change sign inside D, the eigenvalue μ is the first eigenvalue of the Steklov–Dirichlet problem in D. By Proposition 3.1.1 we get $\mu > \lambda_1$, where λ_1 is the first Steklov–Dirichlet eigenvalue of W. This is a contradiction to the assumption $\mu \leq \lambda_1$ of the lemma. Hence, $C \cap B_0 \neq \emptyset$.

Suppose now that $d = 2$. Then C is a curve and, by the argument above, one of its ends belongs to the set B_0. Suppose that the other end of C coincides with one of the endpoints of the interval F'. Then ϕ is an eigenfunction of a mixed Dirichlet–Neumann eigenvalue problem on the domain bounded by B and C with the eigenvalue zero. But then $\phi \equiv$ const, hence $\phi \equiv 0$, and we get a contradiction. This completes the proof of Lemma 1.4.1.

Acknowledgments. The authors would like to thank A. Girouard, N. Kuznetsov, and Yu. Safarov for useful discussions.

References

[1] M. S. Agranovich, *On a mixed Poincare–Steklov type spectral problem in a Lipschitz domain*, Russ. J. Math. Phys. 13(3), (2006), 281-290.

[2] R. Bañuelos, T. Kulczycki, *The Cauchy process and the Steklov problem*, J. Funct. Anal. 211 (2004), no. 2, 355–423.

[3] F. Brock, *An isoperimetric inequality for eigenvalues of the Stekloff Problem*, Z. Angew. Math. Mech. 81 (2001), 69-71.

[4] L. Caffarelli, A. Friedman, *Partial regularity of solutions of linear and superlinear elliptic equations*, J. Diff. Equations 60 (1985), 420-433.

[5] E. B. Davies, *Spectral theory and differential operators*, Cambridge University Press, 1995.

[6] B. Dittmar, A. Yu. Solynin, *The mixed Steklov eigenvalue problem and new extremal properties of the Grötzsch ring*, J. Math. Sci. 115(2) (2003), 2119-2134

[7] L. Euler, *Principia motus fluidorum*, Novi Commentarii Academiae Scientiarum Imperialis Petropolitanae, Tom. VI (1761), 271-311.

[8] N. Filonov, *On an inequality between Dirichlet and Neumann eigenvalues for the Laplace operator*, St. Petersburg Math. J. 16(2) (2005), 413-416.

[9] D. W. Fox, J. R. Kuttler, *Sloshing frequencies*, Z. Angew. Math. Phys. 34 (1983), 668–696.

[10] A. Fraser, R. Schoen, *The first Steklov eigenvalue, conformal geometry, and minimal surfaces*, arXiv:0912:5392.

[11] L. Friedlander, *Some inequalities between Dirichlet and Neumann Eigenvalues*. Arch. Rational Mech. Anal. 116 (1991), 153-160.

[12] A. Girouard, I. Polterovich, *On the Hersch–Payne–Schiffer inequalities for Steklov eigenvalues*, Func. Anal. Appl. 44(2) (2010), 106–117.

[13] P. Henrici, B. Troesch, L. Wuytack, *Sloshing frequencies for a half-space with circular or strip-like aperture*, Z. Angew. Math. Phys. 21 (1970), 285-318.

[14] R. Ibrahim, *Liquid sloshing dynamics. Theory and applications*, Cambridge University Press, 2005.
[15] N. D. Kopachevsky, S. G. Krein, *Operator approach to linear problems of hydrodynamics, Vol. 1: Self-adjoint problems for an ideal fluid*, Birkhäuser, 2001.
[16] V. Kozlov, N. Kuznetsov, *The ice-fishing problem: the fundamental sloshing frequency versus geometry of holes*, Math. Meth. Appl. Sci. 27 (2004), 289–312.
[17] V. Kozlov, N. Kuznetsov, O. Motygin, *On the two-dimensional sloshing problem*, Proc. Roy. Soc. London A. 460 (2004), 2587–2603.
[18] T. Kulczycki and N. Kuznetsov, *'High spots' theorems for sloshing problems*, Bull. Lond. Math. Soc. 41(3) (2009), 495–505.
[19] J. R. Kuttler, *A nodal line theorem for the sloshing problem*, SIAM J. Math. Anal. 15 (1984), No. 6, 1234-1237.
[20] N. Kuznetsov, V. Maz'ya, B. Vainberg, *Linear water waves*, Cambridge University Press, 2002.
[21] N. G. Kuznetsov, O. V. Motygin, *The Steklov problem in a half-plane: the dependence of eigenvalues on a piecewise constant coefficient*, J. Math. Sciences 127 (2005), 2429–2445.
[22] H. Lamb, *Hydrodynamics*, Cambridge University Press, 1932.
[23] H. Levine and H. Weinberger, *Inequalities between Dirichlet and Neumann eigenvalues*, Arch. Rational Mech. Anal. 94 (1986), no. 3, 193–208.
[24] P. McIver, *Sloshing frequencies for cylindrical and spherical containers filled to an arbitrary depth*, J. Fluid Mech. 201 (1989), 243–257.
[25] J. W. Miles, *On the eigenvalue problem for fluid sloshing in a half-space*, Z. Angew. Math. Phys. 23 (1972), 861–869.
[26] N. N. Moiseev, *Introduction to the theory of oscillations of liquid-containing bodies*, Advances in Applied Mechanics, Vol. 8, pp. 233–289, Academic Press, New York (1964).
[27] L. E. Payne, *Inequalities for eigenvalues of plates and membranes*, J. Rational Mech. Anal. 4 (1955), 517-529.
[28] Å. Pleijel, Remarks on Courant's nodal line theorem, Comm. Pure Appl. Math. 9 (1956), 543-550.
[29] G. Pólya, *Remarks on the foregoing paper*, J. Math. Phys. 31 (1952), 55-57.
[30] Yu. Safarov, *On the comparison of the Dirichlet and Neumann counting functions*, Spectral Theory of Differential Operators: M. Sh. Birman 80th Anniversary Collection, Amer. Math. Soc. Transl. (2), Vol. 225 (2008), 191-204.
[31] G. Szegö, *Inequalities for certain eigenvalues of a membrane of a given area*, J. Rational Mech. Anal. 3 (1954), 343-356.
[32] W. Welsh, *Monotonicity for eigenvalues of the Schrödinger operator on unbounded domains*, Arch. Rational Mech. Anal. 49 (1972/73), 129-136.

Department of Mathematics, Purdue University, West Lafayette, IN 47907, USA
E-mail address: banuelos@math.purdue.edu

Institute of Mathematics, Polish Academy of Sciences, ul. Kopernika 18, 51-617 Wrocław, Poland and Institute of Mathematics, Technical University of Wrocław, Wybrzeze Wyspianskiego 27, 50-370 Wrocław, Poland
E-mail address: tkulczycki@impan.pl

Département de mathématiques et de statistique, Université de Montréal, CP 6128 succ. Centre-Ville, Montréal, QC H3C 3J7, Canada
E-mail address: iossif@dms.umontreal.ca

Department of Mathematics, University of Illinois at Urbana-Champaign, 1409 W. Green Street, Urbana, IL 61801, USA
E-mail address: siudeja@illinois.edu

Amer. Math. Soc. Transl.
Volume **231**, 2010

The Analog of the Limiting Absorption Principle in Homogenization of Periodic Elliptic Operators

M. Sh. Birman and T. A. Suslina

Dedicated to the memory of Victor Borisovich Lidskii

M. Sh. Birman died on July 2, 2009, while this volume was in preparation.

ABSTRACT. In $L_2(\mathbb{R}^d)$, we consider a matrix periodic elliptic second-order differential operator $\mathcal{A}_\varepsilon$ with rapidly oscillating coefficients (which depend on $\mathbf{x}/\varepsilon$). We obtain an approximation in the operator norm in $L_2(\mathbb{R}^d)$ for the resolvent $(\mathcal{A}_\varepsilon - \zeta I)^{-1}$ in terms of the resolvent of the effective operator with sharp-order error term $O(\varepsilon)$. The dependence of constants in estimates on a point $\zeta \in \mathbb{C} \setminus [0, \infty)$ is given. Combining these estimates with some facts of scattering theory, we prove an analog of the limiting absorption principle for the operator $\mathcal{A}_\varepsilon$. Namely, it is shown that the sandwiched resolvent $\langle \mathbf{x} \rangle^{-1/2-s}(\mathcal{A}_\varepsilon - (\eta \pm i\varepsilon^\sigma)I)^{-1}\langle \mathbf{x} \rangle^{-1/2-s}$ has a limit in the operator norm in L_2 as $\varepsilon \to 0$, provided that $\eta > 0$, $s > 0$, $0 < \sigma < 1/2$.

§0. Introduction

0.1. The operator $\mathcal{A}_\varepsilon$ and the effective operator $\mathcal{A}^0$. In $L_2(\mathbb{R}^d; \mathbb{C}^n)$, we consider a matrix elliptic second-order operator of the form $\mathcal{A} = b(\mathbf{D})^* g(\mathbf{x}) b(\mathbf{D})$. The matrix of coefficients $g(\mathbf{x})$ (of size $m \times m$) is positive definite, bounded, and periodic with respect to some lattice Γ; $b(\mathbf{D})$ is an $(m \times n)$-matrix homogeneous first-order differential operator. It is assumed that $m \geq n$ and that the symbol $b(\boldsymbol{\xi})$ of the operator $b(\mathbf{D})$ is a matrix of rank n.

Let $\varepsilon > 0$ be a parameter. Consider the family of operators

$$\mathcal{A}_\varepsilon = b(\mathbf{D})^* g(\varepsilon^{-1}\mathbf{x}) b(\mathbf{D}).$$

The coefficients of $\mathcal{A}_\varepsilon$ are rapidly oscillating for small ε. In homogenization theory, properties of such operators in the small period limit (as $\varepsilon \to 0$) are studied. A broad literature is devoted to homogenization problems. First of all, we mention

---

2010 *Mathematics Subject Classification.* Primary 35B27.

Key words and phrases. Periodic differential operators, homogenization, effective operator, limiting absorption principle.

This research was supported by RFBR (grant no. 08-01-00209-a).

the books [**BeLP**], [**BaPa**], [**JKO**]. A typical problem of homogenization theory is to study the behavior of solution $\mathbf{u}_\varepsilon$ of the equation

$$\mathcal{A}_\varepsilon \mathbf{u}_\varepsilon + \mathbf{u}_\varepsilon = \mathbf{F}, \quad \mathbf{F} \in L_2(\mathbb{R}^d; \mathbb{C}^n),$$

for small ε. It is known that, as $\varepsilon \to 0$, the solution $\mathbf{u}_\varepsilon$ tends (in some sense) to the solution $\mathbf{u}_0$ of the "homogenized" equation

$$\mathcal{A}^0 \mathbf{u}_0 + \mathbf{u}_0 = \mathbf{F}.$$

Here $\mathcal{A}^0 = b(\mathbf{D})^* g^0 b(\mathbf{D})$ is the so-called *effective* operator, and g^0 is a constant positive matrix called the *effective matrix.* The definition of g^0 is nontrivial; it is given below in Section 5.2. The character of convergence of $\mathbf{u}_\varepsilon$ to $\mathbf{u}_0$ and error estimates are studied.

0.2. Operator-theoretic approach to homogenization. In [**BSu1, BSu2, BSu3, BSu4, BSu5**] a new operator-theoretic approach to homogenization problems was suggested and developed. By this approach, in [**BSu1, BSu2**] it was proved that

$$\|\mathbf{u}_\varepsilon - \mathbf{u}_0\|_{L_2(\mathbb{R}^d)} \leq C\varepsilon \|\mathbf{F}\|_{L_2(\mathbb{R}^d)}, \quad 0 < \varepsilon \leq 1. \tag{0.1}$$

Estimate (0.1) is order-sharp; the constant C is controlled explicitly. Thus, as $\varepsilon \to 0$, the resolvent $(\mathcal{A}_\varepsilon + I)^{-1}$ tends to the resolvent $(\mathcal{A}^0 + I)^{-1}$ of the effective operator in the operator norm in $L_2(\mathbb{R}^d; \mathbb{C}^n)$, and the following estimate is true:

$$\|(\mathcal{A}_\varepsilon + I)^{-1} - (\mathcal{A}^0 + I)^{-1}\|_{L_2(\mathbb{R}^d) \to L_2(\mathbb{R}^d)} \leq C\varepsilon, \quad 0 < \varepsilon \leq 1. \tag{0.2}$$

The main idea of the suggested approach is that the homogenization procedure can be studied as a *spectral threshold effect* at the bottom of the spectrum of a periodic operator $\mathcal{A}$. By the scaling transformation, estimate (0.2) is equivalent to the inequality

$$\|(\mathcal{A} + \varepsilon^2 I)^{-1} - (\mathcal{A}^0 + \varepsilon^2 I)^{-1}\|_{L_2(\mathbb{R}^d) \to L_2(\mathbb{R}^d)} \leq C\varepsilon^{-1}, \quad 0 < \varepsilon \leq 1. \tag{0.3}$$

The bottom of the spectrum of $\mathcal{A}$ is point $\lambda_0 = 0$. The behavior of the operator $(\mathcal{A}+\varepsilon^2 I)^{-1}$ (the resolvent of $\mathcal{A}$ near the bottom of its spectrum) can be described in terms of the threshold characteristics of the operator $\mathcal{A}$ (spectral characteristics at the bottom of the spectrum). For this, the operator $\mathcal{A}$ is decomposed in the direct integral of operators $\mathcal{A}(\mathbf{k})$ acting in $L_2(\Omega; \mathbb{C}^n)$ (where Ω is the cell of the lattice Γ) and depending on the parameter $\mathbf{k} \in \mathbb{R}^d$ (called the quasimomentum). The operator $\mathcal{A}(\mathbf{k})$ is given by the differential expression $\mathcal{A}(\mathbf{k}) = b(\mathbf{D}+\mathbf{k})^* g(\mathbf{x}) b(\mathbf{D}+\mathbf{k})$ with periodic boundary conditions. The problem is reduced to the approximation of the resolvent $(\mathcal{A}(\mathbf{k})+\varepsilon^2 I)^{-1}$. The operator $\mathcal{A}(\mathbf{k})$ depends on $\mathbf{k}$ analytically and has compact resolvent. We distinguish the one-dimensional parameter $t = |\mathbf{k}|$, putting $\mathbf{k} = t\boldsymbol{\theta}$. The operator family $\mathcal{A}(\mathbf{k}) = A(t, \boldsymbol{\theta})$ is studied by methods of the *analytic perturbation theory* (see [**Ka**]) with respect to the one-dimensional parameter t. A good deal of constructions and estimates is done in the framework of an abstract operator-theoretic method. A crucial role is played by the spectral germ $S(\boldsymbol{\theta})$ of the operator family $A(t, \boldsymbol{\theta})$ at $t = 0$. The germ $S(\boldsymbol{\theta})$ is a selfadjoint operator acting in the finite-dimensional space $\mathfrak{N} = \operatorname{Ker} \mathcal{A}(0)$. The germ is defined in terms of the threshold characteristics of the operator $\mathcal{A}(\mathbf{k})$. It is possible to approximate the resolvent $(\mathcal{A}(\mathbf{k}) + \varepsilon^2 I)^{-1}$ in terms of the resolvent of $t^2 S(\boldsymbol{\theta})$, which is a finite rank operator. It turns out that the effective operator has the same germ as $\mathcal{A}(\mathbf{k})$. Then it is possible to pass from approximation in terms of the germ to approximation in terms of the effective operator, which implies the required estimate (0.3).

0.3. The resolvent of $\mathcal{A}_\varepsilon$ in an arbitrary point. In the present paper, we study the resolvent of $\mathcal{A}_\varepsilon$ in an arbitrary point $\zeta \in \mathbb{C} \setminus \mathbb{R}_+$. We prove that

$$\|(\mathcal{A}_\varepsilon - \zeta I)^{-1} - (\mathcal{A}^0 - \zeta I)^{-1}\|_{L_2(\mathbb{R}^d)\to L_2(\mathbb{R}^d)} \leq C(\zeta)\varepsilon, \quad 0 < \varepsilon \leq 1. \tag{0.4}$$

The dependence of the constant $C(\zeta)$ on the parameter ζ is given. The cases $\operatorname{Re}\zeta < 0$ and $\operatorname{Re}\zeta \geq 0$ are essentially different, because the spectrum of $\mathcal{A}_\varepsilon$ lies on the semiaxis $\mathbb{R}_+$ (i.e., on the half-plane $\operatorname{Re}\zeta \geq 0$).

Though estimate of the form (0.4) can be deduced directly from (0.2) (with the help of some resolvent identities), this way gives a rough result concerning the dependence of $C(\zeta)$ on ζ (we mean the behavior of $C(\zeta)$ for large $|\zeta|$). In order to trace the dependence of $C(\zeta)$ on ζ more accurately, we repeat all the steps of the previous scheme. This gives estimate (0.4) with the constant $C(\zeta) = C_4 + C_3(\eta + |\kappa|)^{-1/2}$ for $\zeta = -\eta + i\kappa$, $\eta > 0$.

If $\zeta = \eta + i\kappa$, $\eta \geq 0$, $0 \neq \kappa \in \mathbb{R}$, then the estimate (0.4) is true with the constant $C(\zeta) = C_5 + C_6(\eta+1)^{3/2}\kappa^{-2}$. It follows that

$$\begin{aligned} &\|(\mathcal{A}_\varepsilon - (\eta \pm i\varepsilon^\sigma)I)^{-1} - (\mathcal{A}^0 - (\eta \pm i\varepsilon^\sigma)I)^{-1}\|_{L_2(\mathbb{R}^d)\to L_2(\mathbb{R}^d)} \\ &\quad \leq (C_5 + C_6(\eta+1)^{3/2})\varepsilon^{1-2\sigma}, \quad \eta > 0,\ 0 < \sigma < 1/2,\ 0 < \varepsilon \leq 1. \end{aligned} \tag{0.5}$$

0.4. The analog of the limiting absorption principle for $\mathcal{A}_\varepsilon$. Now we consider the sandwiched resolvent of the effective operator:

$$\mathcal{B}^0(\eta + i\kappa; s) = \langle \mathbf{x}\rangle^{-1/2-s}(\mathcal{A}^0 - (\eta + i\kappa)I)^{-1}\langle \mathbf{x}\rangle^{-1/2-s}, \tag{0.6}$$

where $\langle \mathbf{x}\rangle = (1+|\mathbf{x}|^2)^{1/2}$, $\eta > 0$, $0 \neq \kappa \in \mathbb{R}$, $s > 0$. In scattering theory, the *limiting absorption principle* is known (see, e.g., [**Ku1**], [**Ya1**]). By this principle, there exist limits of the operators (0.6) in the operator norm in $L_2(\mathbb{R}^d; \mathbb{C}^n)$ as $\kappa \to \pm 0$. These limits are denoted by $\mathcal{B}^0_\pm(\eta; s)$. Moreover, we have

$$\|\mathcal{B}^0(\eta + i\kappa; s) - \mathcal{B}^0_\pm(\eta; s)\|_{L_2(\mathbb{R}^d)\to L_2(\mathbb{R}^d)} \leq C(\eta; s)|\kappa|^\vartheta, \quad \eta > 0,\ \pm\kappa > 0, \tag{0.7}$$

where $\vartheta = s$ for $0 < s < 1$, $\vartheta < 1$ for $s = 1$, and $\vartheta = 1$ for $s > 1$.

Using estimates (0.5) and (0.7), one can prove an analog of the limiting absorption principle for the operator $\mathcal{A}_\varepsilon$. Namely, consider the sandwiched resolvent

$$\mathcal{B}^{(\pm)}_\varepsilon(\eta; \sigma; s) = \langle \mathbf{x}\rangle^{-1/2-s}(\mathcal{A}_\varepsilon - (\eta \pm i\varepsilon^\sigma)I)^{-1}\langle \mathbf{x}\rangle^{-1/2-s}, \quad s > 0,\ \eta > 0,\ 0 < \sigma < 1/2. \tag{0.8}$$

From (0.5) and (0.7) it follows that, as $\varepsilon \to 0$, the operators (0.8) tend to $\mathcal{B}^0_\pm(\eta; s)$ in the operator norm in $L_2(\mathbb{R}^d; \mathbb{C}^n)$. We have

$$\begin{aligned} \|\mathcal{B}^{(\pm)}_\varepsilon(\eta; \sigma; s) - \mathcal{B}^0_\pm(\eta; s)\|_{L_2(\mathbb{R}^d)\to L_2(\mathbb{R}^d)} &\leq (C_5 + C_6(\eta+1)^{3/2})\varepsilon^{1-2\sigma} + C(\eta; s)\varepsilon^{\sigma\vartheta}, \\ &\eta > 0,\ 0 < \sigma < 1/2,\ s > 0,\ 0 < \varepsilon \leq 1. \end{aligned}$$

This fact can be viewed as an analog of the limiting absorption principle for the operator $\mathcal{A}_\varepsilon$.

The results of the paper have been announced in [**BSu6**].

0.5. The structure of the paper. In §1 and §2, an operator family of the form $A(t) = X(t)^*X(t)$, where $X(t) = X_0 + tX_1$, is studied by an abstract operator-theoretic method. An important notion of the spectral germ (see discussion in Section 0.2) is introduced in §1. Theorems 2.1 and 2.7 about approximations of the resolvent $(A(t) - \zeta\varepsilon^2 I)^{-1}$ in terms of the spectral germ are main results of the abstract method. A short §3 contains necessary information about lattices and the Gelfand transformation. In §4, we introduce the class of periodic differential operators $\mathcal{A}$ acting in $L_2(\mathbb{R}^d;\mathbb{C}^n)$. The operator $\mathcal{A}$ is decomposed in the direct integral of operators $\mathcal{A}(\mathbf{k})$ acting in $L_2(\Omega;\mathbb{C}^n)$. In §5, the operators $\mathcal{A}(\mathbf{k})$ are involved in the framework of the abstract method; the effective characteristics are introduced. Using the results of the abstract method, we obtain approximations of the resolvent $(\mathcal{A}(\mathbf{k}) - \zeta\varepsilon^2 I)^{-1}$ (Theorems 5.1 and 5.2). In §6, by the Gelfand transformation, we deduce approximations for the resolvent $(\mathcal{A} - \zeta\varepsilon^2 I)^{-1}$ from the results of §5 (see Theorems 6.1 and 6.2). In §7, by the scaling transformation, main results concerning approximations of the resolvent $(\mathcal{A}_\varepsilon - \zeta I)^{-1}$ are deduced from the results of §6 (Theorems 7.1 and 7.3). Finally, in §8, combining the results of §7 and some facts of scattering theory, we obtain the main result of the paper concerning the limiting absorption principle for $\mathcal{A}_\varepsilon$ (see Theorem 8.2).

0.6. Notation. Let $\mathcal{G}$, $\mathcal{G}_*$ be separable Hilbert spaces. Symbols $(\cdot,\cdot)_{\mathcal{G}}$ and $\|\cdot\|_{\mathcal{G}}$ stand for the inner product and the norm in $\mathcal{G}$; the symbol $\|\cdot\|_{\mathcal{G}\to\mathcal{G}_*}$ means the norm of a linear continuous operator from $\mathcal{G}$ to $\mathcal{G}_*$. Sometimes, we omit indices, if this does not lead to confusion. We denote the identity operator in $\mathcal{G}$ by $I = I_{\mathcal{G}}$. If $A : \mathcal{G} \to \mathcal{G}_*$ is a linear operator, the symbol $\operatorname{Dom} A$ stands for the domain of A. If $\mathfrak{N}$ is a subspace of $\mathcal{G}$, then $\mathfrak{N}^\perp := \mathcal{G} \ominus \mathfrak{N}$. If P is the orthogonal projection of $\mathcal{G}$ onto $\mathfrak{N}$, then $P^\perp$ is the orthogonal projection onto $\mathfrak{N}^\perp$. The symbol $\langle\cdot,\cdot\rangle = \langle\cdot,\cdot\rangle_n$ stands for the standard inner product in $\mathbb{C}^n$, and the symbol $|\cdot|$ stands for the norm of a vector in $\mathbb{C}^n$; $\mathbf{1} = \mathbf{1}_n$ is the unit $(n \times n)$-matrix. For an $(n \times n)$-matrix a, the symbol $|a|$ stands for the norm of the matrix a as a linear operator in $\mathbb{C}^n$.

The L_p-classes of $\mathbb{C}^n$-valued functions in a domain $\mathcal{O} \subset \mathbb{R}^d$ are denoted by $L_p(\mathcal{O};\mathbb{C}^n)$, $1 \le p \le \infty$. The Sobolev classes of order s of $\mathbb{C}^n$-valued functions (in a domain $\mathcal{O} \subseteq \mathbb{R}^d$) are denoted by $H^s(\mathcal{O};\mathbb{C}^n)$, $s \in \mathbb{R}$. For $n = 1$ we write simply $L_p(\mathcal{O})$, $H^s(\mathcal{O})$, etc., but sometimes we use such simplified notation also for classes of vector-valued or matrix-valued functions.

Next, $\mathbf{x} = (x^1,\dots,x^d) \in \mathbb{R}^d$, $iD_j = \partial_j = \partial/\partial x^j$, $j = 1,\dots,d$, $\nabla = \operatorname{grad} = (\partial_1,\dots,\partial_d)$, $\mathbf{D} = -i\nabla = (D_1,\dots,D_d)$.

By C, c (probably, with indices and marks) we denote various constants in estimates. By β (with indices) we denote various *absolute constants*.

0.7. Acknowledgement. The authors are sincerely grateful to D. Yafaev for valuable consultations on the limiting absorption principle.

§1. Abstract operator-theoretic method

1.1. A factorized operator family $A(t)$**.** Let $\mathfrak{H}$ and $\mathfrak{H}_*$ be complex separable Hilbert spaces. Let $X_0 : \mathfrak{H} \to \mathfrak{H}_*$ be a densely defined closed operator, and let $X_1 : \mathfrak{H} \to \mathfrak{H}_*$ be a bounded operator. We put

$$X(t) := X_0 + tX_1, \quad \operatorname{Dom} X(t) = \operatorname{Dom} X_0, \quad t \in \mathbb{R}.$$

The family of selfadjoint operators

$$A(t) := X(t)^* X(t), \quad t \in \mathbb{R}, \tag{1.1}$$

in $\mathfrak{H}$ is our main object. The operator (1.1) is generated by the quadratic form

$$a(t)[u,u] = \|X(t)u\|^2_{\mathfrak{H}_*}, \quad u \in \operatorname{Dom} X_0,$$

which is closed in $\mathfrak{H}$.

We put $A(0) = X_0^* X_0 =: A_0$,

$$\mathfrak{N} := \operatorname{Ker} A_0 = \operatorname{Ker} X_0, \quad \mathfrak{N}_* := \operatorname{Ker} X_0^*.$$

By P and P_* we denote the orthogonal projections of $\mathfrak{H}$ onto $\mathfrak{N}$ and of $\mathfrak{H}_*$ onto $\mathfrak{N}_*$, respectively.

It is assumed that the point $\lambda_0 = 0$ is an isolated point in the spectrum of A_0 (an eigenvalue of A_0) and

$$0 < n := \dim \mathfrak{N} < \infty, \quad n \leq n_* := \dim \mathfrak{N}_* \leq \infty.$$

By d^0 we denote the distance from the point $\lambda_0 = 0$ to the rest of the spectrum of A_0. Let $F(t,\rho)$ be the spectral projection of the operator $A(t)$ for the interval $[0,\rho]$. *We fix a number* $\delta > 0$ *such that* $8\delta < d^0$. Then (see [**BSu2**], Chapter 1, (1.3))

$$F(t,\delta) = F(t,3\delta), \quad \operatorname{rank} F(t,\delta) = n, \quad \text{for } |t| \leq t_0 := \delta^{1/2}\|X_1\|^{-1}. \tag{1.2}$$

Below we write $F(t)$ instead of $F(t,\delta)$.

1.2. The operator R. Let $\mathcal{D} := \operatorname{Dom} X_0 \cap \mathfrak{N}^\perp$, and let $u \in \mathfrak{H}_*$. Consider the following equation for a function $\psi \in \mathcal{D}$ (cf. [**BSu2**], Chapter 1, (1.7)):

$$(X_0\psi, X_0\varphi)_{\mathfrak{H}_*} = (u, X_0\varphi)_{\mathfrak{H}_*}, \quad \forall \varphi \in \mathcal{D}. \tag{1.3}$$

There exists a unique solution ψ of the equation (1.3). This solution satisfies the estimate $\|X_0\psi\|_{\mathfrak{H}_*} \leq \|u\|_{\mathfrak{H}_*}$. Now, let

$$\omega \in \mathfrak{N}, \quad u = -X_1\omega;$$

the corresponding solution of the equation (1.3) is denoted by $\psi(\omega)$. We put $\omega_* = X_0\psi(\omega) + X_1\omega$ and introduce the operator R (see [**BSu2**], Chapter 1, Section 1.2):

$$R : \mathfrak{N} \to \mathfrak{N}_*, \quad R\omega = \omega_* = X_0\psi(\omega) + X_1\omega \in \mathfrak{N}_*.$$

Another description of R is given by the formula $R = P_* X_1|_{\mathfrak{N}}$.

1.3. The spectral germ. A selfadjoint operator

$$S = R^* R : \mathfrak{N} \to \mathfrak{N} \tag{1.4}$$

is called the *spectral germ* of the operator family (1.1) at $t = 0$ (see [**BSu2**], Chapter 1, Section 1.3). The operator (1.4) can be represented as $S = P X_1^* P_* X_1|_{\mathfrak{N}}$, whence

$$\|S\|_{\mathfrak{N}\to\mathfrak{N}} \leq \|X_1\|^2. \tag{1.5}$$

The germ S is called *nondegenerate* if $\operatorname{Ker} S = \{0\}$ or, equivalently, $\operatorname{rank} R = n$.

According to the general analytic perturbation theory (see [**Ka**]), for $|t| \leq t_0$ there exist real-analytic functions $\lambda_l(t)$ and real-analytic $\mathfrak{H}$-valued functions $\varphi_l(t)$ such that

$$A(t)\varphi_l(t) = \lambda_l(t)\varphi_l(t), \quad l = 1, \dots, n, \quad |t| \leq t_0.$$

Herewith, the elements $\varphi_l(t)$, $l = 1, \dots, n$, form an orthonormal basis in $F(t)\mathfrak{H}$. For sufficiently small t_* ($\le t_0$) and $|t| \le t_*$ we have convergent power series expansions:

$$\lambda_l(t) = \gamma_l t^2 + \mu_l t^3 + \cdots, \quad \gamma_l \ge 0, \quad \mu_l \in \mathbb{R}, \quad l = 1, \dots, n, \tag{1.6}$$

$$\varphi_l(t) = \omega_l + t\varphi_l^{(1)} + t^2\varphi_l^{(2)} + \cdots, \quad l = 1, \dots, n.$$

The elements $\omega_l = \varphi_l(0)$, $l = 1, \dots, n$, form an orthonormal basis in $\mathfrak{N}$. According to [**BSu2**], Chapter 1, Section 1.6, the numbers γ_l and the elements ω_l, $l = 1, \dots, n$, are *eigenvalues and eigenvectors of the operator* S:

$$S\omega_l = \gamma_l \omega_l, \quad l = 1, \dots, n. \tag{1.7}$$

1.4. Threshold approximations. In [**BSu2**], Chapter 1, Theorems 4.1 and 4.3, it was shown that

$$\|F(t) - P\|_{\mathfrak{H}\to\mathfrak{H}} \le C_1|t|, \quad |t| \le t_0, \quad C_1 = \beta_1\delta^{-1/2}\|X_1\|; \tag{1.8}$$

$$\|A(t)F(t) - t^2SP\|_{\mathfrak{H}\to\mathfrak{H}} \le C_2|t|^3, \quad |t| \le t_0, \quad C_2 = \beta_2\delta^{-1/2}\|X_1\|^3. \tag{1.9}$$

We recall (see Section 0.6) that β (with indices) denotes various *absolute constants.*

§2. Approximation for the resolvent of the operator $A(t)$

2.1. The nondegeneracy of the spectral germ. Assume that for some $c_* > 0$ we have

$$A(t) \ge c_* t^2 I, \quad |t| \le t_0; \tag{2.1}$$

cf. [**BSu2**], Chapter 1, (5.1). This is equivalent to the fact that the eigenvalues $\lambda_l(t)$ of the operator $A(t)$ satisfy the estimates

$$\lambda_l(t) \ge c_* t^2, \quad |t| \le t_0, \quad l = 1, \dots, n.$$

Then, by (1.6), we have $\gamma_l \ge c_*$, $l = 1, \dots, n$. Combining this with (1.7) we see that the *germ* S *is nondegenerate*:

$$S \ge c_* I_{\mathfrak{N}}. \tag{2.2}$$

Note that (1.5) and (2.2) imply the inequality $c_* \le \|X_1\|^2$.

In [**BSu2**], Chapter 1, Theorem 5.5, on the basis of threshold approximations (1.8), (1.9), an approximation for the resolvent $(A(t) + \varepsilon^2 I)^{-1}$ was obtained under additional assumption (2.1). Now we shall obtain approximation for the resolvent $(A(t) - \zeta\varepsilon^2 I)^{-1}$ in an arbitrary point $\zeta \in \mathbb{C} \setminus [0, \infty)$. Herewith, the cases $\operatorname{Re}\zeta \ge 0$ and $\operatorname{Re}\zeta < 0$ are essentially different.

2.2. Approximation of the operator $(A(t) - \zeta\varepsilon^2 I)^{-1}$ for $\operatorname{Re}\zeta < 0$. We start with the simpler case where $\operatorname{Re}\zeta < 0$. We put

$$\zeta = -\eta + i\kappa, \quad \eta > 0, \quad \kappa \in \mathbb{R}. \tag{2.3}$$

THEOREM 2.1. *Let $A(t)$ be an operator family defined in Section* 1.1. *Let $|t| \le t_0$, where t_0 is defined in* (1.2). *Let ζ be of the form* (2.3), *and let P be the orthogonal projection of $\mathfrak{H}$ onto $\mathfrak{N} = \operatorname{Ker} X_0$. Finally, let S be the spectral germ of the operator family $A(t)$ at $t = 0$. Then for $|t| \le t_0$ and $\varepsilon > 0$ we have*

$$\|(A(t) - \zeta\varepsilon^2 I)^{-1} - (t^2 S - \zeta\varepsilon^2 I_{\mathfrak{N}})^{-1}P\|_{\mathfrak{H}\to\mathfrak{H}} \le (3\delta)^{-1} + C_3(\eta + |\kappa|)^{-1/2}\varepsilon^{-1}, \tag{2.4}$$

where

$$C_3 = \sqrt{2}C_1 c_*^{-1/2} + C_2 c_*^{-3/2}. \tag{2.5}$$

PROOF. By (2.1), for $|t| \le t_0$ we have

$$\|(A(t) - \zeta\varepsilon^2 I)^{-1}\|_{\mathfrak{H}\to\mathfrak{H}} \le \max_{\lambda\ge c_* t^2} |\lambda - \zeta\varepsilon^2|^{-1} = \max_{\lambda\ge c_* t^2} ((\lambda+\eta\varepsilon^2)^2 + \kappa^2\varepsilon^4)^{-1/2} \le ((c_* t^2 + \eta\varepsilon^2)^2 + \kappa^2\varepsilon^4)^{-1/2} \le \sqrt{2}(c_* t^2 + \varepsilon^2(\eta + |\kappa|))^{-1}. \tag{2.6}$$

Similarly, by (2.2), we obtain

$$\|(t^2 S - \zeta\varepsilon^2 I_{\mathfrak{N}})^{-1} P\|_{\mathfrak{H}\to\mathfrak{H}} \le \sqrt{2}(c_* t^2 + \varepsilon^2(\eta + |\kappa|))^{-1}. \tag{2.7}$$

The resolvent of $A(t)$ can be represented as the sum of three terms:

$$(A(t) - \zeta\varepsilon^2 I)^{-1} = (A(t) - \zeta\varepsilon^2 I)^{-1} F(t)^\perp + (A(t) - \zeta\varepsilon^2 I)^{-1} F(t) P^\perp + F(t)(A(t) - \zeta\varepsilon^2 I)^{-1} P. \tag{2.8}$$

The first term in the right-hand side of (2.8) is estimated by using (1.2):

$$\|(A(t) - \zeta\varepsilon^2 I)^{-1} F(t)^\perp\|_{\mathfrak{H}\to\mathfrak{H}} \le \max_{\lambda\ge 3\delta} |\lambda - \zeta\varepsilon^2|^{-1} = \max_{\lambda\ge 3\delta} ((\lambda+\eta\varepsilon^2)^2 + \kappa^2\varepsilon^4)^{-1/2} \le (3\delta)^{-1}, \quad |t| \le t_0. \tag{2.9}$$

Combining (1.8), (2.6) with the identity $F(t)P^\perp = F(t)(F(t) - P)$, we estimate the second term:

$$\|(A(t) - \zeta\varepsilon^2 I)^{-1} F(t) P^\perp\|_{\mathfrak{H}\to\mathfrak{H}} \le C_1\sqrt{2}|t|(c_* t^2 + \varepsilon^2(\eta + |\kappa|))^{-1} \le C_1(2c_*(\eta + |\kappa|))^{-1/2}\varepsilon^{-1}, \quad |t| \le t_0. \tag{2.10}$$

Next, the operator $(t^2 S - \zeta\varepsilon^2 I_{\mathfrak{N}})^{-1} P$ can be represented as

$$(t^2 S - \zeta\varepsilon^2 I_{\mathfrak{N}})^{-1} P = F(t)^\perp (t^2 S - \zeta\varepsilon^2 I_{\mathfrak{N}})^{-1} P + F(t)(t^2 S - \zeta\varepsilon^2 I_{\mathfrak{N}})^{-1} P. \tag{2.11}$$

The first term in the right-hand side of (2.11) is estimated by using (2.7), (1.8), and the identity $F(t)^\perp P = -(F(t) - P)P$:

$$\|F(t)^\perp (t^2 S - \zeta\varepsilon^2 I_{\mathfrak{N}})^{-1} P\|_{\mathfrak{H}\to\mathfrak{H}} \le C_1\sqrt{2}|t|(c_* t^2 + \varepsilon^2(\eta + |\kappa|))^{-1} \le C_1(2c_*(\eta + |\kappa|))^{-1/2}\varepsilon^{-1}, \quad |t| \le t_0. \tag{2.12}$$

Now, consider the operator

$$\Xi(t,\zeta,\varepsilon) := F(t)(A(t) - \zeta\varepsilon^2 I)^{-1} P - F(t)(t^2 S - \zeta\varepsilon^2 I_{\mathfrak{N}})^{-1} P = -F(t)(A(t) - \zeta\varepsilon^2 I)^{-1}(A(t)F(t) - t^2 SP)(t^2 S - \zeta\varepsilon^2 I_{\mathfrak{N}})^{-1} P. \tag{2.13}$$

By (1.9), (2.6), and (2.7), we have

$$\|\Xi(t,\zeta,\varepsilon)\|_{\mathfrak{H}\to\mathfrak{H}} \le 2C_2|t|^3(c_* t^2 + \varepsilon^2(\eta + |\kappa|))^{-2} \le C_2 c_*^{-3/2}(\eta + |\kappa|)^{-1/2}\varepsilon^{-1}, \quad |t| \le t_0. \tag{2.14}$$

As a result, relations (2.8)–(2.14) imply the estimate

$$\|(A(t) - \zeta\varepsilon^2 I)^{-1} - (t^2 S - \zeta\varepsilon^2 I_{\mathfrak{N}})^{-1} P\|_{\mathfrak{H}\to\mathfrak{H}} \le (3\delta)^{-1} + 2C_1(2c_*(\eta + |\kappa|))^{-1/2}\varepsilon^{-1} + C_2 c_*^{-3/2}(\eta + |\kappa|)^{-1/2}\varepsilon^{-1}, \quad |t| \le t_0. \tag{2.15}$$

Estimate (2.15) is equivalent to (2.4). $\square$

Theorem 2.1 gives an approximation of the resolvent $(A(t) - \zeta\varepsilon^2 I)^{-1}$ with $\operatorname{Re}\zeta < 0$ for small ε in terms of the finite rank operator $(t^2 S - \zeta\varepsilon^2 I_{\mathfrak{N}})^{-1}P$. Both operators are of order $O(\varepsilon^{-2})$, while their difference is of order $O(\varepsilon^{-1})$. The estimate (2.4) is order-sharp with respect to ε. The coefficient $C_3(\eta + |\kappa|)^{-1/2}$ at ε^{-1} in (2.4) decays as $|\zeta|^{-1/2}$ if $|\zeta| \to \infty$.

2.3. Approximation of the operator $(A(t) - \zeta\varepsilon^2 I)^{-1}$ **for** $\operatorname{Re}\zeta \geq 0$**.** Now we proceed to the case where $\operatorname{Re}\zeta \geq 0$, which requires more attention, since the spectrum of $A(t)$ lies on $\mathbb{R}_+$ (i.e., on the half-plane $\operatorname{Re}\zeta \geq 0$). We put

$$\zeta = \eta + i\kappa, \quad \eta \geq 0, \quad 0 \neq \kappa \in \mathbb{R}, \tag{2.16}$$

and denote

$$\Sigma(t, \zeta, \varepsilon) := (A(t) - \zeta\varepsilon^2 I)^{-1} - (t^2 S - \zeta\varepsilon^2 I_{\mathfrak{N}})^{-1}P. \tag{2.17}$$

PROPOSITION 2.2. *Let ζ be of the form* (2.16), *and let $|t| \leq t_0$. Then we have*

$$\|(A(t) - \zeta\varepsilon^2 I)^{-1}F(t)^\perp\|_{\mathfrak{H}\to\mathfrak{H}} \leq \delta^{-1}, \quad \eta\varepsilon^2 \leq 2\delta, \tag{2.18}$$

$$\|(A(t) - \zeta\varepsilon^2 I)^{-1}F(t)^\perp\|_{\mathfrak{H}\to\mathfrak{H}} \leq |\kappa|^{-1}\eta^{1/2}(2\delta)^{-1/2}\varepsilon^{-1}, \quad \eta\varepsilon^2 > 2\delta. \tag{2.19}$$

PROOF. By (1.2), we have

$$\|(A(t)-\zeta\varepsilon^2 I)^{-1}F(t)^\perp\|_{\mathfrak{H}\to\mathfrak{H}} \leq \max_{\lambda\geq 3\delta}|\lambda-\eta\varepsilon^2-i\kappa\varepsilon^2|^{-1} = \max_{\lambda\geq 3\delta}((\lambda-\eta\varepsilon^2)^2+\kappa^2\varepsilon^4)^{-1/2}. \tag{2.20}$$

If $\eta\varepsilon^2 \leq 2\delta$, then $\lambda - \eta\varepsilon^2 \geq \delta$ in (2.20), which implies (2.18).

If $\eta\varepsilon^2 > 2\delta$, then $\varepsilon^{-1} < \eta^{1/2}(2\delta)^{-1/2}$, and

$$((\lambda - \eta\varepsilon^2)^2 + \kappa^2\varepsilon^4)^{-1/2} \leq |\kappa|^{-1}\varepsilon^{-2} \leq |\kappa|^{-1}\eta^{1/2}(2\delta)^{-1/2}\varepsilon^{-1}.$$

Together with (2.20), this yields (2.19). □

Note that $c_* t^2 \leq \delta$ for $|t| \leq t_0$, since $c_* \leq \|X_1\|^2$ and $t_0 = \delta^{1/2}\|X_1\|$. Further estimates will be done separately for three cases: (1) $\eta\varepsilon^2 \leq c_* t^2/2$, (2) $\eta\varepsilon^2 > 2\delta$, (3) $c_* t^2/2 < \eta\varepsilon^2 \leq 2\delta$.

PROPOSITION 2.3. *Let ζ be as in* (2.16), *and let $|t| \leq t_0$. Then we have*

$$\|F(t)(A(t) - \zeta\varepsilon^2 I)^{-1}P^\perp\|_{\mathfrak{H}\to\mathfrak{H}} \leq \sqrt{2}C_1 c_*^{-1/2}(\eta/2 + |\kappa|)^{-1/2}\varepsilon^{-1}, \quad \eta\varepsilon^2 \leq c_* t^2/2, \tag{2.21}$$

$$\|F(t)(A(t) - \zeta\varepsilon^2 I)^{-1}P^\perp\|_{\mathfrak{H}\to\mathfrak{H}} \leq \delta^{-1}, \quad \eta\varepsilon^2 > 2\delta, \tag{2.22}$$

$$\|F(t)(A(t) - \zeta\varepsilon^2 I)^{-1}P^\perp\|_{\mathfrak{H}\to\mathfrak{H}} \leq C_1 c_*^{-1/2}|\kappa|^{-1}(2\eta)^{1/2}\varepsilon^{-1}, \quad c_* t^2/2 < \eta\varepsilon^2 \leq 2\delta. \tag{2.23}$$

PROOF. Since $F(t)P^\perp = F(t)(F(t) - P)$, by (1.8) we obtain that

$$\|F(t)P^\perp\|_{\mathfrak{H}\to\mathfrak{H}} \leq C_1|t|, \quad |t| \leq t_0. \tag{2.24}$$

By (2.1), we have

$$\|F(t)(A(t) - \zeta\varepsilon^2 I)^{-1}\|_{\mathfrak{H}\to\mathfrak{H}} \leq \max_{c_* t^2 \leq \lambda \leq \delta}((\lambda - \eta\varepsilon^2)^2 + \kappa^2\varepsilon^4)^{-1/2}, \quad |t| \leq t_0. \tag{2.25}$$

If $\eta\varepsilon^2 \le c_* t^2/2$, then $|\lambda - \eta\varepsilon^2| \ge c_* t^2/2$ in (2.25). Hence, by (2.24) and (2.25),

$$\begin{aligned}
&\|F(t)(A(t) - \zeta\varepsilon^2 I)^{-1}P^\perp\|_{\mathfrak{H}\to\mathfrak{H}} \\
&\quad \le C_1|t|(c_*^2 t^4/4 + \kappa^2\varepsilon^4)^{-1/2} \le \sqrt{2}C_1|t|(c_* t^2/2 + |\kappa|\varepsilon^2)^{-1} \\
&\quad \le \sqrt{2}C_1|t|(c_* t^2/4 + (\eta/2 + |\kappa|)\varepsilon^2)^{-1} \le \sqrt{2}C_1 c_*^{-1/2}(\eta/2 + |\kappa|)^{-1/2}\varepsilon^{-1},
\end{aligned}$$

which proves (2.21).

If $\eta\varepsilon^2 > 2\delta$, then $|\lambda - \eta\varepsilon^2| > \delta$ in (2.25), which implies (2.22).

If $c_* t^2/2 < \eta\varepsilon^2 \le 2\delta$, then $|t|\varepsilon^{-1} < (2\eta)^{1/2}c_*^{-1/2}$. Hence, by (2.24) and (2.25),

$$\|F(t)(A(t) - \zeta\varepsilon^2 I)^{-1}P^\perp\|_{\mathfrak{H}\to\mathfrak{H}} \le C_1|t||\kappa|^{-1}\varepsilon^{-2} \le C_1|\kappa|^{-1}c_*^{-1/2}(2\eta)^{1/2}\varepsilon^{-1},$$

which yields (2.23). □

PROPOSITION 2.4. *Let ζ be as in* (2.16), *and let $|t| \le t_0$. Then we have*

(2.26)
$$\|F(t)^\perp(t^2 S - \zeta\varepsilon^2 I_{\mathfrak{N}})^{-1}P\|_{\mathfrak{H}\to\mathfrak{H}} \le \sqrt{2}C_1 c_*^{-1/2}(\eta/2 + |\kappa|)^{-1/2}\varepsilon^{-1}, \quad \eta\varepsilon^2 \le c_* t^2/2,$$

(2.27)
$$\|F(t)^\perp(t^2 S - \zeta\varepsilon^2 I_{\mathfrak{N}})^{-1}P\|_{\mathfrak{H}\to\mathfrak{H}} \le \delta^{-1}, \quad \eta\varepsilon^2 > 2\delta,$$

(2.28)
$$\|F(t)^\perp(t^2 S - \zeta\varepsilon^2 I_{\mathfrak{N}})^{-1}P\|_{\mathfrak{H}\to\mathfrak{H}} \le C_1 c_*^{-1/2}|\kappa|^{-1}(2\eta)^{1/2}\varepsilon^{-1}, \quad c_* t^2/2 < \eta\varepsilon^2 \le 2\delta.$$

PROOF. Since $F(t)^\perp P = -(F(t) - P)P$, it follows from (1.8) that

(2.29)
$$\|F(t)^\perp P\|_{\mathfrak{H}\to\mathfrak{H}} \le C_1|t|, \quad |t| \le t_0.$$

We have

(2.30)
$$\|(t^2 S - \zeta\varepsilon^2 I_{\mathfrak{N}})^{-1}P\|_{\mathfrak{H}\to\mathfrak{H}} \le \max_{1\le l\le n}((\gamma_l t^2 - \eta\varepsilon^2)^2 + \kappa^2\varepsilon^4)^{-1/2}.$$

Since $\gamma_l \ge c_*$, $l = 1, \dots, n$, for $\eta\varepsilon^2 \le c_* t^2/2$ we have $\gamma_l t^2 - \eta\varepsilon^2 \ge c_* t^2/2$. Combining this with (2.29) and (2.30), we arrive at the estimate

$$\begin{aligned}
&\|F(t)^\perp(t^2 S - \zeta\varepsilon^2 I_{\mathfrak{N}})^{-1}P\|_{\mathfrak{H}\to\mathfrak{H}} \le \sqrt{2}C_1|t|(c_* t^2/2 + |\kappa|\varepsilon^2)^{-1} \\
&\quad \le \sqrt{2}C_1|t|(c_* t^2/4 + (\eta/2 + |\kappa|)\varepsilon^2)^{-1} \le \sqrt{2}C_1 c_*^{-1/2}(\eta/2 + |\kappa|)^{-1/2}\varepsilon^{-1},
\end{aligned}$$

which proves (2.26).

By (1.2) and (1.5),

$$\|t^2 SP\|_{\mathfrak{H}\to\mathfrak{H}} \le t^2\|X_1\|^2 \le \delta, \quad |t| \le t_0.$$

Hence,

(2.31)
$$\gamma_l t^2 \le \delta, \quad l = 1, \dots, n, \quad |t| \le t_0.$$

For $\eta\varepsilon^2 > 2\delta$ relations (2.30) and (2.31) directly imply (2.27).

Finally, for $c_* t^2/2 < \eta\varepsilon^2 \le 2\delta$ we have $|t|\varepsilon^{-1} < (2\eta)^{1/2}c_*^{-1/2}$. In this case it follows from (2.29) and (2.30) that

$$\|F(t)^\perp(t^2 S - \zeta\varepsilon^2 I_{\mathfrak{N}})^{-1}P\|_{\mathfrak{H}\to\mathfrak{H}} \le C_1|t||\kappa|^{-1}\varepsilon^{-2} \le C_1 c_*^{-1/2}|\kappa|^{-1}(2\eta)^{1/2}\varepsilon^{-1},$$

which proves (2.28). □

PROPOSITION 2.5. *Let $\Xi(t,\zeta,\varepsilon)$ be the operator* (2.13), *and let ζ be as in* (2.16). *Then for $|t| \le t_0$ we have*

(2.32)
$$\|\Xi(t,\zeta,\varepsilon)\|_{\mathfrak{H}\to\mathfrak{H}} \le 2\sqrt{2}C_2 c_*^{-3/2}(\eta/2 + |\kappa|)^{-1/2}\varepsilon^{-1}, \quad \eta\varepsilon^2 \le c_* t^2/2,$$

$$\|\Xi(t,\zeta,\varepsilon)\|_{\mathfrak{H}\to\mathfrak{H}} \le \beta_2\delta^{-1}, \quad \eta\varepsilon^2 > 2\delta, \tag{2.33}$$

$$\|\Xi(t,\zeta,\varepsilon)\|_{\mathfrak{H}\to\mathfrak{H}} \le C_2 c_*^{-3/2}\kappa^{-2}(2\eta)^{3/2}\varepsilon^{-1}, \quad c_* t^2/2 < \eta\varepsilon^2 \le 2\delta. \tag{2.34}$$

PROOF. We use identity (2.13):

$$\Xi(t,\zeta,\varepsilon) = -F(t)(A(t)-\zeta\varepsilon^2 I)^{-1}(A(t)F(t) - t^2 SP)(t^2 S - \zeta\varepsilon^2 I_{\mathfrak{N}})^{-1}P.$$

By (1.9), (2.25), and (2.30) (taking (2.31) and the inequalities $\gamma_l \ge c_*$ into account), we obtain

$$\|\Xi(t,\zeta,\varepsilon)\|_{\mathfrak{H}\to\mathfrak{H}} \le C_2|t|^3 \max_{c_* t^2 \le \lambda \le \delta} ((\lambda - \eta\varepsilon^2)^2 + \kappa^2\varepsilon^4)^{-1}, \quad |t| \le t_0. \tag{2.35}$$

For $\eta\varepsilon^2 \le c_* t^2/2$ from (2.35) it follows that

$$\begin{aligned}\|\Xi(t,\zeta,\varepsilon)\|_{\mathfrak{H}\to\mathfrak{H}} &\le C_2|t|^3((c_*t^2/2)^2 + \kappa^2\varepsilon^4)^{-1} \le 2C_2 c_*^{-1}|t|((c_*t^2/2)^2 + \kappa^2\varepsilon^4)^{-1/2} \\ &\le 2\sqrt{2}C_2 c_*^{-1}|t|(c_*t^2/4 + (\eta/2 + |\kappa|)\varepsilon^2)^{-1} \le 2\sqrt{2}C_2 c_*^{-3/2}(\eta/2+|\kappa|)^{-1/2}\varepsilon^{-1},\end{aligned}$$

which proves (2.32).

If $\eta\varepsilon^2 > 2\delta$, then (2.35) directly implies the estimate

$$\|\Xi(t,\zeta,\varepsilon)\|_{\mathfrak{H}\to\mathfrak{H}} \le C_2 t_0^3\delta^{-2}, \quad |t| \le t_0.$$

Recalling expressions for t_0 and C_2 (see (1.2), (1.9)), we arrive at (2.33).

For $c_*t^2/2 < \eta\varepsilon^2 \le 2\delta$ we have $|t|\varepsilon^{-1} \le (2\eta)^{1/2}c_*^{-1/2}$. Therefore, by (2.35),

$$\|\Xi(t,\zeta,\varepsilon)\|_{\mathfrak{H}\to\mathfrak{H}} \le C_2|t|^3\kappa^{-2}\varepsilon^{-4} \le C_2 c_*^{-3/2}(2\eta)^{3/2}\kappa^{-2}\varepsilon^{-1},$$

which proves (2.34). □

Let us summarize the results. The operator (2.17) can be represented as

$$\begin{aligned}\Sigma(t,\zeta,\varepsilon) =& (A(t)-\zeta\varepsilon^2 I)^{-1}F(t)^\perp + F(t)(A(t)-\zeta\varepsilon^2 I)^{-1}P^\perp \\ &- F(t)^\perp(t^2 S - \zeta\varepsilon^2 I_{\mathfrak{N}})^{-1}P + \Xi(t,\zeta,\varepsilon),\end{aligned} \tag{2.36}$$

where $\Xi(t,\zeta,\varepsilon)$ is the operator (2.13). By identity (2.36), Propositions 2.2, 2.3, 2.4, and 2.5 directly imply the following result.

THEOREM 2.6. *Let $A(t)$ be the operator family defined in Section 1.1. Let $|t| \le t_0$, where t_0 is defined in (1.2). Let ζ be as in (2.16), and let P be the orthogonal projection of $\mathfrak{H}$ onto $\mathfrak{N} = \operatorname{Ker} X_0$. Finally, let S be the spectral germ of the operator family $A(t)$ at $t = 0$. Let $\Sigma(t,\zeta,\varepsilon)$ be the operator (2.17). Then we have*

$$\|\Sigma(t,\zeta,\varepsilon)\|_{\mathfrak{H}\to\mathfrak{H}} \le \delta^{-1} + 2\sqrt{2}(C_1 c_*^{-1/2} + C_2 c_*^{-3/2})(\eta/2+|\kappa|)^{-1/2}\varepsilon^{-1}, \quad \eta\varepsilon^2 \le c_* t^2/2, \tag{2.37}$$

$$\|\Sigma(t,\zeta,\varepsilon)\|_{\mathfrak{H}\to\mathfrak{H}} \le (2\delta)^{-1/2}|\kappa|^{-1}\eta^{1/2}\varepsilon^{-1} + (2+\beta_2)\delta^{-1}, \quad \eta\varepsilon^2 > 2\delta, \tag{2.38}$$

$$\begin{aligned}\|\Sigma(t,\zeta,\varepsilon)\|_{\mathfrak{H}\to\mathfrak{H}} \le \delta^{-1} + (2C_1 c_*^{-1/2}|\kappa|^{-1}(2\eta)^{1/2} + C_2 c_*^{-3/2}\kappa^{-2}(2\eta)^{3/2})\varepsilon^{-1}, \\ c_* t^2/2 < \eta\varepsilon^2 \le 2\delta.\end{aligned} \tag{2.39}$$

Roughly estimating and combining (2.37)–(2.39), we arrive at the following result.

THEOREM 2.7. *Under the assumptions of Theorem 2.6, for $|t| \le t_0$ we have*

$$\|\Sigma(t,\zeta,\varepsilon)\|_{\mathfrak{H}\to\mathfrak{H}} \le \widehat{C}_1 + \varepsilon^{-1}\left(\widehat{C}_2(\eta+|\kappa|)^{-1/2} + \widehat{C}_3|\kappa|^{-1}\eta^{1/2} + \widehat{C}_4\kappa^{-2}\eta^{3/2}\right),$$

where

$$\widehat{C}_1 = (2+\beta_2)\delta^{-1}, \tag{2.40}$$

$$\widehat{C}_2 = 4(C_1 c_*^{-1/2} + C_2 c_*^{-3/2}), \tag{2.41}$$

$$\widehat{C}_3 = \max\{(2\delta)^{-1/2}, 2\sqrt{2}C_1 c_*^{-1/2}\}, \tag{2.42}$$

$$\widehat{C}_4 = 2\sqrt{2}C_2 c_*^{-3/2}. \tag{2.43}$$

Theorems 2.6 and 2.7 give approximations for the resolvent $(A(t)-\zeta\varepsilon^2 I)^{-1}$ with $\operatorname{Re}\zeta \geq 0$ for small ε in terms of the finite rank operator $(t^2 S - \zeta\varepsilon^2 I_{\mathfrak{N}})^{-1}P$. The estimates given by Theorems 2.6 and 2.7 are order-sharp with respect to ε. We see that in these estimates the dependence of the coefficient at ε^{-1} on $\zeta = \eta + i\kappa$ is more complicated than for the case $\operatorname{Re}\zeta < 0$. In particular, this coefficient grows as $O(\eta^{3/2})$ if $\eta \to +\infty$ and $\kappa \neq 0$ is fixed, and it grows as $|\kappa|^{-2}$ if $\kappa \to 0$ and η is fixed.

§3. Lattices in $\mathbb{R}^d$. The Gelfand transformation

3.1. Lattices. Let $\mathbf{a}_1, \dots, \mathbf{a}_d$ be a basis in $\mathbb{R}^d$ generating the lattice

$$\Gamma = \{\mathbf{a} = n_1\mathbf{a}_1 + \dots + n_d\mathbf{a}_d, \quad n_j \in \mathbb{Z}, \ j = 1, \dots, d\}.$$

By $\Omega \subset \mathbb{R}^d$ we denote the cell of the lattice Γ:

$$\Omega = \{\mathbf{x} = \tau_1\mathbf{a}_1 + \dots + \tau_d\mathbf{a}_d, \quad \tau_j \in (0,1), \ j = 1, \dots, d\}.$$

The basis $\mathbf{b}_1, \dots, \mathbf{b}_d$ in $\mathbb{R}^d$ dual to the basis $\mathbf{a}_1, \dots, \mathbf{a}_d$ is defined by the relations $\langle \mathbf{b}_j, \mathbf{a}_l \rangle = 2\pi\delta_{jl}$. We introduce the lattice $\widetilde{\Gamma}$ dual to Γ:

$$\widetilde{\Gamma} = \{\mathbf{b} = m_1\mathbf{b}_1 + \dots + m_d\mathbf{b}_d, \quad m_j \in \mathbb{Z}, \ j = 1, \dots, d\}. \tag{3.1}$$

By $\widetilde{\Omega}$ we denote the Brillouin zone of the lattice (3.1):

$$\widetilde{\Omega} = \{\mathbf{k} \in \mathbb{R}^d : \ |\mathbf{k}| < |\mathbf{k} - \mathbf{b}|, \ 0 \neq \mathbf{b} \in \widetilde{\Gamma}\}.$$

Let r_0 be the radius of the ball inscribed in $\operatorname{clos}\widetilde{\Omega}$.

By $\widetilde{H}^s(\Omega; \mathbb{C}^n)$ we denote the subspace formed by the functions in the Sobolev space $H^s(\Omega; \mathbb{C}^n)$ whose Γ-periodic extension to $\mathbb{R}^d$ belongs to $H^s_{\mathrm{loc}}(\mathbb{R}^d; \mathbb{C}^n)$.

3.2. The Gelfand transformation $\mathcal{V}$. First, $\mathcal{V}$ is defined on the functions of the Schwartz class $\mathbf{u} \in \mathcal{S}(\mathbb{R}^d; \mathbb{C}^n)$ by the formula

$$\widetilde{\mathbf{u}}(\mathbf{k}, \mathbf{x}) = (\mathcal{V}\mathbf{u})(\mathbf{k}, \mathbf{x}) = |\widetilde{\Omega}|^{-1/2} \sum_{\mathbf{a} \in \Gamma} \exp(-i\langle \mathbf{k}, \mathbf{x} + \mathbf{a} \rangle)\mathbf{u}(\mathbf{x} + \mathbf{a}), \quad \mathbf{x} \in \Omega, \quad \mathbf{k} \in \widetilde{\Omega}.$$

Herewith,

$$\int_{\widetilde{\Omega}} \int_{\Omega} |\widetilde{\mathbf{u}}(\mathbf{k}, \mathbf{x})|^2 \, d\mathbf{x} d\mathbf{k} = \int_{\mathbb{R}^d} |\mathbf{u}(\mathbf{x})|^2 \, d\mathbf{x}, \quad \widetilde{\mathbf{u}} = \mathcal{V}\mathbf{u},$$

and $\mathcal{V}$ extends by continuity to a *unitary* mapping

$$\mathcal{V} : L_2(\mathbb{R}^d; \mathbb{C}^n) \to \int_{\widetilde{\Omega}} \bigoplus L_2(\Omega; \mathbb{C}^n) \, d\mathbf{k} =: \mathcal{K}.$$

Inclusion $\mathbf{u} \in H^1(\mathbb{R}^d;\mathbb{C}^n)$ is equivalent to the fact that $\widetilde{\mathbf{u}}(\mathbf{k},\cdot) \in \widetilde{H}^1(\Omega;\mathbb{C}^n)$ for a.e. $\mathbf{k} \in \widetilde{\Omega}$ and

$$\int_{\widetilde{\Omega}} \int_{\Omega} \left(|(\mathbf{D}+\mathbf{k})\widetilde{\mathbf{u}}(\mathbf{k},\mathbf{x})|^2 + |\widetilde{\mathbf{u}}(\mathbf{k},\mathbf{x})|^2\right) d\mathbf{x}\, d\mathbf{k} < \infty.$$

§4. Periodic differential operators in $L_2(\mathbb{R}^d;\mathbb{C}^n)$. Direct integral decomposition

4.1. Factorized second-order operators. Let m,n be natural numbers such that $m \geq n$. Assume that $g(\mathbf{x})$ is a measurable $(m\times m)$-matrix-valued function in $\mathbb{R}^d$ which is bounded and positive definite:

$$g, g^{-1} \in L_\infty, \quad g(\mathbf{x}) > 0. \tag{4.1}$$

Assume also that $g(\mathbf{x})$ is periodic with respect to the lattice Γ.

Let $b(\mathbf{D}) : L_2(\mathbb{R}^d;\mathbb{C}^n) \to L_2(\mathbb{R}^d;\mathbb{C}^m)$ be a homogeneous first-order differential operator with constant coefficients. The symbol $b(\boldsymbol{\xi})$, $\boldsymbol{\xi} \in \mathbb{R}^d$, of the operator $b(\mathbf{D})$ is a linear homogeneous $(m \times n)$-matrix-valued function of $\boldsymbol{\xi}$. We assume that

$$\operatorname{rank} b(\boldsymbol{\xi}) = n, \quad 0 \neq \boldsymbol{\xi} \in \mathbb{R}^d. \tag{4.2}$$

Condition (4.2) is equivalent to the inequalities

$$\alpha_0 \mathbf{1}_n \leq b(\boldsymbol{\theta})^* b(\boldsymbol{\theta}) \leq \alpha_1 \mathbf{1}_n, \quad \boldsymbol{\theta} \in \mathbb{S}^{d-1}, \quad 0 < \alpha_0 \leq \alpha_1 < \infty, \tag{4.3}$$

with some positive constants α_0, α_1.

In $L_2(\mathbb{R}^d;\mathbb{C}^n)$, we consider the operator $\mathcal{A}$ formally given by

$$\mathcal{A} = b(\mathbf{D})^* g(\mathbf{x}) b(\mathbf{D}). \tag{4.4}$$

More precisely, $\mathcal{A}$ is the selfadjoint operator in $L_2(\mathbb{R}^d;\mathbb{C}^n)$ generated by the closed quadratic form

$$a[\mathbf{u},\mathbf{u}] = \int_{\mathbb{R}^d} \langle g(\mathbf{x}) b(\mathbf{D})\mathbf{u}, b(\mathbf{D})\mathbf{u}\rangle\, d\mathbf{x}, \quad \mathbf{u} \in H^1(\mathbb{R}^d;\mathbb{C}^n).$$

Using the Fourier transformation and conditions (4.1), (4.3), it is easily seen that

$$c_0 \int_{\mathbb{R}^d} |\mathbf{D}\mathbf{u}|^2\, d\mathbf{x} \leq a[\mathbf{u},\mathbf{u}] \leq c_1 \int_{\mathbb{R}^d} |\mathbf{D}\mathbf{u}|^2\, d\mathbf{x}, \quad \mathbf{u} \in H^1(\mathbb{R}^d;\mathbb{C}^n),$$

where

$$c_0 = \alpha_0 \|g^{-1}\|^{-1}_{L_\infty}, \quad c_1 = \alpha_1 \|g\|_{L_\infty}. \tag{4.5}$$

The operator $\mathcal{A}$ can be written in a factorized form: $\mathcal{A} = \mathcal{X}^*\mathcal{X}$, where the operator $\mathcal{X} : L_2(\mathbb{R}^d;\mathbb{C}^n) \to L_2(\mathbb{R}^d;\mathbb{C}^m)$ is given by the differential expression $\mathcal{X} = g(\mathbf{x})^{1/2} b(\mathbf{D})$ on the domain $\operatorname{Dom} \mathcal{X} = H^1(\mathbb{R}^d;\mathbb{C}^n)$.

4.2. Operators $\mathcal{A}(\mathbf{k})$. In $L_2(\Omega;\mathbb{C}^n)$, we consider the family of operators $\mathcal{A}(\mathbf{k})$ depending on the parameter $\mathbf{k}\in\mathbb{R}^d$ called *quasimomentum*. The operator $\mathcal{A}(\mathbf{k})$ is formally given by the differential expression

$$\mathcal{A}(\mathbf{k}) = b(\mathbf{D}+\mathbf{k})^* g(\mathbf{x}) b(\mathbf{D}+\mathbf{k}) \tag{4.6}$$

with periodic boundary conditions. Precisely, $\mathcal{A}(\mathbf{k})$ is the selfadjoint operator in $L_2(\Omega;\mathbb{C}^n)$ generated by the closed nonnegative quadratic form

$$a(\mathbf{k})[\mathbf{u},\mathbf{u}] = \int_\Omega \langle g(\mathbf{x}) b(\mathbf{D}+\mathbf{k})\mathbf{u}, b(\mathbf{D}+\mathbf{k})\mathbf{u}\rangle\, d\mathbf{x}, \quad \mathbf{u}\in\widetilde{H}^1(\Omega;\mathbb{C}^n).$$

Using the Fourier series expansion for $\mathbf{u}$ and (4.1), (4.3), it is easily seen that

$$c_0 \int_\Omega |(\mathbf{D}+\mathbf{k})\mathbf{u}|^2\, d\mathbf{x} \le a(\mathbf{k})[\mathbf{u},\mathbf{u}] \le c_1 \int_\Omega |(\mathbf{D}+\mathbf{k})\mathbf{u}|^2\, d\mathbf{x}, \quad \mathbf{u}\in\widetilde{H}^1(\Omega;\mathbb{C}^n), \tag{4.7}$$

where c_0, c_1 are defined by (4.5).

The operator $\mathcal{A}(\mathbf{k})$ can be written in a factorized form

$$\mathcal{A}(\mathbf{k}) = \mathcal{X}(\mathbf{k})^*\mathcal{X}(\mathbf{k}), \tag{4.8}$$

where the operator $\mathcal{X}(\mathbf{k}) : L_2(\Omega;\mathbb{C}^n)\to L_2(\Omega;\mathbb{C}^m)$ is defined as follows:

$$\mathcal{X}(\mathbf{k}) = g(\mathbf{x})^{1/2} b(\mathbf{D}+\mathbf{k}), \quad \operatorname{Dom}\mathcal{X}(\mathbf{k}) = \widetilde{H}^1(\Omega;\mathbb{C}^n). \tag{4.9}$$

From (4.7) and compactness of the embedding of $\widetilde{H}^1(\Omega;\mathbb{C}^n)$ in $L_2(\Omega;\mathbb{C}^n)$ it follows that the spectrum of $\mathcal{A}(\mathbf{k})$ is discrete. The resolvent of $\mathcal{A}(\mathbf{k})$ is compact and depends on $\mathbf{k}\in\mathbb{R}^d$ continuously (in the operator norm). We put

$$\mathfrak{N} := \operatorname{Ker}\mathcal{A}(0) = \operatorname{Ker}\mathcal{X}(0). \tag{4.10}$$

Relations (4.7) with $\mathbf{k}=0$ show that the kernel (4.10) consists of constant vector-valued functions

$$\mathfrak{N} = \{\mathbf{u}\in L_2(\Omega;\mathbb{C}^n) : \ \mathbf{u}=\mathbf{c}\in\mathbb{C}^n\}, \quad \dim\mathfrak{N} = n. \tag{4.11}$$

From the lower estimate (4.7) by using the Fourier series expansion, it easily follows that

$$\mathcal{A}(\mathbf{k}) \ge c_*|\mathbf{k}|^2 I, \quad \mathbf{k}\in\widetilde{\Omega}, \quad c_* = c_0 = \alpha_0\|g^{-1}\|^{-1}_{L_\infty}. \tag{4.12}$$

Besides, using (4.7), it is easy to estimate the distance d^0 from the point $\lambda_0 = 0$ to the rest of the spectrum of $\mathcal{A}(0)$ (see [**BSu2**], Chapter 2, (3.6)):

$$d^0 \ge 4c_* r_0^2. \tag{4.13}$$

4.3. Direct integral decomposition for $\mathcal{A}$. By the Gelfand transformation, the operator $\mathcal{A}$ is decomposed in the direct integral of operators $\mathcal{A}(\mathbf{k})$:

$$\mathcal{V}\mathcal{A}\mathcal{V}^{-1} = \int_{\widetilde{\Omega}} \oplus \mathcal{A}(\mathbf{k})\, d\mathbf{k}. \tag{4.14}$$

In more detail, this means the following. Let $\widetilde{\mathbf{u}} = \mathcal{V}\mathbf{u}$, $\mathbf{u}\in\operatorname{Dom} a = H^1(\mathbb{R}^d;\mathbb{C}^n)$. Then

$$\widetilde{\mathbf{u}}(\mathbf{k},\cdot)\in\operatorname{Dom} a(\mathbf{k}) = \widetilde{H}^1(\Omega;\mathbb{C}^n) \quad \text{for a.e. } \mathbf{k}\in\widetilde{\Omega}, \tag{4.15}$$

$$a[\mathbf{u},\mathbf{u}] = \int_{\widetilde{\Omega}} a(\mathbf{k})[\widetilde{\mathbf{u}}(\mathbf{k},\cdot),\widetilde{\mathbf{u}}(\mathbf{k},\cdot)]\, d\mathbf{k}. \tag{4.16}$$

Conversely, if $\widetilde{\mathbf{u}} \in \mathcal{K}$ satisfies (4.15) and the integral in (4.16) is finite, then $\mathbf{u} \in \operatorname{Dom} a$ and (4.16) is true.

§5. Incorporation of the operators $\mathcal{A}(\mathbf{k})$ in the abstract scheme. Effective characteristics. Approximation of the resolvent

5.1. Incorporation of $\mathcal{A}(\mathbf{k})$ in the abstract scheme. For $\mathbf{k} \in \mathbb{R}^d$ we put

$$\mathbf{k} = t\boldsymbol{\theta}, \quad t = |\mathbf{k}|, \quad \boldsymbol{\theta} \in \mathbb{S}^{d-1},$$

and view t as the main perturbation parameter. We shall apply the method of §1. Herewith, all the objects will depend on the additional parameter $\boldsymbol{\theta}$, and we have to make our constructions and estimates uniform in $\boldsymbol{\theta}$. We put

$$\mathfrak{H} = L_2(\Omega; \mathbb{C}^n), \quad \mathfrak{H}_* = L_2(\Omega; \mathbb{C}^m).$$

The role of $X(t)$ is played by the operator $X(t;\boldsymbol{\theta}) = \mathcal{X}(t\boldsymbol{\theta})$. According to (4.9), $X(t;\boldsymbol{\theta}) = X_0 + tX_1(\boldsymbol{\theta})$, where $X_0 = g^{1/2}b(\mathbf{D})$, $\operatorname{Dom} X_0 = \widetilde{H}^1(\Omega;\mathbb{C}^n)$, and $X_1(\boldsymbol{\theta})$ is a bounded operator of multiplication by the matrix $g(\mathbf{x})^{1/2}b(\boldsymbol{\theta})$. Next, the role of $A(t)$ is played by the operator $A(t;\boldsymbol{\theta}) = \mathcal{A}(t\boldsymbol{\theta})$. By (4.8), we have

$$A(t;\boldsymbol{\theta}) = X(t;\boldsymbol{\theta})^* X(t,\boldsymbol{\theta}).$$

According to (4.10), (4.11), the kernel $\mathfrak{N} = \operatorname{Ker} X_0 = \operatorname{Ker} \mathcal{A}(0)$ is n-dimensional and consists of constant vector-valued functions. The orthogonal projection P of $\mathfrak{H}$ onto $\mathfrak{N}$ is the operator of averaging over the cell Ω:

$$P\mathbf{u} = |\Omega|^{-1} \int_\Omega \mathbf{u}(\mathbf{x})\, d\mathbf{x}. \tag{5.1}$$

The number d^0 is estimated in (4.13). The condition $n \le n_* = \dim \operatorname{Ker} X_0^*$ is also satisfied (see [**BSu2**], Chapter 2, §3). Herewith, either $n_* = n$ (if $m = n$) or $n_* = \infty$ (if $m > n$). Thus, all assumptions of the abstract scheme (see Section 1.1) are satisfied.

According to Section 1.1, we should fix a positive number δ such that $8\delta < d^0$. Taking (4.13) into account, we put

$$\delta = \frac{1}{4} c_* r_0^2 = \frac{1}{4} \alpha_0 \|g^{-1}\|_{L_\infty}^{-1} r_0^2. \tag{5.2}$$

Clearly, the operator $X_1(\boldsymbol{\theta})$ satisfies the estimate

$$\|X_1(\boldsymbol{\theta})\| \le \|g\|_{L_\infty}^{1/2} \alpha_1^{1/2}, \tag{5.3}$$

where α_1 is the constant from (4.3). Using (5.2) and (5.3), instead of the precise value $t_0(\boldsymbol{\theta}) = \delta^{1/2}\|X_1(\boldsymbol{\theta})\|^{-1}$ (see (1.2)), we take a smaller number

$$t_0 = \delta^{1/2} \alpha_1^{-1/2} \|g\|_{L_\infty}^{-1/2} = \frac{1}{2} \alpha_0^{1/2} \alpha_1^{-1/2} r_0 \|g\|_{L_\infty}^{-1/2} \|g^{-1}\|_{L_\infty}^{-1/2}, \tag{5.4}$$

which does not depend on $\boldsymbol{\theta}$. Obviously, we have $t_0 \le r_0/2$.

5.2. The spectral germ $S(\boldsymbol{\theta})$. The effective matrix. The spectral germ of the family $A(t;\boldsymbol{\theta})$ depends on $\boldsymbol{\theta}$. As has been shown in [**BSu2**], Chapter 3, §1, the germ $S(\boldsymbol{\theta})$ acts in the subspace $\mathfrak{N}$ (see (4.11)) and admits the representation

$$S(\boldsymbol{\theta}) = b(\boldsymbol{\theta})^* g^0 b(\boldsymbol{\theta}), \quad \boldsymbol{\theta} \in \mathbb{S}^{d-1}, \tag{5.5}$$

where g^0 is a constant positive $(m \times m)$-matrix called the *effective matrix*.

In order to define g^0, we introduce an $(n \times m)$-matrix-valued function $\Lambda(\mathbf{x})$ which is the Γ-periodic solution of the problem

$$b(\mathbf{D})^* g(\mathbf{x})(b(\mathbf{D})\Lambda(\mathbf{x}) + \mathbf{1}_m) = 0.$$

Then

$$g^0 = |\Omega|^{-1} \int_\Omega g(\mathbf{x})(b(\mathbf{D})\Lambda(\mathbf{x}) + \mathbf{1}_m)\, d\mathbf{x}. \tag{5.6}$$

The matrix (5.6) is positive definite and satisfies the estimates

$$\underline{g} \le g^0 \le \overline{g}, \tag{5.7}$$

where

$$\overline{g} := |\Omega|^{-1} \int_\Omega g(\mathbf{x})\, d\mathbf{x}, \quad \underline{g}^{-1} := |\Omega|^{-1} \int_\Omega g(\mathbf{x})^{-1}\, d\mathbf{x}.$$

If $m = n$, then the effective matrix g^0 coincides with $\underline{g}$: $g^0 = \underline{g}$. (See [**BSu2**], Chapter 3, Theorem 1.5.) Estimates (5.7) are known in homogenization theory as the Voight-Reuss bracketing.

According to (2.2), the germ $S(\boldsymbol{\theta})$ satisfies the estimate

$$S(\boldsymbol{\theta}) = b(\boldsymbol{\theta})^* g^0 b(\boldsymbol{\theta}) \ge c_* \mathbf{1}_n. \tag{5.8}$$

We introduce the operator

$$\mathcal{A}^0 = b(\mathbf{D})^* g^0 b(\mathbf{D}), \tag{5.9}$$

called the *effective operator* for $\mathcal{A}$. The corresponding operator family $\mathcal{A}^0(\mathbf{k})$ in $L_2(\Omega;\mathbb{C}^n)$ is given by the expression

$$\mathcal{A}^0(\mathbf{k}) = b(\mathbf{D}+\mathbf{k})^* g^0 b(\mathbf{D}+\mathbf{k}) \tag{5.10}$$

with periodic boundary conditions. Using the Fourier series and (5.8), it is easily seen that

$$\mathcal{A}^0(\mathbf{k}) \ge c_* |\mathbf{k}|^2 I, \quad \mathbf{k} \in \widetilde{\Omega}. \tag{5.11}$$

5.3. Application of the abstract results to $\mathcal{A}(\mathbf{k})$. We shall apply Theorem 2.1 and Theorem 2.7 to the operator family $A(t;\boldsymbol{\theta}) = \mathcal{A}(\mathbf{k})$. Now the operator (2.17) takes the form

$$\Sigma(t,\boldsymbol{\theta};\zeta,\varepsilon) = (A(t;\boldsymbol{\theta}) - \zeta\varepsilon^2 I)^{-1} - (t^2 S(\boldsymbol{\theta}) - \zeta\varepsilon^2 I_{\mathfrak{N}})^{-1} P. \tag{5.12}$$

By (5.5), due to the presence of projection P (see (5.1)), it is clear that the operator $(t^2 S(\boldsymbol{\theta}) - \zeta\varepsilon^2 I_{\mathfrak{N}})^{-1} P$ coincides with $(\mathcal{A}^0(\mathbf{k}) - \zeta\varepsilon^2 I)^{-1} P$, where $\mathcal{A}^0(\mathbf{k})$ is the operator (5.10). Then the operator (5.12) can be rewritten as

$$\Sigma(\mathbf{k};\zeta,\varepsilon) = (\mathcal{A}(\mathbf{k}) - \zeta\varepsilon^2 I)^{-1} - (\mathcal{A}^0(\mathbf{k}) - \zeta\varepsilon^2 I)^{-1} P. \tag{5.13}$$

It remains to realize the constants in the estimates. Instead of the precise values of constants C_1, C_2 (see (1.8), (1.9)) which now depend on $\boldsymbol{\theta}$, we take

$$C_1 = \beta_1 \delta^{-1/2} \alpha_1^{1/2} \|g\|_{L_\infty}^{1/2} = 2\beta_1 r_0^{-1} \alpha_0^{-1/2} \alpha_1^{1/2} \|g\|_{L_\infty}^{1/2} \|g^{-1}\|_{L_\infty}^{1/2}, \tag{5.14}$$

$$C_2 = \beta_2 \delta^{-1/2} \alpha_1^{3/2} \|g\|_{L_\infty}^{3/2} = 2\beta_2 r_0^{-1} \alpha_0^{-1/2} \alpha_1^{3/2} \|g\|_{L_\infty}^{3/2} \|g^{-1}\|_{L_\infty}^{1/2}. \tag{5.15}$$

We have taken (5.3) and (5.2) into account. The constant C_3 (see (2.5)) is given by

$$\begin{aligned} C_3 &= \sqrt{2} C_1 c_*^{-1/2} + C_2 c_*^{-3/2} \\ &= 2\sqrt{2} \beta_1 \alpha_0^{-1} \alpha_1^{1/2} r_0^{-1} \|g\|_{L_\infty}^{1/2} \|g^{-1}\|_{L_\infty} + 2\beta_2 \alpha_0^{-2} \alpha_1^{3/2} r_0^{-1} \|g\|_{L_\infty}^{3/2} \|g^{-1}\|_{L_\infty}^2. \end{aligned} \tag{5.16}$$

Here we have taken the expression for c_* into account (see (4.12)). The constants $\widehat{C}_1$, $\widehat{C}_2$, $\widehat{C}_3$, $\widehat{C}_4$ are defined by (2.40)–(2.43). Using (4.12), (5.2), (5.14), (5.15), we obtain

$$\widehat{C}_1 = 4(2+\beta_2) \alpha_0^{-1} r_0^{-2} \|g^{-1}\|_{L_\infty}, \tag{5.17}$$

$$\widehat{C}_2 = 8 \left(\beta_1 \alpha_0^{-1} \alpha_1^{1/2} r_0^{-1} \|g\|_{L_\infty}^{1/2} \|g^{-1}\|_{L_\infty} + \beta_2 \alpha_0^{-2} \alpha_1^{3/2} r_0^{-1} \|g\|_{L_\infty}^{3/2} \|g^{-1}\|_{L_\infty}^2 \right), \tag{5.18}$$

$$\widehat{C}_3 = \max\{ \sqrt{2} \alpha_0^{-1/2} r_0^{-1} \|g^{-1}\|_{L_\infty}^{1/2}, 2^{5/2} \beta_1 \alpha_0^{-1} \alpha_1^{1/2} r_0^{-1} \|g\|_{L_\infty}^{1/2} \|g^{-1}\|_{L_\infty} \}, \tag{5.19}$$

$$\widehat{C}_4 = 4\sqrt{2} \beta_2 \alpha_0^{-2} \alpha_1^{3/2} r_0^{-1} \|g\|_{L_\infty}^{3/2} \|g^{-1}\|_{L_\infty}^2. \tag{5.20}$$

Clearly, all the constants under consideration are given in terms of r_0, α_0, α_1, $\|g\|_{L_\infty}$, $\|g^{-1}\|_{L_\infty}$ only.

5.4. Approximation of the resolvent $(\mathcal{A}(\mathbf{k}) - \zeta\varepsilon^2 I)^{-1}$ for $\operatorname{Re}\zeta < 0$. Let ζ be of the form (2.3). Applying Theorem 2.1 to the operator family $A(t,\boldsymbol{\theta}) = \mathcal{A}(\mathbf{k})$, we arrive at the following estimate for the operator (5.13):

$$\|\Sigma(\mathbf{k};\zeta,\varepsilon)\|_{L_2(\Omega)\to L_2(\Omega)} \le (3\delta)^{-1} + C_3(\eta + |\kappa|)^{-1/2}\varepsilon^{-1}, \quad |\mathbf{k}| \le t_0, \ \varepsilon > 0. \tag{5.21}$$

Here t_0 is defined by (5.4), δ is defined by (5.2), and C_3 is given by (5.16).

Estimates for $\mathbf{k} \in \widetilde{\Omega}$, $|\mathbf{k}| > t_0$, are trivial. By (4.12), we have

$$\begin{aligned} \|(\mathcal{A}(\mathbf{k}) - \zeta\varepsilon^2 I)^{-1}\|_{L_2(\Omega)\to L_2(\Omega)} &\le \max_{\lambda \ge c_* t_0^2} |(\lambda - \zeta\varepsilon^2)^{-1}| \\ &= \max_{\lambda \ge c_* t_0^2} ((\lambda + \eta\varepsilon^2)^2 + \kappa^2\varepsilon^4)^{-1/2} \le c_*^{-1} t_0^{-2}, \quad \mathbf{k} \in \widetilde{\Omega}, \ |\mathbf{k}| > t_0. \end{aligned} \tag{5.22}$$

Similarly, by (5.11), we obtain

$$\|(\mathcal{A}^0(\mathbf{k}) - \zeta\varepsilon^2 I)^{-1}\|_{L_2(\Omega)\to L_2(\Omega)} \le c_*^{-1} t_0^{-2}, \quad \mathbf{k} \in \widetilde{\Omega}, \ |\mathbf{k}| > t_0. \tag{5.23}$$

Relations (5.22) and (5.23) imply the following estimate for the operator (5.13):

$$\|\Sigma(\mathbf{k};\zeta,\varepsilon)\|_{L_2(\Omega)\to L_2(\Omega)} \le 2c_*^{-1} t_0^{-2}, \quad \mathbf{k} \in \widetilde{\Omega}, \ |\mathbf{k}| > t_0, \ \varepsilon > 0. \tag{5.24}$$

Note that $(3\delta)^{-1} \le 2c_*^{-1} t_0^{-2}$ (since $c_* t_0^2 \le \delta$). Then estimates (5.21) and (5.24) directly imply the following theorem.

THEOREM 5.1. *Let $\mathcal{A}(\mathbf{k})$ be the operator family (4.6) acting in $L_2(\Omega;\mathbb{C}^n)$, and let $\mathcal{A}^0(\mathbf{k})$ be the operator family (5.10). Let P be the orthogonal projection of $L_2(\Omega;\mathbb{C}^n)$ onto the subspace $\mathfrak{N}$ of constant vector-valued functions. Let $\zeta = -\eta + i\kappa$, $\eta > 0$, $\kappa \in \mathbb{R}$. Then for $\mathbf{k} \in \widetilde{\Omega}$, $\varepsilon > 0$, we have*

$$\|(\mathcal{A}(\mathbf{k}) - \zeta\varepsilon^2 I)^{-1} - (\mathcal{A}^0(\mathbf{k}) - \zeta\varepsilon^2 I)^{-1} P\|_{L_2(\Omega)\to L_2(\Omega)} \le 2c_*^{-1} t_0^{-2} + C_3(\eta + |\kappa|)^{-1/2}\varepsilon^{-1}.$$

Here c_* is defined by (4.12), t_0 is defined by (5.4), and C_3 is given by (5.16).

5.5. Approximation of the resolvent $(\mathcal{A}(\mathbf{k}) - \zeta\varepsilon^2 I)^{-1}$ for $\operatorname{Re}\zeta \geq 0$. Let ζ be of the form (2.16). Applying Theorem 2.7 to the operator family $\mathcal{A}(\mathbf{k})$, we arrive at the following estimate for the operator (5.13):

$$(5.25)\quad \begin{aligned} &\|\Sigma(\mathbf{k};\zeta,\varepsilon)\|_{L_2(\Omega)\to L_2(\Omega)} \\ &\leq \widehat{C}_1 + \varepsilon^{-1}\left(\widehat{C}_2(\eta+|\kappa|)^{-1/2} + \widehat{C}_3|\kappa|^{-1}\eta^{1/2} + \widehat{C}_4\kappa^{-2}\eta^{3/2}\right), \quad |\mathbf{k}|\leq t_0,\ \varepsilon>0. \end{aligned}$$

Here constants $\widehat{C}_1$, $\widehat{C}_2$, $\widehat{C}_3$, $\widehat{C}_4$ are defined by (5.17)–(5.20).

Next, let $\mathbf{k}\in\widetilde{\Omega}$ and $|\mathbf{k}|>t_0$. Then, by (4.12), we have

$$(5.26)\qquad \|(\mathcal{A}(\mathbf{k}) - \zeta\varepsilon^2 I)^{-1}\|_{L_2(\Omega)\to L_2(\Omega)} \leq \max_{\lambda\geq c_* t_0^2}((\lambda-\eta\varepsilon^2)^2 + \varepsilon^4\kappa^2)^{-1/2}.$$

Hence, for $\eta\varepsilon^2 \leq c_* t_0^2/2$ we obtain that

$$\|(\mathcal{A}(\mathbf{k}) - \zeta\varepsilon^2 I)^{-1}\|_{L_2(\Omega)\to L_2(\Omega)} \leq 2c_*^{-1}t_0^{-2}.$$

By (5.11), the similar estimate is true for the operator $(\mathcal{A}^0(\mathbf{k}) - \zeta\varepsilon^2 I)^{-1}$. Thus,

$$(5.27)\qquad \|\Sigma(\mathbf{k};\zeta,\varepsilon)\|_{L_2(\Omega)\to L_2(\Omega)} \leq 4c_*^{-1}t_0^{-2}, \quad \mathbf{k}\in\widetilde{\Omega},\quad |\mathbf{k}|>t_0,\quad \eta\varepsilon^2\leq c_* t_0^2/2.$$

If $\eta\varepsilon^2 > c_* t_0^2/2$, then $\varepsilon^{-1} < (2\eta)^{1/2}c_*^{-1/2}t_0^{-1}$, and (5.26) implies that

$$\|(\mathcal{A}(\mathbf{k}) - \zeta\varepsilon^2 I)^{-1}\|_{L_2(\Omega)\to L_2(\Omega)} \leq \varepsilon^{-2}|\kappa|^{-1} \leq \varepsilon^{-1}|\kappa|^{-1}(2\eta)^{1/2}c_*^{-1/2}t_0^{-1}.$$

A similar estimate is true for $(\mathcal{A}^0(\mathbf{k}) - \zeta\varepsilon^2 I)^{-1}$. Hence,

$$(5.28)\qquad \begin{aligned} \|\Sigma(\mathbf{k};\zeta,\varepsilon)\|_{L_2(\Omega)\to L_2(\Omega)} &\leq \varepsilon^{-1}|\kappa|^{-1}2^{3/2}\eta^{1/2}c_*^{-1/2}t_0^{-1}, \\ &\mathbf{k}\in\widetilde{\Omega},\quad |\mathbf{k}|>t_0,\quad \eta\varepsilon^2 > c_* t_0^2/2. \end{aligned}$$

From (5.27) and (5.28) it follows that

$$(5.29)\qquad \begin{aligned} \|\Sigma(\mathbf{k};\zeta,\varepsilon)\|_{L_2(\Omega)\to L_2(\Omega)} &\leq \max\{4c_*^{-1}t_0^{-2}, \varepsilon^{-1}|\kappa|^{-1}2^{3/2}\eta^{1/2}c_*^{-1/2}t_0^{-1}\}, \\ &\mathbf{k}\in\widetilde{\Omega},\quad |\mathbf{k}|>t_0. \end{aligned}$$

Relations (5.25) and (5.29) directly imply the following result.

THEOREM 5.2. *Let $\mathcal{A}(\mathbf{k})$ be the operator family (4.6) acting in $L_2(\Omega;\mathbb{C}^n)$, and let $\mathcal{A}^0(\mathbf{k})$ be the operator family (5.10). Let P be the orthogonal projection of $L_2(\Omega;\mathbb{C}^n)$ onto the subspace $\mathfrak{N}$ of constant vector-valued functions. Let $\zeta=\eta+i\kappa$, $\eta\geq 0$, $0\neq\kappa\in\mathbb{R}$. Then for $\mathbf{k}\in\widetilde{\Omega}$, $\varepsilon>0$, we have*

$$\begin{aligned} &\|(\mathcal{A}(\mathbf{k}) - \zeta\varepsilon^2 I)^{-1} - (\mathcal{A}^0(\mathbf{k}) - \zeta\varepsilon^2 I)^{-1}P\|_{L_2(\Omega)\to L_2(\Omega)} \\ &\leq \widetilde{C}_1 + \varepsilon^{-1}\left(\widehat{C}_2(\eta+|\kappa|)^{-1/2} + \widetilde{C}_3|\kappa|^{-1}\eta^{1/2} + \widehat{C}_4\kappa^{-2}\eta^{3/2}\right). \end{aligned}$$

Here

$$\widetilde{C}_1 = \widehat{C}_1 + 4c_*^{-1}t_0^{-2} = \widehat{C}_1 + 16\alpha_0^{-2}\alpha_1 r_0^{-2}\|g\|_{L_\infty}\|g^{-1}\|_{L_\infty}^2,$$

$$\widetilde{C}_3 = \widehat{C}_3 + 2^{3/2}c_*^{-1/2}t_0^{-1} = \widehat{C}_3 + 2^{5/2}\alpha_0^{-1}\alpha_1^{1/2}r_0^{-1}\|g\|_{L_\infty}^{1/2}\|g^{-1}\|_{L_\infty},$$

and the constants $\widehat{C}_1$, $\widehat{C}_2$, $\widehat{C}_3$, $\widehat{C}_4$ are defined by (5.17)–(5.20).

§6. Approximation for the resolvent of the operator $\mathcal{A}$

6.1. Approximation of the resolvent $(\mathcal{A} - \zeta\varepsilon^2 I)^{-1}$ for $\operatorname{Re}\zeta < 0$. We return to the operator $\mathcal{A} = b(\mathbf{D})^* g(\mathbf{x}) b(\mathbf{D})$ acting in $L_2(\mathbb{R}^d; \mathbb{C}^n)$. Let g^0 be the effective matrix and let $\mathcal{A}^0 = b(\mathbf{D})^* g^0 b(\mathbf{D})$ be the effective operator. Let $\mathcal{V}$ be the Gelfand transformation defined in Section 3.2. Using decomposition (4.14) for the operator $\mathcal{A}$, we represent the resolvent $(\mathcal{A} - \zeta\varepsilon^2 I)^{-1}$ as

$$(\mathcal{A} - \zeta\varepsilon^2 I)^{-1} = \mathcal{V}^{-1} \left(\int_{\widetilde{\Omega}} \bigoplus (\mathcal{A}(\mathbf{k}) - \zeta\varepsilon^2 I)^{-1} \, d\mathbf{k} \right) \mathcal{V}.$$

A similar decomposition is true for the operator $(\mathcal{A}^0 - \zeta\varepsilon^2 I)^{-1}$. Besides, we introduce the operator

$$\Pi := \mathcal{V}^{-1} [P] \mathcal{V},$$

acting in $L_2(\mathbb{R}^d; \mathbb{C}^n)$. Here $[P]$ is the projection in $\mathcal{K}$ that acts in the fibers as P (see (5.1)). It is easily seen (see [**BSu4**], (6.8)) that Π *is a pseudodifferential operator in* $L_2(\mathbb{R}^d; \mathbb{C}^n)$ *with the symbol* $\chi_{\widetilde{\Omega}}(\boldsymbol{\xi})$, where $\chi_{\widetilde{\Omega}}$ is the indicator of the set $\widetilde{\Omega}$. Thus,

$$(\Pi \mathbf{u})(\mathbf{x}) = (2\pi)^{-d/2} \int_{\widetilde{\Omega}} e^{i\langle \mathbf{x}, \boldsymbol{\xi}\rangle} \widehat{\mathbf{u}}(\boldsymbol{\xi}) \, d\boldsymbol{\xi},$$

where $\widehat{\mathbf{u}}$ is the Fourier image of $\mathbf{u}$. The operator Π is smoothing.

Let ζ be of the form (2.3). By the Gelfand transformation, Theorem 5.1 directly implies the estimate

$$\|(\mathcal{A} - \zeta\varepsilon^2 I)^{-1} - (\mathcal{A}^0 - \zeta\varepsilon^2 I)^{-1}\Pi\|_{L_2(\mathbb{R}^d)\to L_2(\mathbb{R}^d)} \le 2c_*^{-1} t_0^{-2} + C_3(\eta + |\kappa|)^{-1/2}\varepsilon^{-1}. \tag{6.1}$$

6.2. Elimination of the smoothing operator Π. Let us show that the estimate (6.1) remains true, if Π is replaced by the identity operator I; only the constant in the estimate will change under such replacement. Let us estimate the norm of the operator $(\mathcal{A}^0 - \zeta\varepsilon^2 I)^{-1}(I - \Pi)$:

$$\|(\mathcal{A}^0 - \zeta\varepsilon^2 I)^{-1}(I - \Pi)\|_{L_2(\mathbb{R}^d)\to L_2(\mathbb{R}^d)} = \max_{\boldsymbol{\xi}\in\mathbb{R}^d\setminus\widetilde{\Omega}} \left| \left(b(\boldsymbol{\xi})^* g^0 b(\boldsymbol{\xi}) + (\eta\varepsilon^2 - i\kappa\varepsilon^2)\mathbf{1}_n \right)^{-1} \right|. \tag{6.2}$$

The symbol of the effective operator $\mathcal{A}^0$ satisfies the estimate

$$b(\boldsymbol{\xi})^* g^0 b(\boldsymbol{\xi}) \ge c_* |\boldsymbol{\xi}|^2 \mathbf{1}_n, \quad \boldsymbol{\xi} \in \mathbb{R}^d, \tag{6.3}$$

which follows from the estimate (5.8) for the germ. For $\boldsymbol{\xi} \in \mathbb{R}^d \setminus \widetilde{\Omega}$, we have $|\boldsymbol{\xi}| \ge r_0$. Recall that r_0 is the radius of the ball inscribed in $\operatorname{clos}\widetilde{\Omega}$. Then from (6.2) it follows that

$$\|(\mathcal{A}^0 - \zeta\varepsilon^2 I)^{-1}(I - \Pi)\|_{L_2(\mathbb{R}^d)\to L_2(\mathbb{R}^d)} \le c_*^{-1} r_0^{-2}. \tag{6.4}$$

Relations (6.1) and (6.4) directly imply the following result.

THEOREM 6.1. *Let $\mathcal{A}$ be the operator (4.4), and let $\mathcal{A}^0$ be the effective operator (5.9). Let $\zeta = -\eta + i\kappa$, $\eta > 0$, $\kappa \in \mathbb{R}$. Then for $\varepsilon > 0$ we have*

$$\|(\mathcal{A} - \zeta\varepsilon^2 I)^{-1} - (\mathcal{A}^0 - \zeta\varepsilon^2 I)^{-1}\|_{L_2(\mathbb{R}^d)\to L_2(\mathbb{R}^d)} \le C_4 + C_3(\eta + |\kappa|)^{-1/2}\varepsilon^{-1}.$$

Here C_3 is given by (5.16). The constant C_4 is defined by

$$C_4 = 2c_*^{-1} t_0^{-2} + c_*^{-1} r_0^{-2} = 8\alpha_1 \alpha_0^{-2} r_0^{-2} \|g\|_{L_\infty} \|g^{-1}\|_{L_\infty}^2 + \alpha_0^{-1} r_0^{-2} \|g^{-1}\|_{L_\infty}. \tag{6.5}$$

6.3. Approximation of the resolvent $(\mathcal{A}-\zeta\varepsilon^2 I)^{-1}$ for $\operatorname{Re}\zeta \geq 0$. Let ζ be as in (2.16). By the Gelfand transformation, Theorem 5.2 implies that

$$\begin{aligned}&\|(\mathcal{A}-\zeta\varepsilon^2 I)^{-1}-(\mathcal{A}^0-\zeta\varepsilon^2 I)^{-1}\Pi\|_{L_2(\mathbb{R}^d)\to L_2(\mathbb{R}^d)}\\ &\leq \widetilde{C}_1+\varepsilon^{-1}\left(\widehat{C}_2(\eta+|\kappa|)^{-1/2}+\widetilde{C}_3|\kappa|^{-1}\eta^{1/2}+\widehat{C}_4\kappa^{-2}\eta^{3/2}\right).\end{aligned}\tag{6.6}$$

Let us show that Π can be replaced by I. We have

$$\|(\mathcal{A}^0-\zeta\varepsilon^2 I)^{-1}(I-\Pi)\|_{L_2(\mathbb{R}^d)\to L_2(\mathbb{R}^d)}=\max_{\boldsymbol{\xi}\in\mathbb{R}^d\setminus\widetilde{\Omega}}|(b(\boldsymbol{\xi})^*g^0b(\boldsymbol{\xi})-(\eta\varepsilon^2+i\kappa\varepsilon^2)\mathbf{1}_n)^{-1}|.\tag{6.7}$$

Using (6.3) and the estimate $|\boldsymbol{\xi}|\geq r_0$ for $\boldsymbol{\xi}\in\mathbb{R}^d\setminus\widetilde{\Omega}$ in the case $\eta\varepsilon^2\leq c_*r_0^2/2$ we obtain

$$\|(\mathcal{A}^0-\zeta\varepsilon^2 I)^{-1}(I-\Pi)\|_{L_2(\mathbb{R}^d)\to L_2(\mathbb{R}^d)}\leq 2c_*^{-1}r_0^{-2},\quad \eta\varepsilon^2\leq c_*r_0^2/2.\tag{6.8}$$

If $\eta\varepsilon^2>c_*r_0^2/2$, then $\varepsilon^{-1}<(2\eta)^{1/2}c_*^{-1/2}r_0^{-1}$. Hence, (6.7) implies that

$$\begin{aligned}\|(\mathcal{A}^0-\zeta\varepsilon^2 I)^{-1}(I-\Pi)\|_{L_2(\mathbb{R}^d)\to L_2(\mathbb{R}^d)}&\leq\varepsilon^{-2}|\kappa|^{-1}\leq\varepsilon^{-1}|\kappa|^{-1}(2\eta)^{1/2}c_*^{-1/2}r_0^{-1},\\ &\qquad \eta\varepsilon^2>c_*r_0^2/2.\end{aligned}\tag{6.9}$$

From (6.8) and (6.9) it follows that

$$\|(\mathcal{A}^0-\zeta\varepsilon^2 I)^{-1}(I-\Pi)\|_{L_2(\mathbb{R}^d)\to L_2(\mathbb{R}^d)}\leq 2c_*^{-1}r_0^{-2}+\varepsilon^{-1}|\kappa|^{-1}(2\eta)^{1/2}c_*^{-1/2}r_0^{-1},\quad\varepsilon>0.\tag{6.10}$$

Relations (6.6) and (6.10) directly imply the following result.

THEOREM 6.2. *Let $\mathcal{A}$ be the operator* (4.4), *and let $\mathcal{A}^0$ be the effective operator* (5.9). *Let $\zeta=\eta+i\kappa$, $\eta\geq 0$, $0\neq\kappa\in\mathbb{R}$. Then for $\varepsilon>0$ we have*

$$\begin{aligned}&\|(\mathcal{A}-\zeta\varepsilon^2 I)^{-1}-(\mathcal{A}^0-\zeta\varepsilon^2 I)^{-1}\|_{L_2(\mathbb{R}^d)\to L_2(\mathbb{R}^d)}\\ &\leq\check{C}_1+\varepsilon^{-1}\left(\widehat{C}_2(\eta+|\kappa|)^{-1/2}+\check{C}_3|\kappa|^{-1}\eta^{1/2}+\widehat{C}_4\kappa^{-2}\eta^{3/2}\right).\end{aligned}$$

Here

$$\check{C}_1=\widetilde{C}_1+2c_*^{-1}r_0^{-2}=\widehat{C}_1+16\alpha_0^{-2}\alpha_1r_0^{-2}\|g\|_{L_\infty}\|g^{-1}\|^2_{L_\infty}+2\alpha_0^{-1}r_0^{-2}\|g^{-1}\|_{L_\infty},\tag{6.11}$$

$$\begin{aligned}\check{C}_3&=\widetilde{C}_3+\sqrt{2}c_*^{-1/2}r_0^{-1}\\ &=\widehat{C}_3+2^{5/2}\alpha_0^{-1}\alpha_1^{1/2}r_0^{-1}\|g\|^{1/2}_{L_\infty}\|g^{-1}\|_{L_\infty}+\sqrt{2}\alpha_0^{-1/2}r_0^{-1}\|g^{-1}\|^{1/2}_{L_\infty}.\end{aligned}\tag{6.12}$$

The constants $\widehat{C}_1$, $\widehat{C}_2$, $\widehat{C}_3$, $\widehat{C}_4$ are defined according to (5.17)–(5.20).

§7. Approximation for the resolvent of the operator $\mathcal{A}_\varepsilon$

7.1. The operator $\mathcal{A}_\varepsilon$. Scaling transformation. We denote $g^\varepsilon(\mathbf{x})=g(\varepsilon^{-1}\mathbf{x})$. In $L_2(\mathbb{R}^d;\mathbb{C}^n)$, consider the operator

$$\mathcal{A}_\varepsilon=b(\mathbf{D})^*g^\varepsilon(\mathbf{x})b(\mathbf{D}),\quad\varepsilon>0.\tag{7.1}$$

More precisely, $\mathcal{A}_\varepsilon$ is defined in terms of the quadratic form

$$a_\varepsilon[\mathbf{u},\mathbf{u}]=\int_{\mathbb{R}^d}\langle g^\varepsilon(\mathbf{x})b(\mathbf{D})\mathbf{u},b(\mathbf{D})\mathbf{u}\rangle\,d\mathbf{x},\quad\mathbf{u}\in H^1(\mathbb{R}^d;\mathbb{C}^n).$$

Let T_ε be the unitary scaling transformation in $L_2(\mathbb{R}^d;\mathbb{C}^n)$:

$$(T_\varepsilon \mathbf{u})(\mathbf{y}) = \varepsilon^{d/2}\mathbf{u}(\varepsilon\mathbf{y}), \quad \mathbf{y}\in\mathbb{R}^d.$$

Then we have

$$\begin{aligned} \mathcal{A}_\varepsilon &= \varepsilon^{-2}T_\varepsilon^* \mathcal{A} T_\varepsilon, \\ (\mathcal{A}_\varepsilon - \zeta I)^{-1} &= \varepsilon^2 T_\varepsilon^*(\mathcal{A} - \zeta\varepsilon^2 I)^{-1}T_\varepsilon, \\ (\mathcal{A}^0 - \zeta I)^{-1} &= \varepsilon^2 T_\varepsilon^*(\mathcal{A}^0 - \zeta\varepsilon^2 I)^{-1}T_\varepsilon. \end{aligned} \tag{7.2}$$

7.2. Approximation of the resolvent $(\mathcal{A}_\varepsilon - \zeta I)^{-1}$ for $\operatorname{Re}\zeta < 0$. Since the operator T_ε is *unitary* in $L_2(\mathbb{R}^d;\mathbb{C}^n)$, combining Theorem 6.1 and (7.2), we arrive at the following result.

THEOREM 7.1. *Let $\mathcal{A}_\varepsilon$ be the operator* (7.1), *and let $\mathcal{A}^0$ be the effective operator* (5.9). *Let $\zeta = -\eta + i\kappa$, $\eta > 0$, $\kappa\in\mathbb{R}$. Then for $\varepsilon > 0$ we have*

$$\|(\mathcal{A}_\varepsilon - \zeta I)^{-1} - (\mathcal{A}^0 - \zeta I)^{-1}\|_{L_2(\mathbb{R}^d)\to L_2(\mathbb{R}^d)} \le C_4\varepsilon^2 + C_3(\eta + |\kappa|)^{-1/2}\varepsilon. \tag{7.3}$$

The constants C_3 and C_4 are defined by (5.16), (6.5) *and depend only on α_0, α_1, $\|g\|_{L_\infty}$, $\|g^{-1}\|_{L_\infty}$ and on parameters of the lattice Γ.*

Roughening the estimate (7.3) for $0 < \varepsilon \le 1$, we obtain the following statement.

COROLLARY 7.2. *Under the assumptions of Theorem 7.1 we have*

$$\|(\mathcal{A}_\varepsilon - \zeta I)^{-1} - (\mathcal{A}^0 - \zeta I)^{-1}\|_{L_2(\mathbb{R}^d)\to L_2(\mathbb{R}^d)} \le (C_4 + C_3(\eta + |\kappa|)^{-1/2})\varepsilon, \quad 0<\varepsilon\le 1.$$

7.3. Approximation of the resolvent $(\mathcal{A}_\varepsilon - \zeta I)^{-1}$ for $\operatorname{Re}\zeta \ge 0$. Similarly to the arguments of Section 7.2, Theorem 6.2 and (7.2) imply the following result.

THEOREM 7.3. *Let $\mathcal{A}_\varepsilon$ be the operator* (7.1), *and let $\mathcal{A}^0$ be the effective operator* (5.9). *Let $\zeta = \eta + i\kappa$, $\eta \ge 0$, $0 \ne \kappa\in\mathbb{R}$. Then for $\varepsilon > 0$ we have*

$$\begin{aligned} &\|(\mathcal{A}_\varepsilon - \zeta I)^{-1} - (\mathcal{A}^0 - \zeta I)^{-1}\|_{L_2(\mathbb{R}^d)\to L_2(\mathbb{R}^d)} \\ &\quad\le \check{C}_1\varepsilon^2 + \left(\widehat{C}_2(\eta + |\kappa|)^{-1/2} + \check{C}_3|\kappa|^{-1}\eta^{1/2} + \widehat{C}_4\kappa^{-2}\eta^{3/2}\right)\varepsilon. \end{aligned} \tag{7.4}$$

Here the constants $\check{C}_1$, $\check{C}_3$ are defined by (6.11), (6.12), *and the constants $\widehat{C}_1$, $\widehat{C}_2$, $\widehat{C}_3$, $\widehat{C}_4$ are defined by* (5.17)–(5.20). *All these constants depend only on α_0, α_1, $\|g\|_{L_\infty}$, $\|g^{-1}\|_{L_\infty}$ and on parameters of the lattice Γ.*

COROLLARY 7.4. *Under the assumptions of Theorem 7.3 we have*

$$\|(\mathcal{A}_\varepsilon - \zeta I)^{-1} - (\mathcal{A}^0 - \zeta I)^{-1}\|_{L_2(\mathbb{R}^d)\to L_2(\mathbb{R}^d)} \le (C_5 + C_6(\eta+1)^{3/2}\kappa^{-2})\varepsilon, \quad 0<\varepsilon\le 1. \tag{7.5}$$

Here

$$C_5 = \check{C}_1 + \frac{3}{4}\widehat{C}_2 + \frac{1}{2}\check{C}_3, \quad C_6 = \frac{1}{4}\widehat{C}_2 + \frac{1}{2}\check{C}_3 + \widehat{C}_4.$$

PROOF. We roughen estimate (7.4). First, $\check{C}_1\varepsilon^2 \le \check{C}_1\varepsilon$ for $0<\varepsilon\le 1$. Second,

$$\begin{aligned} \widehat{C}_2(\eta + |\kappa|)^{-1/2} &\le \widehat{C}_2|\kappa|^{-1/2} \le \frac{1}{2}\widehat{C}_2(1 + |\kappa|^{-1}) \\ &\le \frac{1}{2}\widehat{C}_2 + \frac{1}{4}\widehat{C}_2(1+\kappa^{-2}) \le \frac{3}{4}\widehat{C}_2 + \frac{1}{4}\widehat{C}_2(\eta+1)^{3/2}\kappa^{-2}. \end{aligned}$$

Next,

$$\check{C}_3|\kappa|^{-1}\eta^{1/2} \le \check{C}_3|\kappa|^{-1}(\eta+1)^{1/2} \le \frac{1}{2}\check{C}_3(1+\kappa^{-2}(\eta+1)) \le \frac{1}{2}\check{C}_3 + \frac{1}{2}\check{C}_3(\eta+1)^{3/2}\kappa^{-2}.$$

Finally,

$$\widehat{C}_4\kappa^{-2}\eta^{3/2} \le \widehat{C}_4\kappa^{-2}(\eta+1)^{3/2}.$$

Combining the above arguments with (7.4), we obtain (7.5). □

Putting $\kappa = \pm\varepsilon^\sigma$, $0<\sigma<1/2$, in (7.5), we obtain the following statement.

COROLLARY 7.5. *Under the assumptions of Theorem* 7.3 *for* $0<\sigma<1/2$, $\eta\ge 0$, *and* $0<\varepsilon\le 1$ *we have*

(7.6)
$$\|(\mathcal{A}_\varepsilon-(\eta\pm i\varepsilon^\sigma)I)^{-1}-(\mathcal{A}^0-(\eta\pm i\varepsilon^\sigma)I)^{-1}\|_{L_2(\mathbb{R}^d)\to L_2(\mathbb{R}^d)} \le (C_5+C_6(\eta+1)^{3/2})\varepsilon^{1-2\sigma}.$$

§8. The limiting absorption principle for the operator $\mathcal{A}_\varepsilon$

8.1. The limiting absorption principle for the effective operator. Let $\eta>0$, $0\ne\kappa\in\mathbb{R}$. Consider the "sandwiched" resolvent of the operator $\mathcal{A}^0$:

$$\mathcal{B}^0(\eta+i\kappa;s) = \langle\mathbf{x}\rangle^{-1/2-s}(\mathcal{A}^0-(\eta+i\kappa)I)^{-1}\langle\mathbf{x}\rangle^{-1/2-s},\quad s>0, \tag{8.1}$$

where $\langle\mathbf{x}\rangle = (1+|\mathbf{x}|^2)^{1/2}$. Operators of such a kind are typical for scattering theory.

To simplify references, we first consider the operator (4.4) of the form $\mathcal{A} = \mathbf{D}^* g(\mathbf{x})\mathbf{D}$ (the acoustics operator). In this case $n=1$, $m=d$, $b(\mathbf{D})=\mathbf{D}$. Then the effective operator (5.9) takes the form $\mathcal{A}^0 = \mathbf{D}^* g^0\mathbf{D}$. As is known from scattering theory (see [**Ku1, Ku2**]), in this case, the operators (8.1) have limits in the operator norm in $L_2(\mathbb{R}^d;\mathbb{C}^n)$ as $\kappa\to\pm 0$ provided that $\eta>0$. Denote these limits by $\mathcal{B}^0_\pm(\eta;s)$. Moreover, in [**Ku1, Ku2**], the following statement is proved.

PROPOSITION 8.1. *Let* $\eta>0$, $0\ne\kappa\in\mathbb{R}$. *Let* $\mathcal{A}^0 = \mathbf{D}^* g^0\mathbf{D}$. *Then the operators* $\mathcal{B}^0(\eta+i\kappa;s)$ *defined by* (8.1) *have limits* $\mathcal{B}^0_\pm(\eta;s)$ *in the operator norm in* $L_2(\mathbb{R}^d;\mathbb{C}^n)$ *as* $\kappa\to\pm 0$. *The following estimates are true:*

$$\|\mathcal{B}^0(\eta+i\kappa;s)-\mathcal{B}^0_\pm(\eta;s)\|_{L_2(\mathbb{R}^d)\to L_2(\mathbb{R}^d)} \le C(\eta;s)|\kappa|^\vartheta,\quad \eta>0,\ \pm\kappa>0,$$

where $\vartheta=s$ *for* $0<s<1$, $\vartheta<1$ *for* $s=1$, *and* $\vartheta=1$ *for* $s>1$.

The existence of the limits for the sandwiched resolvent (8.1) as $\kappa\to\pm 0$ is usually refered to as the *limiting absorption principle* (see, e.g., [**Ku1**], [**Ya1**]).

The authors were informed by D. Yafaev that a full analog of Proposition 8.1 is true for the operator $\mathcal{A}^0 = b(\mathbf{D})^* g^0 b(\mathbf{D})$ in the general case as well (see [**Ya2**], Proposition 2.4.1).

8.2. The limiting absorption principle for $\mathcal{A}_\varepsilon$. Let $\eta>0$. We put

$$\mathcal{B}^{(\pm)}_\varepsilon(\eta;\sigma;s) := \langle\mathbf{x}\rangle^{-1/2-s}(\mathcal{A}_\varepsilon-(\eta\pm i\varepsilon^\sigma)I)^{-1}\langle\mathbf{x}\rangle^{-1/2-s}. \tag{8.2}$$

From (7.6), taking the notation (8.1) and (8.2) into account, we obtain

$$\|\mathcal{B}^{(\pm)}_\varepsilon(\eta;\sigma;s)-\mathcal{B}^0(\eta\pm i\varepsilon^\sigma;s)\|_{L_2(\mathbb{R}^d)\to L_2(\mathbb{R}^d)} \le (C_5+C_6(\eta+1)^{3/2})\varepsilon^{1-2\sigma},$$
$$s>0,\quad \eta\ge 0,\quad 0<\sigma<1/2,\quad 0<\varepsilon\le 1.$$

Combining this with what was said in Section 8.1, we obtain the following result.

THEOREM 8.2. *Let* $\mathcal{B}^{(\pm)}_\varepsilon(\eta;\sigma;s)$ *be the operator* (8.2), *and let* $\mathcal{B}^0_\pm(\eta;s)$ *be the limit of operators* (8.1) *as* $\kappa\to\pm 0$. *Assume that* $\eta>0$, $0<\sigma<1/2$, $s>0$. *Then,* *as $\varepsilon \to 0$, the operators $\mathcal{B}_\varepsilon^{(\pm)}(\eta;\sigma;s)$ tend to the operator $\mathcal{B}_\pm^0(\eta;s)$ in the operator norm in $L_2(\mathbb{R}^d;\mathbb{C}^n)$. Moreover, we have*

$$\|\mathcal{B}_\varepsilon^{(\pm)}(\eta;\sigma;s) - \mathcal{B}_\pm^0(\eta;s)\|_{L_2(\mathbb{R}^d)\to L_2(\mathbb{R}^d)} \leq (C_5 + C_6(\eta+1)^{3/2})\varepsilon^{1-2\sigma} + C(\eta;s)\varepsilon^{\sigma\vartheta}, \quad s>0,\quad \eta>0,\quad 0<\sigma<1/2,\quad 0<\varepsilon\leq 1.$$

Here $\vartheta = s$ for $0<s<1$, $\vartheta<1$ for $s=1$, and $\vartheta = 1$ for $s>1$.

Thus, the sandwiched resolvent (8.2) of the operator $\mathcal{A}_\varepsilon$ has limits from above and below on the real half-axis $\eta > 0$. Herewith, the resolvent of the operator $\mathcal{A}_\varepsilon$ (with rapidly oscillating coefficient $g^\varepsilon(\mathbf{x})$) is taken in the point $\zeta = \eta \pm i\varepsilon^\sigma$ (with $0<\sigma<1/2$), which itself depends on ε. The fact established in Theorem 8.2 can be interpreted as an analog of the *limiting absorption principle* for the operator $\mathcal{A}_\varepsilon$.

References

[BaPa] N. S. Bakhvalov, G. P. Panasenko, *Homogenization: averaging processes in periodic media. Mathematical problems in the mechanics of composite materials*, "Nauka", Moscow, 1984; English transl., Math. Appl. (Soviet Ser.), vol. 36, Kluwer Acad. Publ. Group, Dordrecht, 1989.

[BeLP] A. Bensoussan, J. L. Lions, G. Papanicolau, *Asymptotic analysis for periodic structures*, North Holland Publishing Company, Amsterdam, 1978, 700 pp.

[BSu1] M. Sh. Birman, T. A. Suslina, *Threshold effects near the lower edge of the spectrum for periodic differential operators of mathematical physics*, Oper. Theory: Adv. Appl. **129** (2001), Birkhäuser, pp. 71–107.

[BSu2] M. Sh. Birman, T. A. Suslina, *Second order periodic differential operators. Threshold properties and homogenization*, Algebra i Analiz **15** (2003), no. 5, 1–108; English transl., St. Petersburg Math. J. **15** (2004), no. 5, 639–714.

[BSu3] M. Sh. Birman, T. A. Suslina, *Threshold approximations with corrector for the resolvent of a factorized selfadjoint operator family*, Algebra i Analiz **17** (2005), no. 5, 69–90; English transl., St. Petersburg Math. J. **17** (2006), no. 5, 745–762.

[BSu4] M. Sh. Birman, T. A. Suslina, *Homogenization with corrector term for periodic elliptic differential operators*, Algebra i Analiz **17** (2005), no. 6, 1–104; English transl., St. Petersburg Math. J. **17** (2006), no. 6, 897–973.

[BSu5] M. Sh. Birman, T. A. Suslina, *Homogenization with corrector for periodic differential operators. Approximation of solutions in the Sobolev class $H^1(\mathbb{R}^d)$*, Algebra i Analiz **18** (2006), no. 6, 1–130; English transl., St. Petersburg Math. J. **18** (2007), no. 6, 857–955.

[BSu6] M. Sh. Birman, T. A. Suslina, *The limit absorption principle and homogenization procedure for periodic elliptic operators*, Funktsional. Anal. i Prilozhen. **42** (2008), no. 4, 105–108; English transl., Funct. Anal. Appl. **42** (2008), no. 4, 336–339.

[JKO] V. V. Jikov, S. M. Kozlov, O. A. Oleinik, *Homogenization of differential operators*, "Nauka", Moscow, 1993; English transl., Springer-Verlag, Berlin, 1994.

[Ka] T. Kato, *Perturbation theory for linear operators*, Grundlehren Math. Wiss., vol. 132, Springer-Verlag, Berlin–New York, 1972.

[Ku1] S. T. Kuroda, *Scattering theory for differential operators, I, Operator theory*, J. Math. Soc. Japan **25**, no. 1 (1973), 75–104.

[Ku2] S. T. Kuroda, *Scattering theory for differential operators, II, Selfadjoint elliptic operators*, J. Math. Soc. Japan **25**, no. 2 (1973), 222–234.

[Ya1] D. R. Yafaev, *Scattering theory: some old and new problems*, Lecture Notes in Mathematics, vol. 1735, Springer, 2000.

[Ya2] D. R. Yafaev, *Mathematical scattering theory. Analytic theory*, Mathematical Surveys and Monographs, 158. Amer. Math. Soc., Providence, RI, 2010.

Department of Physics, St. Petersburg State University, Ul'yanovskaya 3, Petrodvorets, St. Petersburg, 198504, Russia

Department of Physics, St. Petersburg State University, Ul'yanovskaya 3, Petrodvorets, St. Petersburg, 198504, Russia
E-mail address: suslina@list.ru

Amer. Math. Soc. Transl.
Volume **231**, 2010

On a Class of Nonselfadjoint Periodic Boundary Value Problems with Discrete Real Spectrum

Lyonell Boulton, Michael Levitin, and Marco Marletta

1. Introduction

We study the operator L_{per} defined by

$$L_{\mathrm{per}}u := i\epsilon(f(x)u'(x))' + iu'(x) \tag{1.1}$$

in which f is a given 2π-periodic function having the following properties:

$$f(x+\pi) = -f(x)\,, \qquad f(-x) = -f(x)\,; \tag{1.2}$$

and also

$$f(x) > 0 \qquad \text{for } x \in (0,\pi)\,. \tag{1.3}$$

In particular it follows that $f(\pi\mathbb{Z}) = 0$. We assume that f is continuous and differentiable except possibly at a finite number of points, the points of nondifferentiability excluding $\pi\mathbb{Z}$. We assume that $f'(0) = 2/\pi$ and that $0 < \epsilon < \pi$.

We consider (1.1) on the domain

$$\mathcal{D} = \{u \in L^2(-\pi,\pi) \mid L_{\mathrm{per}}u \in L^2(-\pi,\pi);\ \ u(-\pi) = u(\pi)\}. \tag{1.4}$$

Remark 1. Of course it is not obvious that functions $u \in L^2(-\pi,\pi)$ such that $L_{\mathrm{per}}u \in L^2(-\pi,\pi)$ have boundary values $u(\pm\pi)$; this was proved in [**BLM**], where we showed that if $u \in L^2(-\pi,\pi)$ and $L_{\mathrm{per}}u \in L^2(-\pi,\pi)$, then $u \in H^1(-\pi,\pi)$.

The main results of this paper are

Theorem 2. *The spectrum of operator L_{per} is*

(a) *real,*
(b) *purely discrete, i.e., it consists only of isolated eigenvalues of finite multiplicity with no accumulation points apart from, possibly, infinity.*

Part (a) has been partially proved in [**BLM**], where we showed that all the *eigenvalues* (if they exist) are real. The rest of Theorem 2 follows from

Theorem 3. *The resolvent $(L_{\mathrm{per}} - \lambda)^{-1}$ is a compact operator on $L^2(-\pi,\pi)$ if λ is not an eigenvalue of L_{per}.*

2010 *Mathematics Subject Classification.* Primary 47E05; Secondary 34B24, 76Rxx.
The research of the second author was partially supported by the EPSRC grant EP/D054621.

REMARK 4. The spectrum is always nonempty, as zero is an eigenvalue corresponding to a constant eigenfunction.

In order to prove Theorem 3, we show that when λ is not an eigenvalue of L_{per} (which is guaranteed, for instance, if λ is not real), then the boundary value problem

$$i\epsilon(f(x)u'(x))' + iu'(x) - \lambda u(x) = F(x), \qquad -\pi < x < \pi, \tag{1.5}$$

with periodic boundary conditions $u(-\pi) = u(\pi)$ has a unique solution $u \in \mathcal{D}$ for every $F \in L^2(-\pi, \pi)$.

The compactness of the resolvent is demonstrated by an "explicit" construction of a bounded Green's function $G(x, s)$ such that $u(x) = \int_{-\pi}^{\pi} G(x, s)F(s)\,ds$. The properties of G are established by studying the solutions of an associated homogeneous equation in Sections 2 and 3.

Motivation and scope of the present paper. Our interest in the operator (1.1) with domain determined by condition (1.2) is primarily motivated by [**BeO'BSa**] and [**Da2**], where $f(x) = (2/\pi)\sin x$, and therefore (1.5) takes the form

$$i\tilde{\epsilon}\,(\sin(x)u'(x))' + iu'(x) - \lambda u(x) = F(x), \tag{1.6}$$

with $0 < \tilde{\epsilon} < 2$. This equation arises in fluid dynamics and describes small oscillations of a thin layer of fluid inside a rotating cylinder. From a purely theoretical perspective, the eigenvalue problem associated to L_{per} has recently drawn a substantial amount of attention: see [**ChPe**], [**We1**], [**We2**], [**DaWe**] and [**ChKaPy**]. Despite the fact that (1.6) is highly nonselfadjoint, the spectrum of (1.6) consists exclusively of real eigenvalues of finite multiplicity, it is symmetric with respect to the origin and it accumulates at $\pm\infty$. It is also known that the eigenfunctions do not form an unconditional basis of $L^2(-\pi, \pi)$. Moreover, Davies and Weir [**DaWe**] have recently found explicit asymptotics of the eigenvalues as $\tilde{\epsilon} \to 0$.

In [**BLM**] we established that the eigenvalues of L_{per} are all real and form a symmetric set with respect to the origin. Theorem 2 above rules out completely the possibility of a nonempty essential spectrum and it answers an open question posed in our previous paper.

2. A naïve Frobenius analysis

Let p satisfy the integrating factor equation

$$\frac{p'}{p} = \frac{f'}{f} + \frac{1}{\epsilon f}. \tag{2.1}$$

Then $u(x)$ is a solution of (1.5) iff

$$(p(x)u'(x))' + \frac{i\lambda p}{\epsilon f}u(x) = -\left(\frac{ip}{\epsilon f}F\right)(x), \qquad -\pi < x < \pi. \tag{2.2}$$

For future use, we also recall the homogeneous differential equation

$$(p(x)u'(x))' + \frac{i\lambda p}{\epsilon f}u(x) = 0, \qquad -\pi < x < \pi. \tag{2.2$'$}$$

In order to understand how p behaves near $x = 0$ and $x = \pi$, it is useful to consider a simple model. Suppose that near $x = 0$, the function f satisfies $f(x) = 2x/\pi$—this maintains the normalization $f'(0) = 2/\pi$. Then (2.1) yields $\log p = \log(2x/\pi) +$

$\pi/(2\epsilon)\log(x)$, whence $p(x) = Cx^{1+c/\epsilon}$ where C is an arbitrary nonzero constant and $c = \pi/2$. Similarly, near $x = \pi$, we can consider a simple model $f(x) = (2/\pi)(\pi - x)$ and obtain $p(x) = \tilde{C}(\pi - x)^{1-c/\epsilon}$. In [**BLM**] we proved that, under the minimal assumptions (1.2) on f, the following results hold:

(2.3)

$$p(x) \sim \begin{cases} x^{1+c/\epsilon}, & x \sim 0, \\ (\pi - x)^{1-c/\epsilon}, & x \sim \pi, \end{cases} \quad \text{and} \quad \frac{p(x)}{f(x)} \sim \begin{cases} x^{c/\epsilon}, & x \sim 0, \\ (\pi - x)^{-c/\epsilon}, & x \sim \pi. \end{cases}$$

By considering the model $f(x) = 2x/\pi$, $p(x) = x^{1+c/\epsilon}$ in a neighborhood of the origin and looking for solutions of the differential equation (2.2′) in the form $u(x) = x^{\nu}(1 + a_1 x + a_2 x^2 + \cdots)$, one can establish the asymptotic behavior of solutions for this model in a neighborhood of the origin, and similarly near $x = \pi$. In [**BLM**] we show that, under hypotheses (1.2) on f, there exist solutions u of (2.2′) such that

$$u(x, \lambda) \sim \begin{cases} x^{-c/\epsilon} \text{ or } 1, & x \sim 0, \\ 1 \text{ or } (\pi - x)^{c/\epsilon}, & x \sim \pi, \\ 1 \text{ or } (x + \pi)^{c/\epsilon}, & x \sim -\pi. \end{cases}$$

This implies the existence of a unique solution $\phi(x) = \phi(x, \lambda)$ of (2.2′), such that

$$\phi(x, \lambda) \sim 1, \quad x \to 0. \tag{2.4}$$

The following result is crucial for reducing the problem to the interval $(0, \pi)$; in a sense it plays the same role in the analysis as the orthogonal splitting of the infinite matrix operator in Davies [**Da2**] in $\ell^2(\mathbb{Z})$ into three operators, hence reducing the problem to a problem in $\ell^2(\mathbb{N})$.

LEMMA 5. *The solution ϕ has the symmetry property*

$$\phi(-x, \lambda) = \phi(x, -\lambda). \tag{2.5}$$

PROOF. Define a function $v(x) = \phi(-x, \lambda)$. A direct calculation shows, thanks to the symmetry conditions (1.2), that v satisfies (2.2′) but with λ on the right-hand side replaced by $-\lambda$. It also satisfies $v(0) = 1$. However (2.2′) has only one solution with this property, namely $\phi(x, -\lambda)$. This proves the result. □

We emphasize the important fact that λ is an eigenvalue of (2.2′) if and only if ϕ possesses the additional symmetry property

$$\phi(-\pi, \lambda) = \phi(\pi, \lambda). \tag{2.6}$$

In [**BLM**] we also show that there is also a second solution $\psi(x) = \psi(x, \lambda)$ of (2.2′) satisfying

$$\psi(x, \lambda) \sim \begin{cases} |x|^{-c/\epsilon}, & x \sim 0, \\ |x \mp \pi|^{c/\epsilon}, & x \sim \pm\pi. \end{cases} \tag{2.7}$$

Observe that $\psi(-\pi, \lambda) = \psi(\pi, \lambda) = 0$ and that $\psi(x, \lambda)$ blows up when $x \sim 0$, at least when λ is not an eigenvalue.

Consequently, when λ is not an eigenvalue, we can also normalize $\psi(x, \lambda)$ by the condition

$$p(x)\psi'(x)\phi(x) - p(x)\phi'(x)\psi(x) = 1, \qquad -\pi < x < \pi.$$

The Wronskian in the right-hand side here is obviously a constant, and below we will always assume that $\psi(x, \lambda)$ satisfies this condition.

3. Construction of the Green's function

Now we are back to constructing explicitly the L^2 solution $u(x,\lambda)$ of (2.2) assuming that (2.6) fails. By the standard variation of parameters technique the general solution of (2.2) takes the form

$$\begin{aligned} u(x,\lambda) = \psi(x,\lambda)\int_0^x \phi(s,\lambda)\left(\frac{-\mathrm{i}p}{\epsilon f}F\right)(s)\,\mathrm{d}s \\ + \phi(x,\lambda)\int_x^\pi \psi(s,\lambda)\left(\frac{-\mathrm{i}p}{\epsilon f}F\right)(s)\,\mathrm{d}s \\ + A\phi(x) + B\psi(x)\,, \end{aligned} \tag{3.1}$$

where A and B are arbitrary complex constants.

It remains to check that one can choose constants A and B in such a way that

(a) $u \in L^2(-\pi,\pi)$,
(b) $L_{\mathrm{per}}u \in L^2(-\pi,\pi)$, and
(c) $u(-\pi,\lambda) = u(\pi,\lambda)$.

It is in fact sufficient, and easier, to check that one can choose constants A and B such that

(a′) $u(x,\lambda)$ is continuous at $x=0$,
(b′) $\mathrm{i}\epsilon f u' + \mathrm{i}u$ is continuous at $x=0$,

and (c) all hold. Indeed, (3.1) together with (a′) implies (a), and together with (b′) and (c) implies (b).

Note first that by (a′) and the behavior of ψ near the origin, one is tempted to take $B=0$ in (3.1). We shall show that this choice is indeed the right one by the careful analysis of the remaining terms in (3.1).

By the Cauchy-Schwarz inequality and (2.3), (2.4), (2.7) we have

$$\begin{aligned} &\left|\psi(x,\lambda)\int_0^x \phi(s,\lambda)\left(\frac{-\mathrm{i}p}{\epsilon f}F\right)(s)\,\mathrm{d}s\right| \\ &\le |\psi(x,\lambda)|\left(\int_0^x |\phi(s,\lambda)|^2\left|\frac{-\mathrm{i}p}{\epsilon f}\right|^2\mathrm{d}s\right)^{1/2}\left(\int_0^x |F(s)|^2\,\mathrm{d}s\right)^{1/2} \\ &\le C|x|^{-c/\epsilon}|x|^{c/\epsilon+1/2}\left(\int_0^x |F(s)|^2\,\mathrm{d}s\right)^{1/2} \\ &\le C|x|^{1/2}\|F\|\,. \end{aligned} \tag{3.2}$$

Here, and throughout the rest of this paper, C denotes a generic positive constant; $\|\cdot\|$ is the standard norm in $L^2(-\pi,\pi)$.

Similarly,

$$\begin{aligned} &\left|\phi(x,\lambda)\int_x^\pi \psi(s,\lambda)\left(\frac{-\mathrm{i}p}{\epsilon f}F\right)(s)\,\mathrm{d}s\right| \\ &\le |\phi(x,\lambda)|\left(\int_x^\pi |\psi(s,\lambda)|^2\left|\frac{-\mathrm{i}p}{\epsilon f}\right|^2\mathrm{d}s\right)^{1/2}\left(\int_x^\pi |F(s)|^2\,\mathrm{d}s\right)^{1/2} \\ &\le C\cdot 1\cdot|\pi-x|^{1/2}\left(\int_x^\pi |F(s)|^2\,\mathrm{d}s\right)^{1/2} \\ &\le C|\pi-x|^{1/2}\|F\|\,, \end{aligned} \tag{3.3}$$

so it is bounded. It is also continuous as the integrand is a product of a bounded function $\psi p/f$ and an L^2 function F.

Also, ϕ is continuous at zero, and ψ is not, and therefore (a′) holds if and only if $B = 0$.

To check the condition (b′), it is now sufficient to verify that fu' is continuous at zero. Differentiating (3.1) with respect to x gives

$$\begin{aligned} fu'(x,\lambda) = {} & f\psi'(x,\lambda)\int_0^x \phi(s,\lambda)\left(\frac{-\mathrm{i}p}{\epsilon f}F\right)(s)\,\mathrm{d}s \\ & + f\phi'(x,\lambda)\int_x^\pi \psi(s,\lambda)\left(\frac{-\mathrm{i}p}{\epsilon f}F\right)(s)\,\mathrm{d}s \\ & + Af\phi'(x)\,, \end{aligned} \tag{3.4}$$

as the contributions from differentiating the integrals cancel out.

The last two terms go to zero as $x \to 0$, and we only need to check the continuity of the first term. Similarly to (3.2), we get

$$\begin{aligned} & \left| f(x)\psi'(x,\lambda)\int_0^x \phi(s,\lambda)\left(\frac{-\mathrm{i}p}{\epsilon f}F\right)(s)\,\mathrm{d}s\right| \\ & \le |f(x)\psi'(x,\lambda)|\left(\int_0^x |\phi(s,\lambda)|^2\left|\frac{-\mathrm{i}p}{\epsilon f}\right|^2\mathrm{d}s\right)^{1/2}\left(\int_0^x |F(s)|^2\,\mathrm{d}s\right)^{1/2} \\ & \le C|x||x|^{-c/\epsilon-1}|x|^{c/\epsilon+1/2}\left(\int_0^x |F(s)|^2\,\mathrm{d}s\right)^{1/2} \\ & \le C|x|^{1/2}\|F\|\,, \end{aligned} \tag{3.5}$$

which proves (b′).

Finally, we need to guarantee that we can choose a value of constant A to ensure that condition (c) holds. Direct substitution, again taking account of (2.3), (2.4), (2.7) gives

$$\begin{aligned} u(\pi,\lambda) &= A\phi(\pi,\lambda)\,, \\ u(-\pi,\lambda) &= A\phi(-\pi,\lambda) + \phi(-\pi,\lambda)\int_{-\pi}^{\pi}\psi(s,\lambda)\left(\frac{-\mathrm{i}p}{\epsilon f}F\right)(s)\,\mathrm{d}s, \end{aligned}$$

and as λ is assumed not to be an eigenvalue, and therefore (2.6) is not satisfied, we can choose

$$A = \int_{-\pi}^{\pi}\psi(s,\lambda)\left(\frac{-\mathrm{i}p}{\epsilon f}F\right)(s)\,\mathrm{d}s\Big/\left(\frac{\phi(\pi,\lambda)}{\phi(-\pi,\lambda)} - 1\right). \tag{3.6}$$

This proves the existence of the resolvent of $(L_{\mathrm{per}} - \lambda)^{-1} : L_2(-\pi,\pi) \to L_2(-\pi,\pi)$ for each λ which is not an eigenvalue of L_{per}. Thus, the spectrum of L_{per} is pure point and real.

Now we proceed to writing down the expression for the Green's function $G(x,s)$ of L_{per}. We note that (3.1) can be written, on account of $B = 0$ and (3.6), as

$$u(x,\lambda) = \int_{-\pi}^{\pi} G(x,s)F(s)\,\mathrm{d}s = \int_{-\pi}^{\pi}(G_{\mathrm{I}}(x,s) + G_{\mathrm{II}}(x,s) + G_{\mathrm{III}}(x,s))F(s)\,\mathrm{d}s$$

where we set $G(x,s) := G_{\mathrm{I}}(x,s) + G_{\mathrm{II}}(x,s) + G_{\mathrm{III}}(x,s)$, with

$$G_{\mathrm{I}}(x,s) := \begin{cases} \psi(x,\lambda)\phi(s,\lambda)\left(\frac{-ip(s)}{\epsilon f(s)}\right) & \text{if } |x| \geq |s|, \\ 0 & \text{otherwise;} \end{cases}$$

$$G_{\mathrm{II}}(x,s) := \begin{cases} \phi(x,\lambda)\psi(s,\lambda)\left(\frac{-ip(s)}{\epsilon f(s)}\right) & \text{if } x \leq s, \\ 0 & \text{otherwise;} \end{cases}$$

$$G_{\mathrm{III}}(x,s) := \phi(x,\lambda)\psi(s,\lambda)\left(\frac{-ip(s)}{\epsilon f(s)}\right) \Big/ \left(\frac{\phi(\pi,\lambda)}{\phi(-\pi,\lambda)} - 1\right) \quad \text{for all } -\pi \leq x, s \leq \pi.$$

The functions $G_{\mathrm{II}}(x,s)$ and $G_{\mathrm{III}}(x,s)$ are bounded by (2.3), (2.4), (2.7), so we need to look at $G_{\mathrm{I}}(x,s)$. The only scope for trouble in the expression for G_{I} lies in the fact that $\psi(x,\lambda)$ blows up as $x \to 0$. However if x is small, then, in the region $|x| \geq |s|$ where G_{I} is nonzero, (2.7) and (2.3) yield

$$\left|\psi(x,\lambda)\frac{p(s)}{f(s)}\right| \leq C|x|^{-c/\epsilon}|s|^{c/\epsilon} \leq C.$$

Thus G is bounded and hence is the kernel of a compact operator on $L^2(-\pi,\pi)$. Thus L_{per} has compact resolvent and purely discrete real spectrum.

4. Schatten class properties of the Green's function

We first recall the standard notion of Schatten class operator. Let T be a compact operator and consider its (∞-dimensional) singular value decomposition (see, e.g., [**GoKr**]):

$$T = \sum_{j=0}^{\infty} \alpha_j |v_j\rangle\langle w_j|$$

where the singular values $\alpha_j \geq 0$ and the two sets of vector $\{v_j\}$ and $\{w_j\}$ are orthonormal and not necessarily equal. Here we use the bracket notation: $|v\rangle\langle w|u = \langle u, w\rangle v$. For $p > 0$, we say that T is in the p-Schatten class, $T \in \mathcal{C}_p$, if

$$\|T\|_p := \left(\sum_{j=0}^{\infty} \alpha_j^p\right)^{1/p} < \infty.$$

Note that $\mathcal{C}_1$ are the trace class operators and $\mathcal{C}_2$ are the Hilbert-Schmidt operators.

THEOREM 6. *The resolvent* $(\lambda - L_{\mathrm{per}})^{-1}$ *is in* $\mathcal{C}_p$ *for all* $p > 1$.

PROOF. Let $G_{\mathrm{I}}(x,s), G_{\mathrm{II}}(x,s), G_{\mathrm{III}}(x,s)$ be the components of the Green's function $G(x,s)$. It suffices to show that each of the corresponding integral operators $R_{\mathrm{I}}(\lambda), R_{\mathrm{II}}(\lambda), R_{\mathrm{III}}(\lambda)$ is in $\mathcal{C}_p$.

We start by showing that $R_{\mathrm{III}}(\lambda) \in \mathcal{C}_1$. Indeed, $G_{\mathrm{III}}(x,s) = v(x)w(s)$ where

$$v(x) = \phi(x,\lambda) \quad \text{and} \quad w(s) = \psi(s,\lambda)\left(\frac{-ip(s)}{\epsilon f(s)}\right) \Big/ \left(\frac{\phi(\pi,\lambda)}{\phi(-\pi,\lambda)} - 1\right).$$

By virtue of (2.4), (2.7), and (2.3), both $v, w \in L^2(-\pi,\pi)$. Then $R_{\mathrm{III}}(\lambda) = |v\rangle\langle w|$ and $\|R_{\mathrm{III}}(\lambda)\|_1 = \|v\|_{L^2(-\pi,\pi)}\|w\|_{L^2(-\pi,\pi)} < \infty$, as required.

Let us now show that $R_{\rm II}(\lambda) \in \mathcal{C}_p$ for all $p > 1$. Let $\Omega_{\rm II} = \{(x,s) \in [-\pi,\pi]^2 : x \leq s\}$ be the supports of $G_{\rm II}(x,s)$. Decompose the characteristic function of $\Omega_{\rm II}$ as

$$\mathbb{1}_{\Omega_{\rm II}}(x,s) = \sum_{j=0}^{\infty}\sum_{i=0}^{2^j-1} \mathbb{1}_{I_{2i,j}}(x)\mathbb{1}_{I_{2i+1,j}}(s)$$

where there are 2^{j+1} intervals

$$I_{i,j} = 2\pi\left[\frac{i}{2^{j+1}}, \frac{i+1}{2^{j+1}}\right] - \pi = \left[\frac{i\pi}{2^{j+1}} - \pi, \frac{(i+1)\pi}{2^{j+1}} - \pi\right], \qquad i = 0, \dots, 2^{j+1}-1.$$

Then

$$G_{\rm II}(x,s) = \mathbb{1}_{\Omega_{\rm II}}(x,s)v(x)w(s) = \sum_{j=0}^{\infty}\sum_{i=0}^{2^j-1} v(x)\mathbb{1}_{I_{2i,j}}(x)w(s)\mathbb{1}_{I_{2i+1,j}}(s).$$

Let $S_{i,j} = |v\mathbb{1}_{I_{2i,j}}\rangle\langle w\mathbb{1}_{I_{2i+1,j}}|$. Then

$$\alpha_{i,j} := \|S_{i,j}\| = \|v\|_{L^2(I_{2i,j})}\|w\|_{L^2(I_{2i+1,j})} \leq \frac{m}{2^j}$$

where m is independent of i and j. The constant m depends on λ and it is finite as a consequence of (2.4), (2.7), and (2.3). Let

$$S_j = \sum_{i=0}^{2^j-1} S_{i,j} = \sum_{i=0}^{2^j-1} \alpha_{i,j}\left|\frac{v\mathbb{1}_{I_{2i,j}}}{\|v\|_{L^2(I_{2i,j})}}\right\rangle\left\langle\frac{w\mathbb{1}_{I_{2i+1,j}}}{\|w\|_{L^2(I_{2i+1,j})}}\right|. \tag{4.1}$$

Since the intervals $I_{i,j}$ are pairwise disjoint for a fixed j, the right side of (4.1) is a singular value decomposition for S_j. Then

$$\|S_j\|_p = \left(\sum_{i=0}^{2^j-1}\alpha_i^p\right)^{1/p} \leq \frac{m}{\left(2^{\frac{p-1}{p}}\right)^j}.$$

By the triangle inequality, this ensures that

$$\|R_{\rm II}(\lambda)\|_p < \infty$$

for all $p > 1$ as required.

The fact that $R_{\rm I}(\lambda) \in \mathcal{C}_p$ for all $p > 1$ follows by an analogous decomposition of the support $\Omega_{\rm I} = \{(x,s) \in [-\pi,\pi]^2 : |x| \geq |s|\}$ as the union of disjoint rectangles and a very similar argument. □

REMARK 7. Let λ_n be the eigenvalues of $L_{\rm per}$ and let $\lambda \neq \lambda_n$. By virtue of [**DuSc**, Corollary XI.9.7], the series $\sum_{n=0}^{\infty}(\lambda-\lambda_n)^{-p}$ converges absolutely and

$$\sum_{j=0}^{\infty}|\lambda-\lambda_n|^{-p} \leq \|(\lambda - L_{\rm per})^{-1}\|_p^p \tag{4.2}$$

for all $p > 1$. According to the results of [**We2**] on the case $f(x) = (2/\pi)\sin x$, it is known that $\lambda_n \sim n^2$ as $n \to \infty$. Hence we know that $(\lambda - L_{\rm per})^{-1} \notin \mathcal{C}_{1/2}$. As $L_{\rm per} \neq L_{\rm per}^*$, the inequality in (4.2) cannot generally be reversed for any $p > 0$. The question of whether $(\lambda - L_{\rm per})^{-1} \in \mathcal{C}_p$ for $p > 1/2$ will be addressed in subsequent work.

References

[BeO'BSa] E. S. Benilov, S. B. G. O'Brien, and I. A. Sazonov, *A new type of instability: explosive disturbances in a liquid film inside a rotating horizontal cylinder*, J. Fluid Mech. **497**, 201–224 (2003).

[BLM] L. Boulton, M. Levitin, and M. Marletta, *A PT-symmetric periodic problem with boundary and interior singularities*, J. Diff. Eq., to appear (2010); preprint at `http://arxiv.org/abs/0801.0172`

[ChPe] M. Chugunova and D. Pelinovsky, *Spectrum of a non-self-adjoint operator associated with the periodic heat equation*, J. Math. Anal. Appl. **342**, 970–988 (2008).

[ChKaPy] M. Chugunova, I. Karabash, and S. Pyatkov, *On the nature of ill-posedness of the forward-backward heat equation*, Integral Equations and Operator Theory **65**, 319–344 (2009).

[Da1] E. B. Davies, *Spectral Theory and Differential Operators*, Cambridge University Press, Cambridge, 1995.

[Da2] E. B. Davies, *An indefinite convection-diffusion operator*, LMS J. Comp. Math. **10**, 288–306 (2007).

[DaWe] E. B. Davies and J. Weir, *Convergence of eigenvalues for a highly non-self-adjoint differential operator*, preprint `http://arxiv.org/abs/0809.0787v1` (2008).

[DuSc] N. Dunford and J. Schwartz, *Linear Operators, part II: Spectral Theory*, Interscience, New York (1957).

[GoKr] I. Gohberg and M. Krein, *Introduction to the theory of linear non-selfadjoint operators*, Translation of Mathematical Monographs 18, AMS, Providence (1969).

[We1] J. Weir, *An indefinite convection-diffusion operator with real spectrum*, Appl. Math. Letters **22**, 280–283 (2009).

[We2] J. Weir, *Correspondence of the eigenvalues of a non-self-adjoint operator to those of a self-adjoint operator*, preprint `http://arxiv.org/abs/0801.4959v2` (2008).

Department of Mathematics and Maxwell Institute for Mathematical Sciences, Heriot-Watt University, Riccarton, Edinburgh, EH14 4AS, United Kingdom

E-mail address: `L.Boulton@hw.ac.uk`

Department of Mathematics and Statistics, Reading University, Whiteknights, PO Box 220, Reading RG6 6AX, United Kingdom

E-mail address: `M.Levitin@reading.ac.uk`

Cardiff School of Mathematics and Wales Institute of Mathematical and Computational Sciences, Cardiff University, Senghennydd Road, Cardiff CF24 4AG, United Kingdom

E-mail address: `Marco.Marletta@cs.cardiff.ac.uk`

Amer. Math. Soc. Transl.
Volume **231**, 2010

Spectral Instability of Semiclassical Operators

Nils Dencker

1. Introduction

The spectral instability of nonnormal operators is a phenomenon which has recently attracted interest in applications. It gives an obstruction to the accurate computation of eigenvalues of large nonnormal matrices and has applications in a wide field, from random matrix theory, the stability of flows to things as mundane as brake squeal (see [**70**] for more examples). The standard example of the spectral instability of nonnormal matrices is the following perturbation of the $N \times N$ Jordan matrix

$$\begin{pmatrix} 0 & 1 & 0 & \dots & 0 & 0 \\ 0 & 0 & 1 & \dots & 0 & 0 \\ \vdots & \vdots & \vdots & \ddots & \vdots & \vdots \\ 0 & 0 & 0 & \dots & 1 & 0 \\ 0 & 0 & 0 & \dots & 0 & 1 \\ \varepsilon & 0 & 0 & \dots & 0 & 0 \end{pmatrix}, \qquad \varepsilon > 0,$$

which has characteristic equation $z^N + \varepsilon = 0$ and eigenvalues

$$\lambda_k = \varepsilon^{1/N} e^{i\pi(1+2k)/N}, \qquad k = 1, \dots, N.$$

We find for large N that $|\lambda_k| = \varepsilon^{1/N} \approx 1$. Thus a small perturbation can give a large change of the eigenvalues. This holds in general: Lidskiĭ, in a pioneering work [**52, 56**], showed that small perturbations $A + \varepsilon B$ of an $N \times N$ matrix A could produce "Lidskii circles" of eigenvalues. Davies and Hager [**15**] has gone further, proving that for large N most eigenvalues of random perturbations of the $N \times N$ Jordan matrix will be very close to the unit circle. The spectral instability of Jordan matrices is the worst case, since in general the minimal distance (under permutations) between the spectra of two $N \times N$ matrices is bounded by

$$(\|A\| + \|B\|)^{1-1/N} \|A - B\|^{1/N}$$

where $\|A\|$ is the standard matrix norm; see [**25**]. This spectral instability complicates the mathematical modelling of nonsymmetric problems, since more accurate models usually use larger matrices giving less stability under computational errors. The spectral instability for nonnormal matrices seems to have been discovered and rediscovered at least five times according to [**70**]; see the references there. A simple

2010 *Mathematics Subject Classification.* Primary 35S05; Secondary 35P05, 47G30, 58J40.

way of measuring the spectral stability of matrices is the *ε-pseudospectrum*, given by the following definition.

DEFINITION 1.1. Let A be an $N \times N$ matrix. Then the ε-pseudospectrum $\operatorname{Spec}_\varepsilon(A)$ is the set of $z \in \mathbf{C}$ such that

$$\|(z \operatorname{Id}_N - A)^{-1}\| > \varepsilon^{-1}.$$

Here $(z \operatorname{Id}_N - A)^{-1}$ is the resolvent of A and we define $\|(z \operatorname{Id}_N - A)^{-1}\| = \infty$ if $z \in \operatorname{Spec}(A)$, the spectrum of A. Thus

$$\operatorname{Spec}(A) = \bigcap_{\varepsilon > 0} \operatorname{Spec}_\varepsilon(A).$$

An equivalent definition is that $\operatorname{Spec}_\varepsilon(A)$ is the set of $z \in \mathbf{C}$ such that $z \in \operatorname{Spec}(A+B)$ for some B with $\|B\| < \varepsilon$ (see [**70**] for other equivalent definitions). These eigenvalues are called *ε-pseudoeigenvalues* and the corresponding eigenvectors are called *ε-pseudoeigenvectors* or *ε-pseudomodes.*

For normal matrices we have that

$$\|(z \operatorname{Id}_N - A)^{-1}\| = \operatorname{dist}(z, \operatorname{Spec}(A))^{-1}$$

so that the ε-pseudospectrum is contained in an ε neighborhood of the spectrum. But for nonnormal matrices, the ε-pseudospectrum could be a much larger set, as seen above for the Jordan matrices. Observe that when $|z| \geq \|A\|$ we have for any matrix A the estimate

$$\|(z \operatorname{Id}_N - A)^{-1}\| \leq \cot\left(\frac{\pi}{4N}\right) \operatorname{dist}(z, \operatorname{Spec}(A))^{-1}$$

by Davies and Simon [**16**]. The use of the resolvent norm has given rise to the pseudospectral method in numerical analysis. There are several other ways of measuring spectral stability, for example the *structured ε-pseudospectrum* [**23**] and the *second-order relative spectrum* [**39**]. In the present review article, it will not be possible to make a more thorough treatment of the spectral instability of matrices. For more results, examples, and references we refer the reader to [**70**].

Spectral instability also occurs for nonselfadjoint partial differential operators. Lidskiĭ, in a series of papers [**41**]–[**51**], studied the completeness and summability of eigenfunction expansions of certain types of nonselfadjoint differential operators. The first to study the stability of the spectrum for nonselfadjoint differential operators seems to be Reddy, Schmid, and Henningson [**64**] who studied the complex Airy and the Orr-Sommerfeld equations. When studying the spectral instability of differential operators, it is illuminating (and physically relevant) to study semiclassical operators $P(x, hD_x)$, where the parameter $0 < h \leq 1$ usually is called "Planck's constant". Observe that this parameter could in applications be, for example, the inverse of the Reynolds number, the Péclet number, or the frequency. Then, the spectral instability can be defined as a function of h; see Definition 2.7. Davies studied the semiclassical complex harmonic oscillator [**5**] and the general semiclassical Schrödinger equation with complex potential in one dimension [**6**]. He proved that spectral instability is generic, in the sense that the norm of the resolvent blows up as any power of semiclassical parameter h almost everywhere in the numerical range of the semiclassical principal symbol (see Theorem 2.1 below). Zworski [**72, 73**] made the important observation that the spectral instability of semiclassical differential operators is directly connected with the bracket condition and the solvability question. In these infinite-dimensional problems, the condition numbers of the eigenvectors could grow exponentially; see [**1**]. Another important problem is the behavior of evolution semigroups for nonnormal operators and the relation to pseudospectrum; see for example [**9**].

In this paper, we shall review the spectral instability of nonnormal semiclassical differential operators in several variables, in particular for systems.

2. Semiclassical operators

In this section, we shall consider the spectral instability of semiclassical partial differential operators, following [**21**]. Let us start with a typical (and physically relevant) example, the semiclassical Schrödinger operator:

$$P(h) = h^2\Delta + V(x), \qquad V \in C^\infty(\mathbf{R}^n), \tag{2.1}$$

where $\Delta = -\sum_{j=1}^n \partial_{x_j}^2$ is the positive Laplacian. In [**21**] the *semiclassical pseudospectrum* of $P(h)$ was defined as

$$\Lambda(p) = \overline{\{\xi^2 + V(x) : \ (x,\xi) \in \mathbf{R}^{2n}\,,\ \operatorname{Im}\langle \xi, V'(x)\rangle \neq 0\}}\,. \tag{2.2}$$

Of course, in the analytic case $\Lambda(p)$ is either empty or the closure of the set of all values of $p = \xi^2 + V(x)$. The following result shows that the resolvent blows up almost everywhere inside the semiclassical pseudospectrum.

THEOREM 2.1. *Suppose that* $P(h) = -h^2\Delta + V(x)$, *with* $V \in \mathcal{C}^\infty(\mathbf{R}^n)$. *Then, for any* $z \in \{\xi^2+V(x) : \ (x,\xi) \in \mathbf{R}^{2n}, \operatorname{Im}\langle \xi, V'(x)\rangle \neq 0\}$ *there exists* $u(h) \in L^2(\mathbf{R}^n)$ *with the property that* $\|u(h)\| = 1$ *and*

$$\|(P(h) - z)u(h)\| \leq C_N h^N \quad \forall N, \qquad h \to 0. \tag{2.3}$$

In addition, $u(h)$ *is localized to a point in phase space,* (x,ξ), *with* $p(x,\xi) = z$, *in the sense that* $WF_h(u) = \{(x,\xi)\}$. *Here the semiclassical wave front set,* $WF_h(u)$, *is given by Definition* 2.2. *If the potential* V *is real analytic, then we can replace* h^∞ *by* $\exp(-1/Ch)$ *in* (2.3).

These "almost eigenvectors" are called pseudomodes. Theorem 2.1 was proved by Davies [**6**] for Schrödinger operators in one dimension, but as was pointed out by Zworski [**72, 73**], it follows in general from a simple adaptation of the now classical results of Hörmander [**33, 34**] and Duistermaat-Sjöstrand [**24**] and is connected with the solvability problem by the bracket condition. It follows from (2.3) that, unlike the case of normal operators, the resolvent blows up as any power of h:

$$\|(P(h) - z)^{-1}\| \geq C_N h^{-N} \qquad \forall N, \quad h \to 0,$$

for z in an open set. That is particularly striking when $P(h)$ only has discrete spectrum $\operatorname{Spec}(P(h))$. The Schrödinger equation (2.1) has discrete spectrum if, for example,

$$\begin{aligned} &|\partial_x^\alpha V(x)| \leq C_\alpha (1 + |x|)^{m-|\alpha|}, \\ &(1 + |x|^m + |\xi|^2)/C \leq |\xi^2 + V(x)|, \quad |(x,\xi)| \geq C, \end{aligned} \tag{2.4}$$

where $m > 0$. In the analytic case, it suffices that (2.4) holds for $|\alpha| = 0$ as $|x| \to \infty$ and $|\operatorname{Im} x| < c_0$. The symbol $p = \xi^2 + V(x)$ will then avoid all sufficiently negative values and Fredholm theory guarantees that $P(h)$ has discrete spectrum for h small enough (see Proposition 3.12 below).

If $P(h)-z_1 \operatorname{Id}_N$ is invertible for some z_1, we may consider the following operator which has bounded symbol:

$$(P(h)-z_1)^{-1}(P(h)-z_2), \qquad z_2 \neq z_1;$$

see Remark 3.13. Thus it is sufficient to consider the quantization of functions which have uniform bounds on any derivative:

$$p \in C_{\mathrm{b}}^{\infty}(T^*\mathbf{R}^n) = \{u \in C^{\infty}(T^*\mathbf{R}^n) : \ \partial^\alpha u \in L^\infty(T^*\mathbf{R}^n) \ \ \forall\, \alpha\}.$$

In the analytic case we will assume that $p(x,\xi)$ is bounded and holomorphic in a tubular neighborhood of $T^*\mathbf{R}^n \simeq \mathbf{R}^{2n} \subset \mathbf{C}^{2n}$.

We shall use the Weyl quantization

$$p^w(x,hD_x)u = \frac{1}{(2\pi)^n}\iint p\left(\frac{x+y}{2}, h\xi\right) e^{i\langle x-y,\xi\rangle}u(y)\,dyd\xi \tag{2.5}$$

which for bounded symbols $p \in C_{\mathrm{b}}^{\infty}(T^*\mathbf{R}^n)$ gives operators bounded on $L^2(\mathbf{R}^n)$; see [**22**, Chapter 7]. The advantage of the Weyl quantization is that real symbols give symmetric operators. We shall also consider more general operators,

$$P(h) \sim \sum_{j=0}^{\infty} h^j p_j^w(x,hD),$$

in which case we call p_0 the principal symbol of $P(h)$. Since the results only depend on the principal symbol and different quantizations only differ in the lower-order terms, we could as well have used the Kohn-Nirenberg or other quantizations.

In the case of analytic symbols we assume that p_j are bounded holomorphic functions in a tubular neighborhood of $T^*\mathbf{R}^n$ and that

$$|p_j(z,\zeta)| \le C^j j^j, \qquad |\operatorname{Im}(z,\zeta)| \le 1/C. \tag{2.6}$$

This gives exponentially small errors in the expansions.

Let $\Sigma(p) = \overline{p(T^*\mathbf{R}^n)}$. Then it is well known that the spectrum $\operatorname{Spec}(P(h)) \subseteq \Sigma(p)$ (see Proposition 3.11) and spectral instability can only occur in $\Sigma(p)$. Now, [**21**] generalized the definition (2.2) of the *semiclassical pseudospectrum* to

$$\Lambda(p) \overset{\text{def}}{=} \overline{p(\{(x,\xi) \in T^*\mathbf{R}^n : \ \{p,\bar p\}(x,\xi) \neq 0\})} \subseteq \Sigma(p) \tag{2.7}$$

where we have used the Poisson bracket:

$$\{f,g\} = H_f g = \sum_{j=1}^{n} \partial_{\xi_j} f \partial_{x_j} g - \partial_{x_j} f \partial_{\xi_j} g.$$

The nonvanishing of $\{p,\bar p\}$ is a classical equivalent of the operator not being normal, since this is the principal symbol of the commutator $[P(h),P^*(h)]$. Hörmander showed in [**33**] that the vanishing of $\{p,\bar p\}$ at $p^{-1}(0)$ is a necessary condition for the solvability of $P(h)$. Thus almost all partial differential equations are nonsolvable. Note that in the analytic case we have

$$\Lambda(p) = \emptyset \ \text{ or } \ \Lambda(p) = \Sigma(p).$$

Actually, it is the sign of the Poisson bracket that is important. Therefore, we define

$$\begin{aligned} &\Lambda_\pm(p) = \{p(x,\xi) : \ \pm\{\operatorname{Re} p, \operatorname{Im} p\}(x,\xi) > 0\} \subseteq \Sigma(p), \\ &\Sigma_\infty(p) = \{z : \ \exists\,(x_j,\xi_j) \to \infty \ \ \lim_{j\to\infty} p(x_j,\xi_j) = z\}. \end{aligned} \tag{2.8}$$

Then $\Lambda(p) = \overline{\Lambda_-(p) \bigcup \Lambda_+(p)}$ and $\Sigma_\infty(p)$ is the set of limits of p at infinity. In order to measure the singularities of the solutions, we shall recall the semiclassical wave front sets.

DEFINITION 2.2. For $u \in L^2(\mathbf{R}^n)$ we say that $(x,\xi) \notin \mathrm{WF}_h(u)$ if there exists $a \in C_0^\infty(T^*\mathbf{R}^n)$ such that $a(x,\xi) \neq 0$ and the L^2 norm

$$\|a^w(x,hD)u\| \le C_k h^k \quad \forall\, k, \qquad h \to 0. \tag{2.9}$$

We call $\mathrm{WF}_h(u)$ the semiclassical wave front set of u.

Observe that this definition is equivalent to the definition (2.5) in [**21**] which uses the FBI transform T: $(x_0,\xi_0) \notin \mathrm{WF}_h(u)$ if $\|Tu(x,\xi)\| = \mathcal{O}(h^\infty)$ when $|(x,\xi)-(x_0,\xi_0)| \ll 1$. We may also define the *analytic* semiclassical wave front set by the condition that $\|Tu(x,\xi)\| = \mathcal{O}(e^{-c/h})$ in a neighborhood of (x_0,ξ_0) for some $c>0$; see (2.6) in [**21**]. In this more general setting, Theorem 1.2 in [**21**] gives

THEOREM 2.3. *Suppose that $n \ge 2$, $p \in C_{\mathrm{b}}^\infty(T^*\mathbf{R}^n)$ and that $p^{-1}(z)$ is compact for a dense set of values $z \in \mathbf{C}$. If $P(h)$ has the principal symbol $p(x,\xi)$, then*

$$\Lambda(p) \setminus \Sigma_\infty \subseteq \overline{\Lambda_-(p)}$$

and for every $z \in \Lambda_-(p)$ there exists $u(h) \in L^2(\mathbf{R}^n)$ with the property that $\|u(h)\| = 1$ and

$$\|(P(h)-z)u(h)\| \le C_N h^N \quad \forall\, N, \qquad h \to 0. \tag{2.10}$$

In addition, $u(h)$ is localized to a point in phase space, (x,ξ), with $p(x,\xi)=z$, so that $WF_h(u) = \{(x,\xi)\}$. If in addition p has a bounded holomorphic continuation to $\{(x,\xi) \in \mathbf{C}^{2n},\ |\mathrm{Im}(x,\xi)| \le 1/C\}$, then we obtain that (2.10) holds with h^∞ replaced by $\exp(-c/h)$, $c>0$. If $n=1$, then the same conclusions hold provided that each component of $\mathbf{C} \setminus \Sigma_\infty(p)$ has a nonempty intersection with $\complement\Lambda(p)$.

Theorem 2.3 gives that norm of the resolvent $\|(P(h)-z_0)^{-1}\| \ge C_N h^{-N}$ as $h \to 0$ for any N. In general, one cannot construct an almost solution (or quasimode) $u(h)$ for an arbitrary interior point of $\Lambda(p) \setminus \Sigma_\infty(p)$; see the example in [**21**, Section 3]. However, for many points $\Lambda(p) \setminus \Lambda_-(p)$ quasimodes can exist, since the vanishing of the Poisson bracket $\{\mathrm{Re}\, p, \mathrm{Im}\, p\}$ is not enough to guarantee the absence of a quasimode. In fact, as proved by Pravda-Starov [**57**], a violation of the condition $(\overline{\Psi})$ (see [**36**, Section 26.4]) can produce quasimodes. Condition $(\overline{\Psi})$ is directly connected to the Nirenberg-Treves conjecture, which says that a pseudo-differential operator $P(x,D)$ of principal type is locally solvable if and only if the principal symbol p satisfies condition (Ψ) or, equivalently, $\overline{p}$ satifies condition $(\overline{\Psi})$. Here principal type means that the principal symbol satisfies (2.12) for $z_0=0$. The Nirenberg-Treves conjecture was recently proved in [**19**].

In the case of dimension one the topological condition is necessary, according to the following example from [**21**].

EXAMPLE 2.4. Let

$$p(x,\xi) = \frac{(\xi+ix)^2}{1+x^2+\xi^2}.$$

Then $\{\mathrm{Re}\, p, \mathrm{Im}\, p\}(x,\xi) > 0$ for $(x,\xi) \neq (0,0)$.

For principal symbols arising from differential equations of *even order* in one variable, like the Schrödinger equation, we always have

$$\sum_{(x,\xi)\in p^{-1}(z)} \operatorname{sgn}\{\operatorname{Re} p, \operatorname{Im} p\}(x,\xi) = 0 \tag{2.11}$$

for a dense set of values z. In fact, $p(x,\xi) = p(x,-\xi) = z$,

$$\{\operatorname{Re} p, \operatorname{Im} p\}(x,-\xi) = -\{\operatorname{Re} p, \operatorname{Im} p\}(x,\xi),$$

and the set of values z corresponding to $\xi \neq 0$ is dense in the set of values for which the bracket is nonzero.

In simple one-dimensional examples we can already see that the spectrum, $\sigma(P(h))$, typically lies deep inside the pseudospectrum $\Lambda(p)$; see [**5, 6, 69**] for numerical examples of this phenomenon. Consider for example the following non-selfadjoint operator $P(h) = (hD_x)^2 + i(hD_x) + x^2$. A formal conjugation

$$e^{-x/2h}P(h)e^{x/2h} = (hD_x)^2 + x^2 + \frac{1}{4}$$

shows that the spectrum of $P(h)$ is given by $(2n+1)h + 1/4$, while

$$\Lambda(p) = \{z : \operatorname{Re} z \geq (\operatorname{Im} z)^2\}$$

since $p = \xi^2 + i\xi + x^2$.

To obtain that the spectrum is inside the pseudospectrum for general operators we have to make assumptions on $z_0 \in \partial\Sigma(p)$. One is the principal type condition:

$$p(x,\xi) = z_0 \implies dp(x,\xi) \neq 0, \quad (x,\xi) \in T^*\mathbf{R}^n. \tag{2.12}$$

Then p is of principal type if (2.12) holds for $z_0 = 0$. We also assume a dynamical condition:

(2.13)

$$\exists\, \lambda \in \mathbf{C} \text{ so that } \textit{no} \text{ complete trajectory of } H_{\operatorname{Re}(\lambda p)} \text{ is contained in } p^{-1}(z_0).$$

Recall that the trajectories of $H_{\operatorname{Re}(\lambda p)}$ on $\operatorname{Re}(\lambda p) = 0$ are called bicharacteristics of $\operatorname{Re}(\lambda p)$ and semibicharacteristics of p. Under these conditions we obtain from Theorem 1.3 in [**21**]

THEOREM 2.5. *Suppose that $p \in C_b^\infty(T^*\mathbf{R}^n)$ and that the principal symbol of $P(h)$ is given by $p(x,\xi)$. If $z_0 \in \partial\Sigma(p) \setminus \Sigma_\infty(p)$ satisfies* (2.12) *and* (2.13), *then for any $M > 0$ there exists $h_M > 0$ such that*

$$\{z : |z - z_0| < Mh\log(1/h)\} \cap \sigma(P(h)) = \emptyset, \qquad 0 < h < h_M,$$

and $\|(P(h) - z_0)^{-1}\| \leq C/h$ for $0 < h \ll 1$. If in addition p is a bounded holomorphic function in a complex tubular neighborhood of $\mathbf{R}^n$, then there exists $C_0 > 0$ such that

$$\{z : |z - z_0| < C_0\} \cap \sigma(P(h)) = \emptyset, \qquad 0 < h \ll 1.$$

Observe that we can replace $\Sigma(p)$ in Theorem 2.5 by $\Lambda(p)$. In fact, condition (2.13) gives that $z_0 \in \partial\Lambda(p)$, since the bracket cannot vanish identically on all level sets close to $p^{-1}(z_0)$. If (2.13) is violated, then for a large class of *dissipative operators*, the spectrum lies arbitrarily close (as $h \to 0$) to the boundary of the pseudospectrum; see [**21**, Section 6]. The example in [**21**, Section 4] shows that the C^∞ result of Theorem 2.5 is optimal.

At the boundary of the pseudospectrum we may expect an improved bound on the resolvent when some additional nondegeneracy is assumed. We shall borrow our notation from [**36**, Chapter 27]. If $p = p_1 + ip_2 \in \mathcal{C}^\infty$ with real valued p_j, then we define the repeated Poisson brackets

$$p_I = H_{p_{i_1}} H_{p_{i_2}} \dots H_{p_{i_{k-1}}} p_{i_k}$$

where $I = (i_1, i_2, \dots, i_k) \in \{1, 2\}^k$ and $|I| = k > 1$ is the order of the bracket.

We say that $z_0 \in \Sigma(p) \setminus \Sigma_\infty(p)$ is of *finite type* for p if for any $(x, \xi) \in p^{-1}(z_0)$ there exists $k > 1$ and $I \in \{1, 2\}^k$ such that

$$p_I(x, \xi) \neq 0 \tag{2.14}$$

which implies (2.12). The *order* of p at (x, ξ) is

$$k(x, \xi) = \max\{ j \in \mathbf{Z} :\ p_I(x, \xi) = 0 \ \text{ for } 1 < |I| \leq j \}. \tag{2.15}$$

The order of z_0 is the maximum of the order of p at (x, ξ) for $(x, \xi) \in p^{-1}(z_0)$.

THEOREM 2.6. *Assume that $p \in \mathcal{C}_{\mathrm{b}}^\infty(T^*\mathbf{R}^n)$ and that the principal symbol of $P(h)$ is $p(x, \xi)$. If $z_0 \in \partial\Sigma(p) \setminus \Sigma_\infty(P)$ is of finite type for p of order $k \geq 1$, then k is* even *and $\exists\ h_0,\ C > 0$ so that*

$$\|(P(h) - z_0)^{-1}\| \leq Ch^{-\frac{k}{k+1}}, \qquad 0 < h \leq h_0. \tag{2.16}$$

Also, there exist $h_1,\ c_0 > 0$ such that

$$\left\{ z :\ |z - z_0| \leq c_0 (h \log h^{-1})^{\frac{k}{k+1}} \right\} \cap \operatorname{Spec}(P(h)) = \emptyset, \qquad 0 < h \leq h_1. \tag{2.17}$$

We obtain (2.16) from [**21**, Theorem 1.4] and (2.17) from [**66**]. In one dimension, the resolvent estimate was proved in [**71**] and in some special cases by Boulton [**3**] who also showed that the bounds are optimal. As was demonstrated by Trefethen [**69**], this is also easy to see numerically.

We have the following simple higher-dimensional example from [**21**] to which we can apply Theorem 2.6. Let $W \in \mathcal{C}_{\mathrm{b}}^\infty(\mathbf{R}^2)$ be a nonnegative function, vanishing on the circle $x_1^2 + x_2^2 = 1$ and consider

$$P(h) = -h^2\Delta + iW(x) + i(x_1^2 + x_2^2 - 1)^m \quad \text{with } m \text{ even.}$$

Then the estimate (2.16) holds for $z_0 > 0$ uniformly on compact subsets of $(0, \infty)$, with $k = 2m$. This is due to the (simple) tangency of some bicharacteristics of the real part to the set where the imaginary part vanishes.

One could also define the semiclassical pseudospectrum of $P(h)$ as the closure of the set of values z for which (2.10) holds; actually this was an alternative definition in [**21**]. Pravda-Starov [**59**] introduced the following refined definitions.

DEFINITION 2.7. Let $P(h)$, $0 < h \leq 1$, be a semiclassical family of operators on $L^2(\mathbf{R}^n)$ with domain D. For $\mu > 0$ we define the *pseudospectrum of index* μ as the set

$$\begin{aligned}\Lambda_\mu^{\mathrm{sc}}(P(h)) = \{z \in \mathbf{C} :\ &\forall C > 0,\ \forall h_0 > 0, \\ &\exists 0 < h < h_0,\ \|(P(h) - z\,\mathrm{Id}_N)^{-1}\| \geq Ch^{-\mu}\}\end{aligned}$$

and the *injectivity pseudospectrum of index* μ as

$$\begin{aligned}\lambda_\mu^{\mathrm{sc}}(P(h)) = \{z \in \mathbf{C} :\ &\forall C > 0,\ \forall h_0 > 0, \\ &\exists 0 < h < h_0,\ \exists u \in D,\ \|u\| = 1,\ \|(P(h) - z\,\mathrm{Id}_N)u\| \leq Ch^\mu\}.\end{aligned}$$

We define the *pseudospectrum of infinite index* as $\Lambda_\infty^{\mathrm{sc}}(P(h)) = \bigcap_\mu \Lambda_\mu^{\mathrm{sc}}(P(h))$ and correspondingly the *injectivity pseudospectrum of infinite index*.

With these definitions, we find that the semiclassical harmonic oscillator has pseudospectrum $\lambda_\infty^{\mathrm{sc}} = \Lambda_\infty^{\mathrm{sc}} = \overline{\mathbf{R}_+}$, since the eigenvalues are $(2n+1)h$, $n \in \mathbf{N}$. One way of avoiding this is to have $\exists\, h_0 > 0 \;\; \forall\, 0 < h < h_0$ instead in the definitions.

Observe that we have the obvious inclusion $\lambda_\mu^{\mathrm{sc}}(P(h)) \subseteq \Lambda_\mu^{\mathrm{sc}}(P(h))$, $\forall\, \mu$. We get equality if, for example, $P(h)$ is Fredholm of index ≥ 0. Theorem 2.3 and Proposition 3.11 show that

$$\Lambda_-(p) \subseteq \lambda_\infty^{\mathrm{sc}}(P(h)) \subseteq \Lambda_\infty^{\mathrm{sc}}(P(h)) \subseteq \Sigma(p).$$

The interplay between classical properties of symbols and the existence of localized quasimodes can also be observed in the Berezin-Toeplitz quantization of compact symplectic Kähler manifolds. See Chapman and Trefethen [**4**] for the case of the torus and Borthwick and Uribe [**2**] for the general C^∞ case.

For nonprincipal type scalar semiclassical operators, there are still many open questions about the pseudospectrum. Pravda-Starov has studied the pseudospectrum for nonprincipal type operators in one dimension [**60**], in the case when the Hessian of the principal symbol is elliptic and nonnormal. He has also studied the pseudospectrum for general elliptic nonnormal quadratic Weyl operators [**61**].

Sjöstrand and Hager have proved that for certain random perturbations of pseudodifferential operators, the spectrum will satisfy an asymptotic Weyl law; see [**30**]. It is interesting to compare this to the recent proof by Tao and Vu [**68**] of the *circular law* for random matrices, for which the spectrum is uniformly distributed in a disk.

3. Systems of semiclassical operators

In this section, we will show how the results for semiclassical scalar operators generalize to systems, following [**20**]. We shall consider $N \times N$ systems of semiclassical pseudodifferential operators of the form

$$P(h) \sim \sum_{j=0}^{\infty} h^j P_j^w(x, hD) \tag{3.1}$$

with $P_j(x,\xi) \in C_{\mathrm{b}}^\infty(T^*\mathbf{R}^n, \mathcal{L}(\mathbf{C}^N, \mathbf{C}^N))$ using the Weyl quantization given by (2.5). As before, C_{b}^∞ is the set of C^∞ functions having all derivatives in L^∞ and $P_0 = \sigma(P(h))$ is the principal symbol of $P(h)$. The operator is said to be elliptic if the principal symbol P_0 is invertible, and of principal type if P_0 vanishes of first order on the kernel; see Definition 4.1. We shall also consider operators with analytic symbols. Then we shall assume that $P_j(x,\xi)$ are bounded and holomorphic in a tubular neighborhood of $T^*\mathbf{R}^n$ satisfying (2.6), which gives exponentially small errors in the calculus. In the following, we shall use the notation $w = (x,\xi) \in T^*\mathbf{R}^n$.

We shall consider the spectrum $\mathrm{Spec}(P(h))$ which is the set of values $z \in \mathbf{C}$ such that the resolvent $(P(h) - z\,\mathrm{Id}_N)^{-1}$ is *not* a bounded operator; here Id_N is the identity in $\mathbf{C}^N$. The spectrum of $P(h)$ is essentially contained in the spectrum of the principal symbol $\mathrm{Spec}(P(w))$, which is given by

$$|P(w) - z\,\mathrm{Id}_N| = 0$$

where $|A|$ is the determinant of the matrix A. For example, if $P(w) = \sigma(P(h))$ is bounded and z is not an eigenvalue of $P(w)$ for any $w = (x,\xi)$ (or a limit of eigenvalues at infinity), then $P(h) - z\,\mathrm{Id}_N$ is invertible by Proposition 3.11 below. When $P(w)$ is an unbounded symbol, one needs additional conditions; see for example

Proposition 3.12. As before, we shall restrict our study to bounded symbols, but we can reduce to this case if $P(h) - z_1 \operatorname{Id}_N$ is invertible for some z_1 by considering

$$(P(h) - z_1 \operatorname{Id}_N)^{-1}(P(h) - z_2 \operatorname{Id}_N), \qquad z_2 \neq z_1;$$

see Remark 3.13. But unless we have conditions on the eigenvalues at infinity, this does not always give a bounded operator.

EXAMPLE 3.1. Let

$$P(\xi) = \begin{pmatrix} 0 & \xi \\ 0 & 0 \end{pmatrix}, \qquad \xi \in \mathbf{R}.$$

Then 0 is the only eigenvalue of $P(\xi)$ but

$$(P(\xi) - z \operatorname{Id}_N)^{-1} = -1/z \begin{pmatrix} 1 & \xi/z \\ 0 & 1 \end{pmatrix} \tag{3.2}$$

and $(P^w - z \operatorname{Id}_N)^{-1} P^w = -z^{-1} P^w$ is unbounded for any $z \neq 0$.

DEFINITION 3.2. Let $P \in C^\infty(T^*\mathbf{R}^n)$ be an $N \times N$ system. We denote the closure of the set of eigenvalues of P by

$$\Sigma(P) = \overline{\{\, \lambda \in \mathbf{C} : \exists\, w \in T^*\mathbf{R}^n,\ |P(w) - \lambda \operatorname{Id}_N| = 0 \,\}} \tag{3.3}$$

and the eigenvalues at infinity by

$$\Sigma_\infty(P) = \{\lambda \in \mathbf{C} :\ \exists\, w_j \to \infty,\ \exists\, u_j \in \mathbf{C}^N,\ |u_j| = 1, \\ |P(w_j)u_j - \lambda u_j| \to 0,\ j \to \infty \,\} \tag{3.4}$$

which is closed in $\mathbf{C}$.

In fact, that $\Sigma_\infty(P)$ is closed follows by taking a suitable diagonal sequence. Observe that, as in the scalar case, we could have $\Sigma_\infty(P) = \Sigma(P)$, for example if $P(w)$ is constant in one direction. It follows from the definition that $\lambda \notin \Sigma_\infty(P)$ if and only if the resolvent is defined and bounded when $|w|$ is large enough:

$$\|(P(w) - \lambda \operatorname{Id}_N)^{-1}\| \le C, \qquad |w| \gg 1, \tag{3.5}$$

where as before $\|A\|$ is the norm of the matrix A.

It is clear from the definition that $\Sigma_\infty(P)$ contains all finite limits of eigenvalues of P at infinity. In fact, if $P(w_j)u_j = \lambda_j u_j$, $|u_j| = 1$, $w_j \to \infty$, and $\lambda_j \to \lambda$, then

$$P(w_j)u_j - \lambda u_j = (\lambda_j - \lambda)u_j \to 0.$$

Example 3.1 shows that in general $\Sigma_\infty(P)$ could be a larger set; in fact, then $\Sigma(P) = \{\,0\,\}$ but $\Sigma_\infty(P) = \mathbf{C}$ by (3.2) and (3.5). But for bounded $N \times N$ symbols $P \in C_{\mathrm{b}}^\infty(T^*\mathbf{R}^n)$ we obtain from Proposition 2.4 in [**20**] that $\Sigma_\infty(P)$ is the set of all limits of the eigenvalues of P at infinity.

One problem with studying systems $P(w)$ is that the eigenvalues are not regular in the parameter w. In general they depend only continuously on w when the multiplicity is not constant.

DEFINITION 3.3. For an $N \times N$ system $P \in C^\infty(T^*\mathbf{R}^n)$ we define

$$\kappa_P(w, \lambda) = \operatorname{Dim} \operatorname{Ker}(P(w) - \lambda \operatorname{Id}_N)$$

and

$$K_P(w, \lambda) = \max\left\{\, k :\ \partial_\lambda^j p(w, \lambda) = 0 \text{ for } j < k \,\right\}$$

where $p(w,\lambda) = |P(w) - \lambda \operatorname{Id}_N|$ is the characteristic polynomial. Let

$$\Omega_k(P) = \{ (w,\lambda) \in T^*\mathbf{R}^n \times \mathbf{C} : \ K_P(w,\lambda) \ge k \}, \qquad k \ge 1.$$

Then $\emptyset = \Omega_{N+1}(P) \subseteq \Omega_N(P) \subseteq \cdots \subseteq \Omega_1(P)$ and we may define

$$\Xi(P) = \bigcup_{j>1} \partial\Omega_j(P) \tag{3.6}$$

where $\partial\Omega_j(P)$ is the boundary of $\Omega_j(P)$ in the relative topology of $\Omega_1(P)$.

Clearly, $\Omega_j(P)$ is a closed set for any $j \ge 1$. We have $\kappa_P \le K_P$ with equality for symmetric systems but in general we need not have equality; see Example 3.1 where $\kappa_P < K_P = 2$ when $\lambda = 0$ and $\xi \ne 0$. By the definition we find that the multiplicity $K_P(w,\lambda)$ of the zeros of $|P(w) - \lambda \operatorname{Id}_N|$ is locally constant on $\Omega_1(P) \setminus \Xi(P)$. If $P(w)$ is symmetric, then κ_P is also constant on $\Omega_1(P) \setminus \Xi(P)$, but in general this is not true; see Example 3.1 where κ_P is discontinuous when $\xi = 0$.

REMARK 3.4. *We find that $\Xi(P)$ is closed and nowhere dense in the relative topology of $\Omega_1(P)$ since it is the union of boundaries of closed sets. We also find that*

$$(w,\lambda) \in \Xi(P) \Leftrightarrow (w,\overline{\lambda}) \in \Xi(P^*)$$

since $|P^* - \overline{\lambda} \operatorname{Id}_N| = \overline{|P - \lambda \operatorname{Id}_N|}$.

EXAMPLE 3.5. Let

$$P(t) = \begin{pmatrix} 0 & 1 \\ t & 0 \end{pmatrix}, \qquad t \in \mathbf{R}.$$

Then $P(t)$ has the eigenvalues $\pm\sqrt{t}$, $\kappa_P \equiv 1$ on $\Omega_1(P)$, and $\Omega_2(P) = \{ (0,0) \}$.

EXAMPLE 3.6. Let

$$P = \begin{pmatrix} w_1 + w_2 & w_3 \\ w_3 & w_1 - w_2 \end{pmatrix}.$$

Then

$$\Omega_1(P) = \left\{ (w;\lambda_j) : \ \lambda_j = w_1 + (-1)^j \sqrt{w_2^2 + w_3^2}, \ j = 1,\ 2 \right\}$$

and $\Omega_2(P) = \{ (w_1, 0, 0; w_1) : \ w_1 \in \mathbf{R} \}$.

DEFINITION 3.7. Let π_j be the projections

$$\pi_1(w,\lambda) = w, \qquad \pi_2(w,\lambda) = \lambda.$$

Then we define for $\lambda \in \mathbf{C}$ the closed sets

$$\Sigma_\lambda(P) = \pi_1\left(\Omega_1(P) \bigcap \pi_2^{-1}(\lambda)\right) = \{ w : \ |P(w) - \lambda \operatorname{Id}_N| = 0 \}$$

and

$$X(P) = \pi_1(\Xi(P)) \subseteq T^*\mathbf{R}^n.$$

Observe that $X(P)$ is nowhere dense in $T^*\mathbf{R}^n$ and that $P(w)$ has constant characteristics near $w_0 \notin X(P)$. This means that $|P(w) - \lambda \operatorname{Id}_N| = 0$ if and only if $\lambda = \lambda_j(w)$ for $j = 1, \dots l$, where the eigenvalues $\lambda_j(w) \ne \lambda_k(w)$ for $j \ne k$ when $|w - w_0| \ll 1$.

DEFINITION 3.8. For an $N \times N$ system $P \in C^\infty(T^*\mathbf{R}^n)$ we define the *weakly singular eigenvalue set*

$$\Sigma_{ws}(P) = \pi_2\,(\Xi(P)) \subseteq \mathbf{C} \tag{3.7}$$

and the *strongly singular eigenvalue set*

$$\Sigma_{ss}(P) = \left\{\, \lambda :\ \pi_2^{-1}(\lambda) \bigcap \Omega_1(P) \subseteq \Xi(P) \,\right\}. \tag{3.8}$$

It follows from the definition that $\Sigma_{ss}(P) \subseteq \Sigma_{ws}(P)$. Clearly $\Sigma_{ws}(P) \bigcup \Sigma_\infty(P)$ and $\Sigma_{ss}(P) \bigcup \Sigma_\infty(P)$ are closed, and $\Sigma_{ss}(P)$ is nowhere dense. On the other hand, it is possible that $\Sigma_{ws}(P) = \Sigma(P)$; for instance in Example 3.6 we have $\Sigma_{ws}(P) = \Sigma(P) = \mathbf{R}$ and $\Sigma_{ss}(P) = \emptyset$. In fact, the eigenvalues then coincide only when $w_2 = w_3 = 0$ but the eigenvalue $\lambda = w_1$ is also attained at some point where $w_2 \neq 0$. When we have constant characteristics, the Implicit Function Theorem immediately gives the following result.

REMARK 3.9. *Let $P \in C^\infty(T^*\mathbf{R}^n)$ be an $N \times N$ system. If $(w_0, \lambda_0) \in \Omega_1(P) \setminus \Xi(P)$, then there exists a unique C^∞ function $\lambda(w)$ so that $(w, \lambda) \in \Omega_1(P)$ if and only if $\lambda = \lambda(w)$ in a neighborhood of (w_0, λ_0). If $\lambda_0 \in \Sigma(P) \setminus (\Sigma_{ws}(P) \bigcup \Sigma_\infty(P))$, then $\exists\ \lambda(w) \in C^\infty$ such that $(w, \lambda) \in \Omega_1(P)$ if and only if $\lambda = \lambda(w)$ in a neighborhood of $\Sigma_{\lambda_0}(P)$.*

By Remark 3.9 we find that $\Omega_1(P) \setminus \Xi(P)$ is locally given as a C^∞ manifold over $T^*\mathbf{R}^n$ and that the eigenvalues $\lambda_j(w) \in C^\infty$ outside $X(P)$. This is not true if we instead assume that κ_P is constant on $\Omega_1(P)$; see Example 3.5.

DEFINITION 3.10. A C^∞ function $\lambda(w)$ is called a *germ of eigenvalues* at w_0 for the $N \times N$ system $P \in C^\infty(T^*\mathbf{R}^n)$ if

$$|P(w) - \lambda(w)\,\mathrm{Id}_N\,| \equiv 0 \qquad \text{in a neighborhood of } w_0. \tag{3.9}$$

If this holds in a neighborhood of every point in $\omega \Subset T^*\mathbf{R}^n$, then we say that $\lambda(w)$ is a germ of eigenvalues for P on ω.

If $\lambda_0 \in \Sigma(P) \setminus (\Sigma_{ss}(P) \bigcup \Sigma_\infty(P))$, then there exists $w_0 \in \Sigma_{\lambda_0}(P)$ so that $(w_0, \lambda_0) \in \Omega_1(P) \setminus \Xi(P)$. By Remark 3.9 there exists a C^∞ germ $\lambda(w)$ of eigenvalues at w_0 for P such that $\lambda(w_0) = \lambda_0$. If $\lambda_0 \in \Sigma(P) \setminus (\Sigma_{ws}(P) \bigcup \Sigma_\infty(P))$, then there exists a C^∞ germ $\lambda(w)$ of eigenvalues on $\Sigma_{\lambda_0}(P)$.

As in the scalar case we obtain that the spectrum is essentially discrete outside $\Sigma_\infty(P)$ by the following result, which is Proposition 2.19 in [**20**].

PROPOSITION 3.11. *Assume that the $N \times N$ system $P(h)$ is given by (3.1) with principal symbol $P \in C_{\mathrm{b}}^\infty(T^*\mathbf{R}^n)$. Let Ω be an open connected set, satisfying*

$$\overline{\Omega} \cap \Sigma_\infty(P) = \emptyset \quad \text{and} \quad \Omega \cap \complement\Sigma(P) \neq \emptyset.$$

Then $(P(h) - \lambda\,\mathrm{Id}_N)^{-1}$, $0 < h \ll 1$, $\lambda \in \Omega$, is a meromorphic family of operators with poles of finite rank. In particular, for h sufficiently small, the spectrum of $P(h)$ is discrete in any such set. When $\Omega \cap \Sigma(P) = \emptyset$, we find that Ω contains no spectrum of $P(h)$.

Proposition 3.11 shows that $\Lambda_\mu^{\mathrm{sc}}(P(h)) \subseteq \Sigma(P)$ for any $\mu > 0$. We shall show how the reduction to the case of bounded operator can be done in the systems case, following [**20, 21**]. Let $m(w)$ be a positive function on $T^*\mathbf{R}^n$ satisfying

$$1 \le m(w) \le C(1 + |w - w_0|)^M m(w_0) \qquad \forall\ w,\ w_0 \in T^*\mathbf{R}^n$$

for some C and M. Then m is an admissible weight function and we can define the symbol classes $P \in S(m)$ by

$$\|\partial_w^\alpha P(w)\| \leq C_\alpha m(w) \qquad \forall \alpha.$$

Following [**22**], we can then define the semiclassical operator $P(h) = P^w(x, hD)$. In the analytic case we require that the symbol estimates hold in a tubular neighborhood of $T^*\mathbf{R}^n$:

$$\|\partial_w^\alpha P(w)\| \leq C_\alpha m(\operatorname{Re} w) \qquad \text{for} \quad |\operatorname{Im} w| \leq 1/C \qquad \forall \alpha. \tag{3.10}$$

One typical example of an admissible weight function is $m(x, \xi) = (\langle \xi \rangle^2 + \langle x \rangle^p)$. We shall make the ellipticity assumption

$$\|P^{-1}(w)\| \leq C_0 m^{-1}(w), \qquad |w| \gg 1, \tag{3.11}$$

and in the analytic case we assume this in a tubular neighborhood of $T^*\mathbf{R}^n$ as in (3.10). We then obtain the following result from Proposition 2.20 in [**20**].

PROPOSITION 3.12. *Assume that $P \in S(m)$ is an $N \times N$ system satisfying* (3.11) *and that $z \notin \Sigma(P) \cup \Sigma_\infty(P)$. Then we find that $P^w(x, hD) - z \operatorname{Id}_N$ is invertible for small enough h.*

When m is bounded, (3.11) can be replaced by $z \notin \Sigma_\infty(P)$ and then the result follows from Proposition 3.11. In the reduction to operators with bounded symbols we may use the following result.

REMARK 3.13. *If $z_1 \notin \operatorname{Spec}(P)$, we may define the operator*

$$Q = (P - z_1 \operatorname{Id}_N)^{-1}(P - z_2 \operatorname{Id}_N), \qquad z_2 \neq z_1.$$

Then the resolvents of Q and P are related by

$$(Q - \zeta \operatorname{Id}_N)^{-1} = (1 - \zeta)^{-1}(P - z_1 \operatorname{Id}_N)\left(P - \frac{\zeta z_1 - z_2}{\zeta - 1} \operatorname{Id}_N\right)^{-1}, \qquad \zeta \neq 1,$$

when $\frac{\zeta z_1 - z_2}{\zeta - 1} \notin \operatorname{Spec}(P)$.

EXAMPLE 3.14. Let

$$P(x, \xi) = |\xi|^2 \operatorname{Id}_N + iK(x)$$

where $0 \leq K(x) \in C_{\mathrm{b}}^\infty$. Then we find that $P \in S(m)$ with $m(x, \xi) = |\xi|^2 + 1$. If $0 \notin \Sigma_\infty(K)$, then $K(x)$ is invertible for $|x| \gg 1$, so $P^{-1} \in S(m^{-1})$ at infinity. Since $\operatorname{Re} z \geq 0$ in $\Sigma(P)$, we find from Proposition 3.12 that $P^w(x, hD) + \operatorname{Id}_N$ is invertible for small enough h and $P^w(x, hD)(P^w(x, hD) + \operatorname{Id}_N)^{-1}$ is bounded in L^2 with principal symbol $P(w)(P(w) + \operatorname{Id}_N)^{-1} \in C_{\mathrm{b}}^\infty$.

Observe that if $u = (u_1, \ldots, u_N) \in L^2(\mathbf{R}^n, \mathbf{C}^N)$, we may define $\operatorname{WF}_h(u) = \bigcup_j \operatorname{WF}_h(u_j)$, but this gives no information about which components of u are singular. Therefore we shall define the corresponding vector-valued semiclassical polarization sets.

DEFINITION 3.15. For $u \in L^2(\mathbf{R}^n, \mathbf{C}^N)$, we say that $(w_0, z_0) \notin \operatorname{WF}_h^{pol}(u) \subseteq T^*\mathbf{R}^n \times \mathbf{C}^N$ if there exists $A(h)$ given by (3.1) with principal symbol $A(w)$ such that $A(w_0)z_0 \neq 0$ and $\|a^w(x, hD)u\| \leq C_k h^k$, $\forall k$. We call $\operatorname{WF}_h^{pol}(u)$ the semiclassical polarization set of u.

We could similarly define the *analytic* semiclassical polarization set by using the FBI transform and analytic pseudodifferential operators; see (2.6) in [**21**].

REMARK 3.16. *The semiclassical polarization sets are closed, linear in the fiber, and have the functorial properties of the C^∞ polarization sets in* [**17**]. *In particular, we find that*

$$\pi(\mathrm{WF}_h^{pol}(u) \setminus 0) = \mathrm{WF}_h(u) = \bigcup_j \mathrm{WF}_h(u_j)$$

if π is the projection along the fiber variables: $\pi : T^*\mathbf{R}^n \times \mathbf{C}^N \mapsto T^*\mathbf{R}^n$. We also find that

$$A(\mathrm{WF}_h^{pol}(u)) = \left\{ (w, A(w)z) : \ (w,z) \in \mathrm{WF}_h^{pol}(u) \right\} \subseteq \mathrm{WF}_h^{pol}(A(h)u)$$

if $A(w)$ is the principal symbol of $A(h)$, which gives that $\mathrm{WF}_h^{pol}(Au) = A(\mathrm{WF}_h^{pol}(u))$ when $A(h)$ is elliptic.

Remark 3.16 follows from the proofs of Propositions 2.5 and 2.7 in [**17**].

EXAMPLE 3.17. Let $u = (u_1, \dots, u_N) \in L^2(\mathbf{R}^n, \mathbf{C}^N)$ where $\mathrm{WF}_h(u_1) = \{ w_0 \}$ and $\mathrm{WF}_h(u_j) = \emptyset$ for $j > 1$. Then

$$\mathrm{WF}_h^{pol}(u) = \{ (w_0, (z, 0, \dots)) : \ z \in \mathbf{C} \}$$

since $w_0 \in \mathrm{WF}_h(u)$ and $\|A^w(x, hD)u\| = \mathcal{O}(h^\infty)$ if $A^w u = \sum_{j>1} A_j^w u_j$. By taking a suitable invertible E we obtain

$$\mathrm{WF}_h^{pol}(Eu) = \{ (w_0, zv) : \ z \in \mathbf{C} \}$$

for any $0 \neq v \in \mathbf{C}^N$.

4. Systems of principal type

Recall that the scalar symbol $p(x,\xi) \in C^\infty(T^*\mathbf{R}^n)$ is of *principal type* if $dp \neq 0$ when $p = 0$ by (2.12). In the following we let $\partial_\nu P(w) = \langle \nu, dP(w) \rangle$ for $P \in C^1(T^*\mathbf{R}^n)$ and $\nu \in T^*\mathbf{R}^n$. We shall use the following definition of systems of principal type; in fact, most of the systems we consider will be of this type. We shall denote by $\operatorname{Ker} P$ and $\operatorname{Ran} P$ the kernel and range of the matrix P.

DEFINITION 4.1. The $N \times N$ system $P(w) \in C^\infty(T^*\mathbf{R}^n)$ is of *principal type* at w_0 if

$$\operatorname{Ker} P(w_0) \ni u \mapsto \partial_\nu P(w_0)u \in \operatorname{Coker} P(w_0) = \mathbf{C}^N / \operatorname{Ran} P(w_0) \tag{4.1}$$

is bijective for some $\nu \in T_{w_0}(T^*\mathbf{R}^n)$. The operator $P(h)$ given by (3.1) is of principal type if the principal symbol $P = \sigma(P(h))$ is of principal type.

REMARK 4.2. *If $P(w) \in C^\infty$ is of principal type and $A(w)$, $B(w) \in C^\infty$ are invertible, then APB is of principal type. We have that $P(w)$ is of principal type if and only if the adjoint P^* is of principal type.*

In fact, by Leibniz's rule we have

$$\partial(APB) = (\partial A)PB + A(\partial P)B + AP\partial B \tag{4.2}$$

and $\operatorname{Ran}(APB) = A(\operatorname{Ran} P)$ and $\operatorname{Ker}(APB) = B^{-1}(\operatorname{Ker} P)$ when A and B are invertible, which gives the invariance under left and right multiplication. Since $\operatorname{Ker} P^*(w_0) = \operatorname{Ran} P(w_0)^\perp$, we find that P satisfies (4.1) if and only if

$$\operatorname{Ker} P(w_0) \times \operatorname{Ker} P^*(w_0) \ni (u, v) \mapsto \langle \partial_\nu P(w_0)u, v \rangle \tag{4.3}$$

is a nondegenerate bilinear form. Since $\langle \partial_\nu P^* v, u \rangle = \overline{\langle \partial_\nu P u, v \rangle}$, we find that P^* is of principal type if and only if P is.

Observe that if P has only one vanishing eigenvalue λ (with multiplicity one), then the condition that P is of principal type reduces to the condition in the scalar case: $d\lambda \neq 0$. In fact, by using the spectral projection one can find invertible systems A and B so that

$$APB = \begin{pmatrix} \lambda & 0 \\ 0 & E \end{pmatrix}$$

with E invertible $(N-1)\times(N-1)$ system, and this is of principal type if and only if $d\lambda \neq 0$.

EXAMPLE 4.3. Consider the system

$$P(w) = \begin{pmatrix} \lambda_1(w) & 1 \\ 0 & \lambda_2(w) \end{pmatrix}$$

where $\lambda_j(w) \in C^\infty$, $j = 1, 2$. We find that $P(w) - \lambda \operatorname{Id}_2$ is not of principal type when $\lambda = \lambda_1(w) = \lambda_2(w)$ since $\operatorname{Ker}(P(w) - \lambda \operatorname{Id}_2) = \operatorname{Ran}(P(w) - \lambda \operatorname{Id}_2) = \mathbf{C} \times \{ 0 \}$ is invariant under ∂P.

Observe that the property of being of principal type is not stable under C^1 perturbation, not even when $P = P^*$ is symmetric, by the following example.

EXAMPLE 4.4. The system

$$P(w) = \begin{pmatrix} w_1 - w_2 & w_2 \\ w_2 & -w_1 - w_2 \end{pmatrix} = P^*(w), \qquad w = (w_1, w_2),$$

is of principal type when $w_1 = w_2 = 0$, but it is *not* of principal type when $w_2 \neq 0$ and $w_1 = 0$. In fact,

$$\partial_{w_1} P = \begin{pmatrix} 1 & 0 \\ 0 & -1 \end{pmatrix}$$

is invertible, and when $w_2 \neq 0$, we have that

$$\operatorname{Ker} P(0, w_2) = \operatorname{Ker} \partial_{w_2} P(0, w_2) = \{ z(1, 1) : z \in \mathbf{C} \}$$

which is mapped to $\operatorname{Ran} P(0, w_2) = \{ z(1, -1) : z \in \mathbf{C} \}$ by $\partial_{w_1} P$.

We obtain from Proposition 3.5 in [**20**] a simple characterization of systems of principal type. For that we recall κ_P, K_P, and $\Xi(P)$ from Definition 3.3.

PROPOSITION 4.5. *Assume $P(w) \in C^\infty$ is an $N \times N$ system and that $(w_0, \lambda_0) \in \Omega_1(P) \setminus \Xi(P)$. Then $P(w) - \lambda_0 \operatorname{Id}_N$ is of principal type at w_0 if and only if $\kappa_P \equiv K_P$ at (w_0, λ_0) and $d\lambda(w_0) \neq 0$ for the C^∞ germ of eigenvalues for P at w_0 satisfying $\lambda(w_0) = \lambda_0$.*

Here, the germs of eigenvalues are given by Definition 3.10. Now, for symmetric systems we have $\kappa_P \equiv K_P$ and the differential $d\lambda \neq 0$ almost everywhere on $\Omega_1(P) \setminus \Xi(P)$. Proposition 4.5 shows that for a *symmetric* system the property to be of principal type is stable outside $\Xi(P)$: if the symmetric system $P(w) - \lambda \operatorname{Id}_N$ is of principal type at a point $(w_0, \lambda_0) \notin \Xi(P)$, then it is in a neighborhood. It follows from the Sard Theorem that symmetric systems $P(w) - \lambda \operatorname{Id}_N$ are of principal type almost everywhere on $\Omega_1(P)$. For eigenvalues $\lambda_0 \notin \Sigma_{ss}(P)$ we can define the following bracket condition.

DEFINITION 4.6. Let $P \in C^\infty(T^*\mathbf{R}^n)$ be an $N \times N$ system. Then we define

$$\Lambda(P) = \overline{\Lambda_-(P) \cup \Lambda_+(P)}$$

where $\Lambda_\pm(P)$ is the set of $\lambda_0 \in \Sigma(P)$ such that there exists $w_0 \in \Sigma_{\lambda_0}(P)$ so that $(w_0, \lambda_0) \notin \Xi(P)$ and

$$\pm \{ \operatorname{Re}\lambda, \operatorname{Im}\lambda \} (w_0) > 0 \tag{4.4}$$

for the unique C^∞ germ $\lambda(w)$ of eigenvalues at w_0 for P such that $\lambda(w_0) = \lambda_0$.

Observe that $\Lambda_\pm(P) \cap \Sigma_{ss}(P) = \emptyset$, and it follows from Proposition 4.5 that $P(w) - \lambda_0 \operatorname{Id}_N$ is of principal type at $w_0 \in \Lambda_\pm(P)$ if and only if $\kappa_P = K_P$ at (w_0, λ_0), since $d\lambda(w_0) \neq 0$ when (4.4) holds. Because of the bracket condition (4.4) we find that $\Lambda_\pm(P)$ is contained in the interior of $\Sigma(P)$. The following result generalizes Theorem 2.3 to systems; for a proof see Theorem 3.10 in [**20**].

THEOREM 4.7. *Let $P \in C^\infty(T^*\mathbf{R}^n)$ be an $N \times N$ system. Then we have that*

$$\Lambda(P) \setminus (\Sigma_{ws}(P) \cup \Sigma_\infty(P)) \subseteq \overline{\Lambda_-(P)} \tag{4.5}$$

when $n \geq 2$. Assume that $P(h)$ is given by (3.1) with principal symbol $P \in C_{\mathrm{b}}^\infty(T^*\mathbf{R}^n)$ and that $\lambda_0 \in \Lambda_-(P)$, $0 \neq u_0 \in \operatorname{Ker}(P(w_0) - \lambda_0 \operatorname{Id}_N)$ and $P(w) - \lambda \operatorname{Id}_N$ is of principal type on $\Sigma_\lambda(P)$ near w_0 for $|\lambda - \lambda_0| \ll 1$, for the $w_0 \in \Sigma_{\lambda_0}(P)$ in Definition 4.6. Then there exists $u(h) \in L^2(\mathbf{R}^n)$ so that $\|u(h)\| \leq 1$,

$$\|(P(h) - \lambda_0 \operatorname{Id}_N)u(h)\| \leq C_M h^M \quad \forall M, \qquad h \to 0, \tag{4.6}$$

and $\mathrm{WF}_h^{pol}(u(h)) = \{ (w_0, u_0) \}$. There also exists a dense subset of values $\lambda_0 \in \Lambda(P)$ so that

$$\|(P(h) - \lambda_0 \operatorname{Id}_N)^{-1}\| \geq C'_M h^{-M} \quad \forall M, \qquad h \to 0. \tag{4.7}$$

If all the terms P_j in the expansion (3.1) are analytic satisfying (2.6), then $h^{\pm M}$ may be replaced by $\exp(\mp c/h)$ in (4.6)–(4.7).

Theorem 4.7 together with Proposition 3.11 gives that

$$\Lambda_-(P) \subseteq \lambda_\infty^{\mathrm{sc}}(P(h)) \subseteq \Lambda_\infty^{\mathrm{sc}}(P(h)) \subseteq \Sigma(P)$$

as in the scalar case.

EXAMPLE 4.8. Let

$$P(x,\xi) = |\xi|^2 \operatorname{Id} + iK(x), \qquad (x,\xi) \in T^*\mathbf{R}^n,$$

where $K(x) \in C^\infty(\mathbf{R}^n)$ is symmetric for all x. Then we find from Sard's Theorem that

$$\overline{\Lambda_-(P)} = \Lambda(P) = \left\{ \operatorname{Re} z \geq 0 \wedge \operatorname{Im} z \in \overline{\Sigma(K) \setminus (\Sigma_{ss}(K) \cup \Sigma_\infty(K))} \right\}.$$

As in the scalar case, when the dimension is equal to one we have to add some conditions in order to get the inclusion (4.5).

LEMMA 4.9. *Let $P(w) \in C^\infty(T^*\mathbf{R})$ be an $N \times N$ system. Then for every component Ω of $\mathbf{C} \setminus (\Sigma_{ws}(P) \cup \Sigma_\infty(P))$ which has nonempty intersection with $\complement\Lambda(P)$ we have*

$$\Lambda(P) \bigcap \Omega \subseteq \overline{\Lambda_-(P)}. \tag{4.8}$$

For a proof, see the proof of [**20**, Lemma 3.15] (which has an error in the formulation). Recall that the topological condition in Lemma 4.9 is necessary even in the scalar case by Example 2.4.

5. Quasisymmetrizable systems

Next, we shall study the behavior at the boundary $\partial\Sigma(P)$ of the eigenvalues. First we note that if the system $P(w) - z\,\mathrm{Id}_N$ is of principal type near $\Sigma_z(P)$ for z close to $\lambda \in \partial\Sigma(P) \setminus (\Sigma_{ws}(P) \cup \Sigma_\infty(P))$ and $\Sigma_\lambda(P)$ has no closed bicharacteristics, then one can generalize Theorem 1.3 in [**21**] to obtain

$$\|(P(h) - \lambda\,\mathrm{Id}_N)^{-1}\| \le C/h, \qquad h \to 0. \tag{5.1}$$

In fact, by using the reduction in the proof of [**20**, Theorem 3.10], this follows from the scalar case; see Example 5.7. But in this case the eigenvalues close to λ have constant multiplicity.

Generically, we have that the eigenvalues of the principal symbol P have constant multiplicity almost everywhere since $\Xi(P)$ is nowhere dense. But at the boundary $\partial\Sigma(P)$ this needs not be the case. For example, if

$$P(t,\tau) = \tau\,\mathrm{Id} + iK(t)$$

where $C^\infty \ni K \ge 0$ is unbounded and $0 \in \Sigma_{ss}(K)$, then $\mathbf{R} = \partial\Sigma(P) \subseteq \Sigma_{ss}(P)$.

When the multiplicity of the eigenvalues of the principal symbol is not constant, the situation is more complicated. Then the following example shows that it is not sufficient to have conditions only on the eigenvalues in order to obtain the estimate (5.1), not even in the principal type case.

Example 5.1. Let $a(t) \in C^\infty(\mathbf{R})$ be real valued and let

$$P^w(t,hD_t) = \begin{pmatrix} hD_t + a(t) & t - ia(t) \\ t + ia(t) & -hD_t + a(t) \end{pmatrix} = P^w(t,hD_t)^*.$$

Then the eigenvalues of $P(t,\tau)$ are

$$\lambda = a(t) \pm \sqrt{\tau^2 + t^2 + a^2(t)} \in \mathbf{R}$$

which coincide if and only if $\tau = t = a(t) = 0$. We have that

$$\frac{1}{2}\begin{pmatrix} 1 & i \\ 1 & -i \end{pmatrix} P^w \begin{pmatrix} 1 & 1 \\ i & -i \end{pmatrix} = \begin{pmatrix} hD_t + it & 0 \\ 2a(t) & hD_t - it \end{pmatrix} = \widetilde{P}(h).$$

Thus we can construct pseudomodes $u_h(t) = {}^t(0, u_2(t))$ so that $\|u_h\| = 1$ and $\widetilde{P}(h)u_h = \mathcal{O}(\exp(-c/h))$ by Theorem 2.3. By the invariance, we see that P is of principal type at $t = \tau = 0$ if and only if $a(0) = 0$. If $a(0) = 0$, then $\Sigma_{ws}(P) = \{0\}$, and when $a \neq 0$, we have that P^w is a selfadjoint diagonalizable system. In the case $a \equiv 0$ the eigenvalues of P^w are $\pm\sqrt{2kh}$, $k \in \mathbf{N}$; see the proof of Proposition 3.6.1 in [**32**].

Of course, the problem is that the eigenvalues are not invariant under multiplication with elliptic systems. To obtain the estimate (5.1) for operators that are *not* of principal type, it is not even sufficient that the principal symbol is C^∞ diagonalizable with real eigenvalues with constant multiplicity one, according to the following example.

Example 5.2. Let $a(t) \in C^\infty(\mathbf{R})$, $a(0) = 0$, $a'(0) > 0$, and

$$P(h) = \begin{pmatrix} 1 & hD_t \\ h & iha(t) \end{pmatrix}$$

with principal symbol $\begin{pmatrix} 1 & \tau \\ 0 & 0 \end{pmatrix}$ having eigenvalues 0 and 1. Thus $\Xi(P) = \emptyset$. Since

$$\begin{pmatrix} 1 & 0 \\ -h & 1 \end{pmatrix} P(h) \begin{pmatrix} 1 & -hD_t \\ 0 & 1 \end{pmatrix} = \begin{pmatrix} 1 & 0 \\ 0 & -h \end{pmatrix} \begin{pmatrix} 1 & 0 \\ 0 & hD_t - ia(t) \end{pmatrix},$$

we obtain from by Theorem 2.3 that $\|P(h)^{-1}\| \ge C_M h^{-M}$ when $h \to 0$, $\forall\, M$, and for analytic $a(t)$ we obtain $\|P(h)^{-1}\| \ge Ce^{c/h}$, $h \to 0$. Now $\partial_\tau P$ maps $\operatorname{Ker} P(0)$ into $\operatorname{Ran} P(0)$ so the system is not of principal type. Observe that this property is not preserved under the multiplications above, since not all the systems are elliptic.

Instead of using properties of the eigenvalues of the principal symbol, we shall use properties that are invariant under multiplication with invertible systems. We recall the following normal form for scalar symbols of principal type near the boundary $\partial\Sigma(P)$.

EXAMPLE 5.3. Assume that $p(x,\xi) \in C^\infty(T^*\mathbf{R}^n)$ is scalar of principal type and $0 \in \partial\Sigma(p) \setminus \Sigma_\infty(p)$. Then we find from the proof of Lemma 4.1 in [**21**] that we can choose symplectic coordinates so that

$$p(x,\xi) = e(x,\xi)(\xi_1 + if(x,\xi')), \qquad \xi = (\xi_1, \xi'), \tag{5.2}$$

in a neighborhood of $w_0 \in \Sigma_0(p)$, where $e \ne 0$ and $f \ge 0$. If there are no closed semibicharacteristics of p, we obtain (5.2) in a neighborhood of $\Sigma_0(P)$.

Recall that a *semibicharacteristic* of p is a nontrivial bicharacteristic of $\operatorname{Re} qp$, for some $q \ne 0$. The normal form (5.2) in the scalar case motivates the following definition, which is Definition 4.5 in [**20**].

DEFINITION 5.4. We say that the $N \times N$ system $P(w) \in C^\infty(T^*\mathbf{R}^n)$ is *quasisymmetrizable* with respect to the real C^∞ vector field V in $\Omega \subseteq T^*\mathbf{R}^n$ if $\exists\, N \times N$ system $M(w) \in C^\infty(T^*\mathbf{R}^n)$ and c, $C > 0$ so that

$$\begin{cases} \operatorname{Re}\langle M(VP)u, u\rangle \ge c\|u\|^2 - C\|Pu\|^2, \\ \operatorname{Im}\langle MPu, u\rangle \ge -C\|Pu\|^2, \end{cases} \quad \text{in } \Omega \tag{5.3}$$

for any $u \in \mathbf{C}^N$. The system M is called a *symmetrizer* for P.

The definition is clearly independent of the choice of coordinates in $T^*\mathbf{R}^n$ and the choice of base in $\mathbf{C}^N$. When P is elliptic, we may take $M = iP^*$ as multiplier. Then P is quasisymmetrizable with respect to any vector field because $\|Pu\| \cong \|u\|$. We see from Example 5.3 that the scalar symbol p of principal type is quasisymmetrizable in the neighborhood of any point at $\partial\Sigma(p) \setminus \Sigma_\infty(p)$. We obtain the following result from Propositions 4.7 and 4.10 in [**20**].

PROPOSITION 5.5. *If $P(w) \in C^\infty(T^*\mathbf{R}^n)$ is a quasisymmetrizable $N \times N$ system, then P is of principal type and P^* is quasisymmetrizable. If $A(w)$ and $B(w) \in C^\infty(T^*\mathbf{R}^n)$ are invertible $N \times N$ systems, then BPA is quasisymmetrizable.*

EXAMPLE 5.6. The $N \times N$ system

$$P(w) = w_1 \operatorname{Id}_N + iF(w), \qquad F \ge 0,$$

is quasisymmetrizable with respect to ∂_{w_1}.

EXAMPLE 5.7. Assume $P(w) \in C^\infty$ is $N\times N$ and $z \in \Sigma(P)\setminus(\Sigma_{ws}(P)\cap\Sigma_\infty(P))$ such that $P(w) - \lambda \operatorname{Id}_N$ is of principal type when $|\lambda - z| \ll 1$. Then we can make a base change $B(w) \in C^\infty$ so that

$$P(w) = B^{-1}(w)\begin{pmatrix}\lambda(w)\operatorname{Id}_K & 0\\ 0 & P_{22}(w)\end{pmatrix}B(w) \tag{5.4}$$

near $w_0 \in \Sigma_z(P)$, where $|P_{22}(w) - \lambda(w)\operatorname{Id}| \neq 0$. We find from Proposition 4.5 that $d\lambda \neq 0$ when $\lambda = z$, so $\lambda - z$ is of principal type. Proposition 5.5 and Example 5.3 give that $P - z\operatorname{Id}_N$ is quasisymmetrizable near any $w_0 \in \Sigma_z(P)$ if $z \in \partial\Sigma(\lambda)$. If there are no closed semibicharacteristics of $\lambda - z$, then we also find from Example 5.3 that $P - z\operatorname{Id}_N$ is quasisymmetrizable in a neighborhood of $\Sigma_z(P)$.

EXAMPLE 5.8. Let

$$P(x,\xi) = |\xi|^2 \operatorname{Id}_N + iK(x)$$

where $0 \le K(x) \in C^\infty$. When $\lambda > 0$, we find that $P - \lambda\operatorname{Id}_N$ is quasisymmetrizable in a neighborhood of $\Sigma_\lambda(P)$ with respect to the exterior normal $\langle \xi, \partial_\xi\rangle$ to $\Sigma_\lambda(P) = \{\, |\xi|^2 = \lambda\,\}$.

It is interesting to note that the operator in Example 5.8 is of the type considered by Lidskiĭ in [**43, 44, 46**]. For scalar symbols, we find that $0 \in \partial\Sigma(p)$ if and only if p is quasisymmetrizable; see Example 5.3. But in the system case, this needs not be the case according to the following example from [**20**].

EXAMPLE 5.9. Let

$$P(w) = \begin{pmatrix} w_2 + iw_3 & w_1\\ w_1 & w_2 - iw_3\end{pmatrix}, \qquad w = (w_1, w_2, w_3).$$

Then P is quasisymmetrizable with respect to ∂_{w_1} with symmetrizer $M = \begin{pmatrix}0 & 1\\ 1 & 0\end{pmatrix}$. In fact, $\partial_{w_1} MP = \operatorname{Id}_2$ and

$$MP(w) = \begin{pmatrix} w_1 & w_2 - iw_3\\ w_2 + iw_3 & w_1\end{pmatrix} = (MP(w))^*.$$

Since eigenvalues of $P(w)$ are $w_2 \pm \sqrt{w_1^2 - w_3^2}$, we find that $\Sigma(P) = \mathbf{C}$. Thus $0 \in \overset{\circ}{\Sigma}(P)$ is an interior point of the eigenvalues.

For quasisymmetrizable systems we obtain the following result from Theorem 4.15 in [**20**], which generalizes Theorem 2.5 to systems.

THEOREM 5.10. *Let the $N \times N$ system $P(h)$ be given by (3.1) with principal symbol $P \in C_{\mathrm{b}}^\infty(T^*\mathbf{R}^n)$. Assume that $z \notin \Sigma_\infty(P)$ and that there exists a real-valued function $T(w) \in C^\infty$ such that $P(w) - z\operatorname{Id}_N$ is quasisymmetrizable with respect to the Hamilton vector field $H_T(w)$ in a neighborhood of $\Sigma_z(P)$. Then for any $K > 0$ we have*

$$\{\zeta \in \mathbf{C} :\ |\zeta - z| < Kh\log(1/h)\} \cap \operatorname{Spec}(P(h)) = \emptyset, \qquad 0 < h \ll 1, \tag{5.5}$$

and

$$\|(P(h) - z)^{-1}\| \le C/h, \qquad 0 < h \ll 1. \tag{5.6}$$

If P is analytic in a tubular neighborhood of $T^*\mathbf{R}^n$, then $\exists\, c_0 > 0$ such that

$$\{\zeta \in \mathbf{C} :\ |\zeta - z| < c_0\} \cap \operatorname{Spec}(P(h)) = \emptyset, \qquad 0 < h \ll 1. \tag{5.7}$$

Condition (5.6) means that $z \notin \Lambda_1^{\mathrm{sc}}(P)$, which is the pseudospectrum of index 1 by Definition 2.7. The conditions in Theorem 5.10 give some geometrical information on the bicharacteristic flow of the eigenvalues according to the following remark.

REMARK 5.11. *The conditions in Theorem* 5.10 *imply that the limit set at* $\Sigma_\lambda(P)$ *of the nontrivial semibicharacteristics of the eigenvalues close to zero of* $Q = M(P - \lambda \operatorname{Id}_N)$ *is a union of compact curves on which* T *is strictly monotone. Thus they cannot form closed orbits.*

EXAMPLE 5.12. Consider the system in Example 5.8:

$$P(x,\xi) = |\xi|^2 \operatorname{Id}_N + iK(x)$$

where $0 \le K(x) \in C^\infty$. Then for $\lambda > 0$ we find that $P - \lambda \operatorname{Id}_N$ is quasisymmetric in a neighborhood of $\Sigma_\lambda(P)$ with respect to $V = H_T$, for $T(x,\xi) = -\langle \xi, x\rangle$. If $K(x) \in C_{\mathrm{b}}^\infty$ with $0 \notin \Sigma_\infty(K)$, then we obtain from Proposition 3.12, Remark 3.13, Example 3.14, and Theorem 5.10 that

$$\|(P^w(x,hD) - \lambda)^{-1}\| \le C/h, \qquad 0 < h \ll 1,$$

since $0 < \lambda \notin \Sigma_\infty(P)$.

One can also generalize the improved bounds of finite type boundary points given by Theorem 2.6 to systems. The situation is more complicated in the systems case; see Definition 3.15 and Theorem 5.20 in [**20**].

References

[1] A. Aslanyan, E. B. Davies, *Spectral instability for some Schrödinger operators*, Numer. Math. **85** (2000), 525–552.

[2] D. Borthwick and A. Uribe, *On the pseudospectra of Berezin-Toeplitz operators*, Methods Appl. Anal. **10** (2003), 31–65.

[3] L. S. Boulton, *Non-self-adjoint harmonic oscillator, compact semigroups and pseudospectra,* J. Operator Theory **47** (2002), 413–429.

[4] S. J. Chapman and L. N. Trefethen, *Wave packet pseudomodes of twisted Toeplitz matrices,* Comm. Pure Appl. Math. **57** (2004), 1233–1264.

[5] E. B. Davies *Pseudo-spectra, the harmonic oscillator and complex resonances.* R. Soc. Lond. Proc. Ser. A Math. Phys. Eng. Sci. **455** (1999), 585–599.

[6] ———, *Semi-classical states for non-self-adjoint Schrödinger operators,* Comm. Math. Phys. **200** (1999), 35–41.

[7] ———, *Non-self-adjoint differential operators*, Bull. London Math. Soc. **34** (2002), 513-532.

[8] ———, *Eigenvalues of an elliptic system*, Math. Z. **243** (2003), 719–743.

[9] ———, *Computing the decay of a simple reversible sub-Markov semigroup*, LMS J. Comput. Math. **7** (2004), 1–20.

[10] ———, *Spectral bounds using higher-order numerical ranges*, LMS J. Comput. Math. **8** (2005), 17–45.

[11] ———, *Semigroup growth bounds*, J. Operator Theory **53** (2005), 225–249.

[12] ———, *Semi-classical analysis and pseudo-spectra*, J. Differential Equations **216** (2005), 153–187.

[13] ———, *Approximate diagonalization*, SIAM J. Matrix Anal. Appl. **29** (2007), 1051–1064.

[14] E. B. Davies and A. B. J. Kuijlaars, *Spectral asymptotics of the non-self-adjoint harmonic oscillator,* J. London Math. Soc. (2) **70** (2004), 420–426.

[15] E. B. Davies and M. Hager, *Perturbations of Jordan matrices*, J. Approx. Theory **156** (2009), 82–94.

[16] E. B. Davies and B. Simon, *Eigenvalue estimates for non-normal matrices and the zeros of random orthogonal polynomials on the unit circle*, J. Approx. Theory **141** (2006), 189-213.

[17] N. Dencker, *On the propagation of polarization sets for systems of real principal type*, J. Funct. Anal. **46** (1982), 351–372.

[18] ———, *Preparation theorems for matrix valued functions*, Ann. Inst. Fourier (Grenoble) **43** (1993), 865–892.
[19] ———, *The resolution of the Nirenberg-Treves conjecture*, Ann. of Math. **163** (2006), 405–444.
[20] ———, *The pseudospectrum of systems of semiclassical operators*, Anal. PDE **1** (2008), 323–373.
[21] N. Dencker, J. Sjöstrand, and M. Zworski, *Pseudospectra of semiclassical (pseudo-) differential operators*, Comm. Pure Appl. Math. **57** (2004), 384–415.
[22] M. Dimassi and J. Sjöstrand, *Spectral Asymptotics in the Semi-classical Limit,* Cambridge University Press, 1999.
[23] K. Du and Y. Wei, *Structured pseudospectra and structured sensitivity of eigenvalues*, J. Comput. Appl. Math. **197** (2006), 502–519.
[24] J. J. Duistermaat and J. Sjöstrand, *A global construction for pseudo-differential operators with non-involutive characteristics,* Inv. Math. **20** (1973), 209-225.
[25] L. Elsner, *An optimal bound for the spectral variation of two matrices*, Linear Algebra Appl. **71** (1985), 77–80.
[26] I. C. Gohberg and M. G. Krein, *Introduction to the Theory of Linear Non-self-adjoint operators,* Translations of Mathematical Monographs **18**, American Mathematical Society, Providence, 1969.
[27] M. Hager, *Instabilité spectrale semiclassique pour des opérateurs non-autoadjoints. I. Un modèle*, Ann. Fac. Sci. Toulouse Math. (6) **15** (2006), 243–280.
[28] ———, *Instabilité spectrale semiclassique d'opérateurs non-autoadjoints. II*, Ann. Henri Poincaré **7** (2006), 1035–1064.
[29] ———, *Bound on the number of eigenvalues near the boundary of the pseudospectrum*, Proc. Amer. Math. Soc. **135** (2007), 3867–3873.
[30] M. Hager and J. Sjöstrand, *Eigenvalue asymptotics for randomly perturbed non-selfadjoint operators*, Math. Ann. **42** (2008), 177–243.
[31] B. Helffer and J. Sjöstrand, *Résonances en limite semi-classique*, Mém. Soc. Math. France (N.S.) **114** (1986).
[32] ———, *Analyse semi-classique pour l'équation de Harper. II. Comportement semi-classique près d'un rationnel*, Mém. Soc. Math. France (N.S.) **40** (1990).
[33] L. Hörmander, *Differential operators of principal type,* Math. Ann. **140** (1960), 124-146.
[34] ———, *Differential equations without solutions,* Math. Ann. **140** (1960), 169-173.
[35] ———, *On the solvability of pseudodifferential equations*, in *Structure of solutions of differential equations (Katata/Kyoto, 1995),* 183–213, World Sci. Publ., River Edge, NJ, 1996.
[36] ———, *The analysis of linear partial differential operators*, vols. I–IV, Springer-Verlag, Berlin, 1983–1985.
[37] M. Kashiwara and T. Kawai, *Microhyperbolic pseudo-differential operators. I,* J. Math. Soc. Japan **27** (1975), 359-404.
[38] T. Kato, *Perturbation theory for linear operators*, Die Grundlehren der mathematischen Wissenschaften, Band 132, Springer-Verlag New York, Inc., New York, 1966.
[39] M. Levitin and E. Shargorodsky, *Spectral pollution and second-order relative spectra for self-adjoint operators*, IMA J. Numer. Anal. **24** (2004), 393–416.
[40] H. Lewy, *An example of a smooth linear partial differential equation without solution,* Ann. Math. **66** (1957), 155-158.
[41] V. B. Lidskiĭ, *On the completeness of the system of eigen- and associated functions of a non-self-adjoint differential operator*, (Russian) Dokl. Akad. Nauk SSSR (N.S.) **110** (1956), 172–175.
[42] ———, *A theorem on the spectra of a perturbed differential operator*, (Russian) Dokl. Akad. Nauk SSSR (N.S.) **112** (1957), 994–997.
[43] ———, *On the completeness of a system of eigen elements and adjoint elements of a compact operator*, (Russian) Dokl. Akad. Nauk SSSR (N.S.) **115** (1957) 234–236.
[44] ———, *Theorems on the completeness of a system of characteristic and adjoint elements of operators having a discrete spectrum*, (Russian) Dokl. Akad. Nauk SSSR **119** (1958) 1088–1091.
[45] ———, *Non-selfadjoint operators with a trace*, (Russian) Dokl. Akad. Nauk SSSR **125** (1959), 485–487.

[46] ———, *Conditions for completeness of a system of root subspaces for non-selfadjoint operators with discrete spectrum*, (Russian) Trudy Moskov. Mat. Obšč. **8** (1959), 83–120.

[47] ———, *Summation of series over the principal vectors of non-selfadjoint operators*, Dokl. Akad. Nauk SSSR **132** (1960), 275–278 (Russian); translated as Soviet Math. Dokl. **1** (1960), 540–543.

[48] ———, *A non-selfadjoint operator of Sturm-Liouville type with a discrete spectrum*, (Russian) Trudy Moskov. Mat. Obšč. **9** (1960), 45–79.

[49] ———, *The Fourier series expansion in terms of the principal functions of a non-selfadjoint elliptic operator*, (Russian) Mat. Sb. (N.S.) **57** (99) (1962), 137–150.

[50] ———, *Summability of series in terms of the principal vectors of non-selfadjoint operators*, (Russian) Trudy Moskov. Mat. Obšč. **11** (1962), 3–35.

[51] ———, *On the theory of perturbations of nonselfadjoint operators*, (Russian) Ž. Vyčisl. Mat. i Mat. Fiz. **6** (1966), 52–60.

[52] ———, *On the theory of perturbations of nonselfadjoint operators*, (Russian) Ž. Vyčisl. Mat. i Mat. Fiz. **6** (1966), 52–60. English translation in Comput. Math. Phys. **1** (1966), 73-85.

[53] A. Markus, *Introduction to the spectral theory of polynomial operator pencils*, Translations of Mathematical Monographs, **71**. American Mathematical Society, Providence, RI, 1988.

[54] A. Martinez, *An introduction to semiclassical and microlocal analysis*, Universitext, Springer-Verlag, New York, 2002.

[55] A. Melin and J. Sjöstrand, *Determinants of pseudodifferential operators and complex deformations of phase space*, Methods and Applications of Analysis **9** (2002), 177–237.

[56] J. Moro, J. V. Burke, and M. L. Overton, *On the Lidskii-Vishik-Lyusternik perturbation theory for eigenvalues of matrices with arbitrary Jordan structure*, SIAM J. Matrix Anal. Appl. **18** (1997), 793–817.

[57] K. Pravda-Starov, *A general result about pseudospectrum for Schrödinger operators*, R. Soc. Lond. Proc. Ser. A Math. Phys. Eng. Sci. **460** (2004), 471-477.

[58] ———, *Etude du pseudo-spectre d'opérateurs non auto-adjoints*, Ph.D. thesis, Université de Rennes I, 2006.

[59] ———, *A complete study of the pseudo-spectrum for the rotated harmonic oscillator*, J. London Math. Soc. (2) **73** (2006), 745–761.

[60] ———, *Boundary pseudospectral behaviour for semiclassical operators in one dimension*, Int. Math. Res. Not. **2007:9**.

[61] ———, *On the pseudospectrum of elliptic quadratic differential operators*, Duke Math. J. **145** (2008), 249–279.

[62] ———, *Contraction semigroups of elliptic quadratic differential operators*, Math. Z. **259** (2008), 363–391.

[63] ———, *Pseudo-spectrum for a class of semi-classical operators*, Bull. Soc. Math. France **136** (2008), 329–372.

[64] S. C. Reddy, P. J. Schmid, and D. S. Henningson, *Pseudospectra of the Orr-Sommerfeld operator*, SIAM J. Appl. Math. **53** (1993), 15–47.

[65] J. Sjöstrand, *Singularités analytiques microlocales*, Astérisque, **95** (1982), 95–149.

[66] ———, *Resolvent estimates for non-self-adjoint operators via semi-groups*, arXiv:0906.0094.

[67] J. Sjöstrand and M. Zworski, *Elementary linear algebra for advanced spectral problems*, Ann. Inst. Fourier (Grenoble) **57** (2007), 2095–2141.

[68] T. Tao and V. Vu, *Universality of the ESD and the Circular law (with an appendix by M. Krishnapur)*, arXiv:0807.4898.

[69] L. N. Trefethen, *Pseudospectra of linear operators*, SIAM Review, **39** (1997), 383-400.

[70] L. N. Trefethen and M. Embree, *Spectra and Pseudospectra*, Princeton University Press, Princeton, N.J., 2005.

[71] M. Zworski, *Pseudospectra of semi-classical operators*, talk at King's College, May 2001, unpublished.

[72] ———, *A remark on a paper by E.B. Davies*, Proc. AMS **129** (2001), 2955-2957.

[73] ———, *Numerical linear algebra and solvability of partial differential equations*, Comm. Math. Phys. **229** (2002), 293-307.

Centre for Mathematical Sciences, University of Lund, Box 118, SE-221 00 Lund, Sweden

E-mail address: dencker@maths.lth.se

Amer. Math. Soc. Transl.
Volume **231**, 2010

Estimates of Solutions to Some Weighted Systems Defined on $\mathbb{R}^N$

Jacqueline Fleckinger and Na Wei

Abstract. We study the sign of the solution of a 2×2 system defined on $\mathbb{R}^N$, $N \geq 3$,

$$(S_1) \qquad -\Delta u_1 = \lambda\rho(x)(au_1 + bu_2) + f_1, \quad x \in \mathbb{R}^N,$$

$$(S_2) \qquad -\Delta u_2 = \lambda\rho(x)(cu_1 + du_2) + f_2, \quad x \in \mathbb{R}^N,$$

$$(S_3) \qquad u_1 \longrightarrow 0\,,\ u_2 \longrightarrow 0 \text{ as } |x| \longrightarrow \infty,$$

where a, b, c, d are real numbers and ρ, the weight function which is allowed to change sign, has a positive part which tends to 0 fast enough at infinity. Denoting by $(\lambda_\rho, \varphi_\rho)$ the "principal" eigenpair which satisfies

$$(EV) \qquad -\Delta\varphi_\rho = \lambda_\rho\rho(x)\varphi_\rho,\ \varphi_\rho > 0;\ \lim_{|x|\to+\infty} \varphi_\rho(x) = 0,\ x \in \mathbb{R}^N,$$

we compare u_1 and u_2 (locally) to the groundstate φ_ρ for $\lambda \in \mathbb{R}$ close enough to the "principal" eigenvalue λ_ρ and we determine their sign.

1. Introduction

We are concerned with the sign and the behavior of the solutions of the following 2×2 system defined on $\mathbb{R}^N$, $N \geq 3$, with an "indefinite" weight function ρ (that is, a function which is allowed to change sign):

$$(S) \qquad -\Delta U = \lambda\rho(x)MU + F;\ U \to 0 \text{ as } |x| \to \infty;$$

here λ is a real parameter; U and F are column vectors with components u_i and f_i, respectively, $i = 1$ or 2. The functions f_i are measurable and bounded. The positive part of the weight $\rho^+ = \max(\rho, 0)$ tends to 0 fast enough at infinity. The matrix $M = \begin{pmatrix} a & b \\ c & d \end{pmatrix}$ has real constant coefficients and real eigenvalues $\sigma_1 > \sigma_2$ with $\sigma_1 > 0$.

We derive our results from results on the equation

$$(E) \qquad -\Delta u = \lambda\rho(x)u + f, \quad x \in \mathbb{R}^N, \quad \lim_{|x|\to\infty} u(x) = 0, N \geq 3.$$

We study first the associated eigenvalue problem $-\Delta u = \lambda\rho u$ in $\mathbb{R}^N$ and we recall

2010 *Mathematics Subject Classification.* Primary 35-XX, 35J05, 35P05.

in particular results on the existence of a positive "principal eigenvalue", that is, an eigenvalue which is associated to an eigenfunction which does not change sign. For this equation with indefinite weight, the existence of a principal eigenvalue—under suitable hypotheses—has been established first in [**6**]; indeed there are many papers on this topic since they are involved in several studies arising, e.g., from biology, optics, geology, etc. The case of ρ^+ with compact support is studied in [**8**], $\rho^+ \in L^{N/2}$ in [**1**]. For completeness, let us recall that the case of the p-Laplacian was also studied ([**12**], [**17**], [**20**], etc.).

When the existence of a principal eigenpair $(\lambda_\rho, \varphi_\rho)$ is known, comparison of the $u := u(\lambda, f)$ solution of (E) with φ_ρ, the groundstate satisfying (EV), is a natural question. Under a suitable hypothesis on λ, ρ, M, and f, we show that, locally (on any ball), there exists a positive constant k such that $|u| > k\varphi_\rho$.

For a bounded domain this result is classical. It is more difficult for $\mathbb{R}^N$. For the Schrödinger operator $L := -\Delta + q$ defined, as here, on $\mathbb{R}^N$ with a positive potential tending to ∞ sufficiently fast, several results have been established in [**3**], [**4**], [**5**], in particular the "Maximum Principle" and its improvement, "Fundamental Positivity", the "Antimaximum Principle" and its improvement, "Fundamental Negativity".

The (strong) maximum principle says that any solution to $Lu = f \geq 0, \not\equiv 0$ is positive. Following Alziary and coauthors [**3**] we say that u is "fundamentally positive" (resp. "fundamentally negative") if there exists a positive constant c such that $u > c\varphi_\rho$ (resp. $u < -c\varphi_\rho$). The antimaximum principle has been shown first by Clément and Peletier ([**10**]) for a linear equation $-\Delta u = \lambda u + f$ defined on a smooth bounded domain with Dirichlet boundary condition.

Proposition (Antimaximum Principle). *Assume $f \geq 0, \not\equiv 0$ and denote by λ^* the principal eigenvalue of the Dirichlet Laplacian defined on a (smooth) domain Ω. There exists $\delta > 0$ depending on f such that*

$$\lambda^* < \lambda < \lambda^* + \delta \;\Rightarrow\; u < 0 \text{ on } \Omega\,;\; \partial u/\partial n|_{\partial\Omega} > 0.$$

Since the maximum principle holds for $\lambda < \lambda^*$, the sign of the solution is reversed as λ crosses the first eigenvalue. A local antimaximum principle for the p-Laplacian defined on $\mathbb{R}^N$ is shown in [**13**].

Here we prove local "fundamental positivity" ($u \geq k\varphi_\rho > 0$) or local "fundamental negativity" ($u \leq -k\varphi_\rho < 0$) on balls for equation (E). In particular we show that, for λ tending to λ_ρ, u tends to $\pm\infty$. These estimates for equations are established in order to study systems.

There are many results dealing with systems, essentially cooperative ones. A system is said to be "cooperative" when the terms outside the diagonal in matrix M are positive (e.g., [**11**], [**18**], [**16**], [**17**], [**15**], etc.). Here we study systems which are not necessarily cooperative, and under a suitable hypothesis, we are able to obtain local fundamental positivity or negativity for u_1 and u_2 satisfying (S).

Analogous results for not necessarily cooperative systems defined on unbounded domains have been shown in [**9**] or [**19**] for systems of Schrödinger operators.

This paper is organized as follows: First we give results for ODE. Then we study a 2×2 system; we prove our results first for a positive ρ, then for an "indefinite" weight.

2. The case of one equation

We study, for $\lim_{|x|\to\infty} \rho^+(x) = 0$, the equation

$$(E) \qquad -\Delta u = \lambda\rho(x)u + f, \ \lim_{|x|\to\infty} u(x) = 0, \ x \in \mathbb{R}^N, \ N \geq 3.$$

Here $f \in L^\infty(\mathbb{R}^N) \cap L^2_{1/\rho}$ and $\rho^+(x) := \max\{\rho(x), 0\}$ decreases faster than $1/|x|^2$ at infinity. More precisely ρ satisfies

(H_ρ) ρ is smooth and bounded and $\rho^+(x) \leq K(1+|x|^2)^{-\alpha}$, for some constants $K > 0$ and $\alpha > 1$. Moreover $\Omega_+ := \{x \in \mathbb{R}^N \,/\, \rho(x) > 0\}$ has positive Lebesgue measure.

Note that ρ, which is positive somewhere, is allowed to be positive everywhere.

We work in the space $\mathcal{D}^{1,2}$, the closure of $\mathcal{C}_0^\infty(\mathbb{R}^N)$ w.r.t.

$$\|u\|^2_{\mathcal{D}^{1,2}} = \int_{\mathbb{R}^N} |\nabla u|^2.$$

REMARK 1. *For $\rho \in L^{N/2}$, $\exists \gamma > 0$ s.t. $\int_{\mathbb{R}^N} |\nabla u|^2 \geq \gamma \int_{\mathbb{R}^N} \rho u^2$, $\forall\, u \in \mathcal{D}^{1,2}$. Moreover the imbedding $\mathcal{D}^{1,2} \hookrightarrow L^2_\rho$ is continuous and compact.*

Considering equation (E), we seek weak solutions, that is, $u \in \mathcal{D}^{1,2}$ such that

$$\int_{\mathbb{R}^N} (\nabla u \cdot \nabla \zeta) = \lambda \int_{\mathbb{R}^N} \rho(x) u \zeta + \int_{\mathbb{R}^N} f\zeta, \ \ \forall \zeta \in \mathcal{D}^{1,2}.$$

2.1. The Principal Eigenpair.

THEOREM 1 ([**6**] Section 4; [**17**], Theorem 2.6). *Under hypothesis (H_ρ) there exists a positive principal eigenvalue λ_ρ satisfying*

$$(1) \qquad -\Delta u = \lambda\rho(x)u, \ \lim_{|x|\to\infty} u(x) = 0; \ u(x) > 0, \ x \in \mathbb{R}^N.$$

λ_ρ is the smallest positive eigenvalue of (1); it is simple with multiplicity 1. The associated eigenfunction does not change sign.

Denote by φ_ρ the associated eigenfunction (the "groundstate") which is in $\mathcal{D}^{1,2}$; for simplicity we choose it to be positive and with norm 1 in L^2_ρ:

$$(2) \qquad -\Delta\varphi_\rho = \lambda_\rho \rho(x)\varphi_\rho, \ x \in \mathbb{R}^N; \ \varphi_\rho(x) > 0, \ \forall x \in \mathbb{R}^N; \ \int_{\mathbb{R}^N} \rho\varphi_\rho^2 = 1.$$

THEOREM 2. *Assume (H_ρ). The positive principal eigenvalue λ_ρ is given by*

$$(3) \qquad 0 < \lambda_\rho = \inf_{\{u \in \mathcal{D}^{1,2} / \int_{\mathbb{R}^N} \rho u^2 = 1\}} \|u\|^2_{\mathcal{D}^{1,2}}.$$

The equality in (3) holds if and only if u is proportional to φ_ρ.

The existence of this eigenpair, $(\lambda_\rho, \varphi_\rho)$, follows from the compactness of the operator T defined in $\mathcal{D}^{1,2}$ by

$$A(Tu, v) = B(u, v), \ \forall\, u, v \in \mathcal{D}^{1,2}$$

where (see, e.g., [**6**], Section 4, [**15**], Lemma 4)

$$A(u, v) = \int_{\mathbb{R}^N} \nabla u \cdot \nabla v; \ \ B(u, v) = \int_{\mathbb{R}^N} \rho uv;$$

hence $\lambda_\rho = \inf_{\{u / B(u,u)=1\}} A(u, u)$. This implies (3).

REMARK 2. *The eigenvalue problem* $(-\Delta+w(x))u = \lambda\rho(x)u$ *where* $0 < w \le k\rho$ *with* $k > 0$ *can be treated the same way by replacing* $A(u,u)$ *by* $\int(|\nabla u|^2 + w(x)u^2)$. *The principal eigenvalue, denoted by* $\lambda(w,\rho)$, *is*

$$\lambda(w,\rho) = \inf_{\{u/\int \rho u^2=1\}} \int (|\nabla u|^2 + w(x)u^2). \tag{3'}$$

REMARK 3. *It is important to notice that* $\lambda_\rho > 0$. *A nonpositive principal eigenvalue may exist according to the hypothesis on* ρ. *Since* $\lambda\rho = (-\lambda)(-\rho)$, *we limit our study to the case of positive* λ. *The case of a nonpositive eigenvalue could be derived by changing* ρ *to* $-\rho$.

2.2. Results for equation (E)**.** We derive now the sign of the solution u and obtain some estimates when the parameter λ varies around λ_ρ. For completeness we recall first a classical result of the maximum principle (see, e.g., [**15**]). We also state a (local) antimaximum established (for the p-Laplacian) in [**20**] or [**13**]. We improve these results by proving local fundamental positivity or negativity.

Proposition (MP): (Maximum Principle for one equation). *We suppose that* ρ *satisfies* (H_ρ) *and that* $f := \rho h$ *satisfies*

$$f = \rho h \in L^\infty \cap L^2_{1/\rho} \;\Leftrightarrow\; h \in L^\infty \cap L^2_\rho. \tag{H_f}$$

A necessary and sufficient condition for having a (strong) maximum principle, that is, $\int_{\mathbb{R}^N} f\varphi_\rho \ge 0 \Rightarrow u > 0$, *is* $\lambda < \lambda_\rho$ *with* u *satisfying* (E). *Moreover if* $\lambda < \lambda_\rho$, *there exists a solution and this solution is unique and positive.*

Proposition (LAM): (Local Antimaximum Principle) ([**20**], Theorem 5.2 or [**13**], Theorem 3.3). *Assume* (H_ρ) *and* $f \ge 0, \not\equiv 0$, $f \in L^\infty$. *Let* $R > 0$ *be given; there exists* $\delta > 0$ *depending on* R, f, *and* ρ *such that for* $\lambda_\rho < \lambda < \lambda_\rho + \delta$, $u < 0$ *on the ball* $B_R = \{x \in \mathbb{R}^N / |x| < R\}$.

Proposition (LFP): (Local Fundamental Positivity). *We suppose that* (H_ρ) *and* (H_f) *are satisfied; moreover there exist positive constants* c' *and* c'' *such that*

$$0 < c'\rho\varphi_\rho \le f \le c''\rho\varphi_\rho \;\Leftrightarrow\; 0 < c'\varphi_\rho \le h \le c''\varphi_\rho. \tag{4}$$

Then, for any $R > 0$, *for* $\lambda < \lambda_\rho$, *and for* $\lambda_\rho - \lambda$ *small enough, there exist constants* $k' > 0$ *and* k'', *depending on* f *and* R , *such that on* $B_R = \{x \in \mathbb{R}^N \,/\, |x| < R\}$,

$$0 < \frac{k'}{(\lambda_\rho - \lambda)}\varphi_\rho \le u \le \frac{k''}{(\lambda_\rho - \lambda)}\varphi_\rho. \tag{5}$$

Proposition (LFN): (Local Fundamental Negativity). *Assume that* ρ *satisfies* (H_ρ) *and that* $f = \rho h$ *satisfies* (H_f) *and* (4). *Let* $R > 0$ *be given. There exists* $\delta > 0$ *(depending on* R, f, *and* ρ*) such that for* $\lambda_\rho < \lambda < \lambda_\rho + \delta$, *on the ball* $B_R = \{x \in \mathbb{R}^N/|x| < R\}$, *there are positive constants* K', K'', *depending on* f, φ_ρ, *and* R *such that on* B_R

$$-\frac{K''}{(\lambda - \lambda_\rho)}\varphi_\rho \le u \le -\frac{K'}{(\lambda - \lambda_\rho)}\varphi_\rho < 0. \tag{6}$$

REMARK 4. *Inequalities* (5) *and* (6) *play an important role when studying systems since they show that* $|u|/\varphi_\rho \to +\infty$ *as* $\lambda \to \lambda_\rho$.

2.3. Proofs for equation (E).

2.3.1. *Recall: Propositions (MP) and (LAM) for* $\rho > 0$. We study now equation (E) with $\rho > 0$. Equation (E) is defined in $\mathbb{R}^N$ with $N \geq 3$; we suppose that ρ satisfies (H_ρ) and that $f := \rho h$ satisfies (H_f). The maximum principle is shown in [**15**], Propositions 1 and 2 with $f \geq 0, f \not\equiv 0$ in place of $\int f\varphi_\rho > 0$. The proofs are the same. We notice that in this proposition, λ is allowed to be negative.

The local antimaximum principle is shown in [**20**], Theorem 5.2 or [**13**], Theorem 3.3 for the p-Laplacian.

2.3.2. *Proof of Propositions (LFP) and (LFN) for a positive weight.* We decompose u. Set $u = u^*\varphi_\rho + u^\perp$ where $\int u^\perp \rho\varphi_\rho = 0$. Hence, by (2), $u^* = \int \rho u \varphi_\rho$. Analogously consider h such that $f = \rho h$ and decompose h: $h = h^*\varphi_\rho + h^\perp$ with $\int h^\perp \rho\varphi_\rho = 0$. We first notice that $h^* > 0$. Indeed $h^* = \int \rho h \varphi_\rho = \int f\varphi_\rho > c' > 0$, by (4) and (2). Equation (E) becomes

$$(\lambda_\rho - \lambda)\rho u^*\varphi_\rho - \Delta u^\perp - \lambda\rho u^\perp = h^*\rho\varphi_\rho + \rho h^\perp.$$

Multiply by φ_ρ and integrate. We obtain $(\lambda_\rho - \lambda)u^* = h^*$. Hence

$$u = \frac{h^*}{\lambda_\rho - \lambda}\varphi_\rho + u^\perp. \tag{7}$$

Also $u^\perp$ solves $-\Delta u^\perp = \lambda\rho u^\perp + \rho h^\perp$. We know that $\int |\nabla u^\perp|^2 \geq \lambda_2 \int \rho(u^\perp)^2$, λ_2 being the second eigenvalue of our problem. The parameter λ stays away from λ_2 since $\lambda < \lambda_\rho < \lambda_2$; choose, e.g., $\frac{\lambda_\rho}{2} < \lambda < \frac{(\lambda_\rho + \lambda_2)}{2}$.

$$\lambda_2 \int_{\mathbb{R}^N} \rho(u^\perp)^2 \leq \int_{\mathbb{R}^N} |\nabla u^\perp|^2 = \lambda \int_{\mathbb{R}^N} \rho(u^\perp)^2 + \int_{\mathbb{R}^N} f^\perp u^\perp$$

$$\Rightarrow (\lambda_2 - \lambda) \int_{\mathbb{R}^N} \rho(u^\perp)^2 \leq [\int_{\mathbb{R}^N} \rho(u^\perp)^2 \int_{\mathbb{R}^N} \frac{1}{\rho}(f^\perp)^2]^{1/2}.$$

λ varies around λ_ρ as above; hence

$$\frac{(\lambda_2 - \lambda_\rho)}{2}(\int_{\mathbb{R}^N} \rho(u^\perp)^2)^{1/2} \leq [\int_{\mathbb{R}^N} \frac{1}{\rho}(f^\perp)^2]^{1/2}.$$

$u^\perp$ is bounded in $L^2_\rho(\mathbb{R}^N)$; hence it is bounded in $L^2_\rho(B_R)$. By bootstrap arguments (as [**7**], p. 198 or [**14**]) $u^\perp$ is bounded in $\mathcal{C}^2(B_R)$ by a constant which does not depend on λ.

On B_R, the groundstate φ_ρ is bounded below; denote by $\kappa(R)$ this minimum. It is possible to choose $\lambda_\rho - \lambda > 0$ small enough so that on B_R, $|u^\perp| < (1/4)\frac{h^*}{\lambda_\rho - \lambda}\kappa(R) < (1/4)\frac{h^*}{\lambda_\rho - \lambda}\varphi_\rho$. We derive (5).

Proposition (LFN) is shown exactly as Proposition (LFP); it is an improvement of the "local antimaximum principle".

2.3.3. *Proofs for an indefinite weight.* Consider now equation (E) with indefinite weight; we still assume (H_ρ), (H_f), and (4).

$$-\Delta u = \lambda\rho(x)u + f. \tag{E}$$

Setting $\rho = \rho^+ - \rho^-$, (E) can be rewritten as

$$-\Delta u + \lambda\rho^-(x)u = \lambda\rho^+(x)u + f.$$

This equation with positive weight can be treated as (E) by decomposition of u: $u = u^*\varphi_\rho + u^\perp$ where $\int u^\perp\rho\varphi_\rho = 0$. Also $u^* = \int \rho u\varphi_\rho$. Analogously set $f = \rho h$

and decompose h: $h = h^*\varphi_\rho + h^\perp$. We still have $h^* > 0$. Equation (7) is still valid. We get similar results.

3. 2×2 system with constant coefficients

We consider a 2×2 system with constant coefficients:

$$(S) \qquad -\Delta U = \lambda\rho(x)MU + F\,;\; U \to 0 \text{ as } |x| \to \infty.$$

(S) can be written

$$(S_1) \qquad -\Delta u_1 = \lambda\rho(x)(au_1 + bu_2) + f_1, \quad x \in \mathbb{R}^N,$$

$$(S_2) \qquad -\Delta u_2 = \lambda\rho(x)(cu_1 + du_2) + f_2, \quad x \in \mathbb{R}^N,$$

$$(S_3) \qquad u_1 \longrightarrow 0\,,\; u_2 \longrightarrow 0 \text{ as } |x| \longrightarrow \infty.$$

As for one equation, we seek weak solutions in $\mathcal{D}^{1,2}$.

3.1. Hypothesis and results.

We assume
- $a > d$. (Of course it is possible to change the order of the equations.)
- ρ satisfies (H_ρ).
- Matrix M is such that

(H_M): M has only real eigenvalues $\sigma_1 > \sigma_2 \neq 0$, that is,

$$(8) \qquad (a-d)^2 + 4bc > 0.$$

Moreover the largest eigenvalue σ_1 is positive and algebraically and geometrically simple.

REMARK 5. *Since σ_1 is simple, so is σ_2.*

We assume that both f_1 and f_2 satisfy (H_f); more precisely

$$(H_F) \qquad f_1\,,\; f_2 \in L^\infty \cap L^2_{\frac{1}{\rho}}(\mathbb{R}^N).$$

Our purpose here is to obtain the sign of the solutions. Under a suitable hypothesis on F we show that if $|\lambda_\rho - \lambda\sigma_1|$ is small enough, for $\lambda_\rho - \lambda\sigma_1 > 0$ (resp. < 0) $bu_1 > 0$ (resp. < 0) and $bcu_2 > 0$ (resp. < 0).

(H'_F) We suppose that there exists positive constants, $\gamma'_k > 0$, $\gamma''_k > 0$ such that

$$(9) \qquad 0 < \gamma'_k\rho\varphi_\rho \leq \hat{f}_k \leq \gamma''_k\rho\varphi_\rho, \; k = 1, 2,$$

with

$$(9') \qquad \hat{f}_1 = \frac{1}{bc}[cf_1 + (\sigma_1 - a)f_2]\,;\; \hat{f}_2 = \frac{1}{bc}[(a-\sigma_1)f_1 + bf_2].$$

THEOREM 3. *Assume that (H_ρ), (H_M), (H_F), and (H'_F) are satisfied. For a given $R > 0$ there exists $\delta > 0$ (depending on R, M, ρ, f_k), such that, on B_R,*

** for $0 < \lambda_\rho - \lambda\sigma_1 < \delta$, then, u_1 has the sign of b and u_2 has the sign of bc.*

** for $0 < \lambda\sigma_1 - \lambda_\rho < \delta$, u_1 has the sign of $-b$ and u_2 has the sign of $-bc$.*

Moreover there exists a positive constant k such that $|u_1| > k\varphi_\rho$, $|u_2| > k\varphi_\rho$.

REMARK 6. *The functions $\hat{f}_k$ seem to be artificial, but they will appear naturally later.*

3.2. Maximum principle for cooperative systems. Assume that (S) is a cooperative system, that is, the terms outside the diagonal in M are positive:

$$b,\ c > 0.$$

REMARK 7. *Of course* (8) *is satisfied for a cooperative system!*

The case $\lambda = 1$ is studied in [**15**]. We derive Theorem 3 from [**15**] (with a change of constants):

THEOREM 4. *Assume* (H_F), (H_ρ), *and* (H_M) *are satisfied;* $b > 0$, $c > 0$. *A necessary and sufficient condition for having the maximum principle is*

$$\lambda a < \lambda_\rho;\ \lambda d < \lambda_\rho;\ (\lambda_\rho - \lambda a)(\lambda_\rho - \lambda d) > \lambda^2 bc. \tag{10}$$

REMARK 8. *Indeed* (10) *has first been introduced in* [**11**] *which studies a system defined on a bounded domain.*

3.3. Proofs for a general system with positive weight. Now we consider a system which is not necessarily cooperative; that means that b and c have no given sign. We prove Theorem 3, that is, (local) fundamental positivity or negativity for u_1 or u_2 when $\rho > 0$. The proof is done in several steps.

(i) **Some computations on σ_1 and σ_2.** Assume (H_M). The eigenvalues of M, σ_1 and σ_2, are the roots of the associated characteristic polynomial

$$Q(\sigma) = \sigma^2 - (a+d)\sigma + ad - bc. \tag{11}$$

Since $a + d = \sigma_1 + \sigma_2$, we derive $a - \sigma_1 = \sigma_2 - d$. By computation, we get

$$\sigma_1 = (1/2)(a + d + \sqrt{(a-d)^2 + 4bc});\quad \sigma_2 = (1/2)(a + d - \sqrt{(a-d)^2 + 4bc}).$$

Also, by (H_M), $\sigma_1 > 0$. If $a + d > 0$, a simple calculation shows that $\sigma_1 > 0 \Rightarrow ad - bc < 0$. But $ad - bc = \det M = \sigma_1\sigma_2$. Hence $\operatorname{Det} M < 0$ and therefore $\sigma_2 < 0$.

REMARK 9. $Q(\sigma) > 0 \Leftrightarrow \sigma > \sigma_1\ (> 0)$ *or* $\sigma < \sigma_2$.

From now on set

$$S = \sigma_1 - a = d - \sigma_2. \tag{12}$$

By definition of σ_1 and σ_2, we derive the following:
- If $bc > 0$, then $(a-d)^2 + 4bc > (a-d)^2$; hence $\sigma_1 > a > d > \sigma_2$, and $S > 0$.
- If $bc < 0$, $(a-d)^2 + 4bc < (a-d)^2$; hence $a > \sigma_1 > \sigma_2 > d$, and $S < 0$.

(ii) **Diagonalization of the system.** Since the eigenvalues are simple, there exists a matrix P such that $P^{-1}MP = D$ with D a diagonal matrix. The terms on the diagonal are the eigenvalues σ_k, $k = 1, 2$. Set

(13) $U = PV$, $\tilde{F} = P^{-1}F$, with components $\tilde{f}_1$ and $\tilde{f}_2$.

System (S) can be rewritten as

$$-\Delta V = \lambda\rho(x)DV + \tilde{F},$$

that is,

$$\begin{aligned} -\Delta v_1 &= \lambda\rho(x)\sigma_1 v_1 + \tilde{f}_1; \\ -\Delta v_2 &= \lambda\rho(x)\sigma_2 v_2 + \tilde{f}_2. \end{aligned} \tag{14}$$

(iii) **Computations on the change of basis matrix P.** We compute now P, P^{-1}. Denote by $|P|$ the determinant of P.

$$P = \begin{pmatrix} b & \sigma_2 - d \\ \sigma_1 - a & c \end{pmatrix} = \begin{pmatrix} b & -S \\ S & c \end{pmatrix}. \tag{15}$$

$$P^{-1} = \frac{1}{|P|}\begin{pmatrix} c & d - \sigma_2 \\ a - \sigma_1 & b \end{pmatrix} = \frac{1}{|P|}\begin{pmatrix} c & S \\ -S & b \end{pmatrix}. \tag{15'}$$

LEMMA 3.1. *$|P|$ and bc have the same sign.*

PROOF.

$$\begin{aligned} |P| &= bc - (\sigma_1 - a)(\sigma_2 - d) = bc + S^2 \\ &= bc + \sigma_1 d + \sigma_2 a - \sigma_1\sigma_2 - ad \\ &= 2(bc - ad) + \frac{(a+d)^2}{2} + \frac{(d-a)}{2}\sqrt{(a-d)^2 + 4bc} \\ &= \frac{4bc + (a-d)^2}{2} - [\frac{(a-d)}{2}\sqrt{(a-d)^2 + 4bc}]. \end{aligned}$$

Recall that $a > d$ and $4bc + (a-d)^2 > 0$ by (8). Hence multiply $|P|$ by the positive quantity $X = 4bc + (a-d)^2 + (a-d)\sqrt{(a-d)^2 + 4bc}$. We have

$$2X|P| = [4bc + (a-d)^2]^2 - (a-d)^2[(a-d)^2 + 4bc] = 4bc[(a-d)^2 + 4bc].$$

Since by (8) the last factor is positive as well as X, $|P|$ and bc have the same sign. □

(iv) **The sign of the solutions.** We suppose that there exist positive constants $c_k' > 0$, $c_k'' > 0$ such that

$$0 < c_k' \rho\varphi_\rho \le \tilde{f}_k \le c_k'' \rho\varphi_\rho, \ k = 1, 2, \tag{16}$$

or, denoting as above $\rho\tilde{h}_k = \tilde{f}_k$,

$$0 < c_k' \varphi_\rho \le \tilde{h}_k \le c_k'' \varphi_\rho, \ k = 1, 2. \tag{16'}$$

By Lemma 3.1, (16) is exactly (9) in (H_F') since $bc\hat{f}_k = |P|\tilde{f}_k$.

Case 1. $\lambda\sigma_1 < \lambda_\rho$. It follows from the first equation—by the maximum principle for one equation—that for $0 < \lambda\sigma_1 < \lambda_\rho$, v_1 has the sign of $\tilde{f}_1$: $v_1 > 0$. Moreover, it follows from (16) and (5) that, in any B_R, for $\lambda_\rho - \lambda\sigma_1$ small enough, there exist positive constants K', K'' such that

$$0 < \frac{K'}{(\lambda_\rho - \sigma_1\lambda)}\varphi_\rho \le v_1 \le \frac{K''}{(\lambda_\rho - \sigma_1\lambda)}\varphi_\rho. \tag{17}$$

LEMMA 3.2. *$v_1/\varphi_\rho \to +\infty$ as $0 < \lambda\sigma_1 \nearrow \lambda_\rho$.*

Consider now v_2. If $\lambda\sigma_1 < \lambda_\rho$, then $\lambda\sigma_2 < \lambda\sigma_1 < \lambda_\rho$ so that $\lambda\sigma_2$ stays away from λ_ρ and v_2/φ_ρ stays bounded as $\lambda\sigma_1 \to \lambda_\rho$. Of course v_2 is also bounded.

Finally let us go back to (u_1, u_2). From $U = PV$ we derive

$$u_1 = bv_1 - Sv_2, \quad u_2 = Sv_1 + cv_2. \tag{18}$$

We apply Remark 4 to v_1. It follows that, on B_R, for $\lambda_\rho - \lambda\sigma_1 > 0$ small enough, that is, for v_1/φ_ρ large enough, u_1 has the sign of bv_1, that is, the sign of b since $v_1 > 0$ by (17). Also u_2 has the sign of S.

Of course the existence of solutions for system (S) follows from the existence of solutions for equations (14).

Case 2. $\lambda\sigma_1 > \lambda_\rho$. If $\lambda_\rho < \lambda\sigma_1 < \lambda_\rho + \delta$, we can apply the (local) antimaximum principle and (6); hence on B_R

$$0 < c'\varphi_\rho \leq \tilde{h}_1 \leq c''\varphi_\rho$$
$$\Rightarrow -\frac{K''}{\lambda\sigma_1 - \lambda_\rho}\varphi_\rho \leq v_1 \leq -\frac{K'}{\lambda\sigma_1 - \lambda_\rho}\varphi_\rho < 0.$$

REMARK 10. *For a cooperative system, the condition* $\lambda_\rho - \lambda\sigma_1 > 0$ *is equivalent to* (10).

3.4. Indefinite weight. We study $-\Delta U = \lambda\rho(x)MU + F$ which after diagonalization of the matrix M is written as $-\Delta V = \lambda\rho(x)DV + P^{-1}F$. We are led again to equations (14):

$$-\Delta v_1 = \lambda\sigma_1\rho v_1 + \tilde{f}_1, \; x \in \mathbb{R}^N,$$
$$-\Delta v_2 = \lambda\sigma_2\rho v_2 + \tilde{f}_2, \; x \in \mathbb{R}^N.$$

We also get (18),

$$u_1 = bv_1 - Sv_2, \quad u_2 = Sv_1 + cv_2.$$

We study now the sign of the solutions. We suppose that there exist positive constants $c'_k > 0$, $c''_k > 0$ such that

$$(16') \qquad 0 < c'_k\varphi_\rho \leq \tilde{h}_k \leq c''_k\varphi_\rho, \; k = 1, 2.$$

Case 1. $0 < \lambda\sigma_1 < \lambda_\rho$. It follows from the first equation by the maximum principle for one equation with an indefinite weight that for $0 < \sigma_1\lambda < \lambda_\rho$, $v_1 > 0$ since $\tilde{f}_1 > 0$. Similar discussions as in Section 3.3 lead to the result.

Case 2. $\lambda\sigma_1 > \lambda_\rho$. If $\lambda_\rho < \lambda\sigma_1 < \lambda_\rho + \delta$, we can apply the (local) antimaximum principle and (6): $v_1 < -K'\varphi_\rho < 0$ on some ball B_R if $\tilde{f}_1 > c\varphi_\rho > 0$, and the result follows as above.

References

[1] W. Allegretto, *Principal eigenvalues for indefinite weight elliptic problems in* $\mathbb{R}^n$. Proc. Amer. Math. Soc., 116, 1992, pp. 701-706.

[2] B. Alziary, J. Fleckinger, M.-H. Lécureux, *Principal eigenvalue and maximum principle for some elliptic systems defined on general domains with refined Dirichlet boundary condition.* Comm. Math. Anal., 7, 2, 2009, pp. 1-11.

[3] B. Alziary, J. Fleckinger, P. Takáč, *Positivity and negativity of solutions to a Schrödinger equation in* $\mathbb{R}^N$. Positivity, 5, 4, 2001, pp. 359-382.

[4] B. Alziary, P. Takáč, *A pointwise lower bound for positive solutions of a Schrödinger equation in* $\mathbb{R}^N$. J. Diff. Equ., 133, 2, 1997, pp. 280-295.

[5] B. Alziary, P. Takáč, *Compactness for a Schrödinger operator in the ground-state space over* $\mathbb{R}^N$. Electron. J. Diff. Eqns., 16, 2007, pp. 35-58.

[6] K. J. Brown, C. Cosner, J. Fleckinger, *Principal eigenvalues for problems with indefinite weight function on* $\mathbb{R}^N$. Proc. Amer. Math. Soc., 109, 1, 1990, pp. 147-155.

[7] H. Brezis, *Analyse Fonctionnelle*, Masson, Paris.

[8] K. J. Brown, A. Tertikas, *The existence of principal eigenvalues for problems with indefinite weight function on* $\mathbb{R}^k$. Proc. Royal Soc. Edinburgh, 123A, 1993, pp. 561-569.

[9] L. Cardoulis, *Existence et estimatations de solutions pour des systèmes d'équations aux dérivées partielles avec des opérateurs de Schrödinger sur* $\mathbb{R}^N$. HDR Université de Toulouse 1, 2008.
[10] P. Clément, L. Peletier, *An anti-maximum principle for second order elliptic operators.* J. Diff. Equ., 34, 1979, pp. 218-229.
[11] D. de Figueiredo, E.Mitidieri, *Maximum principles for cooperative elliptic systems.* C. R. Acad. Sci. Paris 310, Serie I, 1990, pp. 49-52.
[12] J. Fleckinger, J. P. Gossez, P. Takáč, F. de Thélin, *Nonexistence of solutions and an anti-maximum principle for cooperative systems with the p-Laplacian.* Math. Nachr., 194, 1998, pp. 49-78.
[13] J. Fleckinger, J. P. Gossez, F. de Thélin, *Antimaximum principle in* $\mathbb{R}^N$*: local versus global.* J. Diff. Equ., 8, 7B, 2004, pp. 159-188.
[14] J. Fleckinger, J. P. Gossez, F. de Thélin, *Maximum and antimaximum principles near the second eigenvalue* (submitted).
[15] J. Fleckinger-Pellé, H. Serag, *Semilinear cooperative elliptic systems on* $\mathbb{R}^n$, Rend. Mat., S.VII, V15, 1995, pp. 89-108.
[16] J. Fleckinger, J. Hernandez, F. de Thélin, *On Maximum principles and existence of positive solutions for some cooperative elliptic systems.*, Diff. Int. Eq., 8, 12, 1995, pp. 69-85.
[17] J. Fleckinger, R. Manasevich, N. Stavrakakis, F. de Thélin, *Principal eigenvalues for some quasilinear elliptic systems on* $\mathbb{R}^n$. Advances in Diff. Eq., 2, 6, 1997, pp. 981-1003.
[18] P. Hess, *On the eigenvalue problem for weakly coupled elliptic systems.* Arch. Rat. Mech. Anal., 81, 1983, pp. 151-159.
[19] M.-H. Lécureux, *Comparison with groundstate for solutions of noncooperative systems of Schrödinger operator on* $\mathbb{R}^N$ (submitted to Rostock Math. Koll.).
[20] N. Stavrakakis, F. de Thélin, *Principal eigenvalues and anti-maximum principle for some quasilinear elliptic equations on* $\mathbb{R}^n$. Math. Nachr., 212, 1, 2000, pp. 155-171.

Institut de Mathématique et Ceremath-UT1, Université de Toulouse, 31042 Toulouse Cedex, France
E-mail address: jfleckinger@gmail.com

Department of Applied Mathematics, Northwestern Polytechnical University, 710072 Xi'an, Shaanxi, People's Republic of China

Amer. Math. Soc. Transl.
Volume **231**, 2010

On Guided Electromagnetic Waves in Photonic Crystal Waveguides

Peter Kuchment and Beng-Seong Ong

Dedicated to the memory of Professor V. B. Lidskii

ABSTRACT. This paper addresses the issue of existence and confinement of electromagnetic modes guided by linear defects in photonic crystals. Sufficient conditions are provided for the existence of such waves near a given spectral location. Confinement to the guide is achieved due to a photonic band gap in the bulk dielectric medium.

1. Introduction

A photonic crystal, also called photonic band gap (PBG) material, is a periodic medium which plays the role of an optical analog of a semiconductor. Such a medium has a gap in the frequency spectrum of electromagnetic (EM) waves. The idea of a photonic crystal was first suggested in 1987 [**17**, **31**] and has since been intensively studied experimentally and theoretically (see, e.g., the recent books [**16**,**18**,**26**,**27**,**32**], the mathematical survey [**21**], the online bibliography [**24**], and references therein). This interest has been triggered by the numerous promising applications of PBG materials, one of which is using photonic crystal for manufacturing highly efficient optical waveguides. The idea is to introduce a linear "defect" into a PBG material and to guide through it EM waves of a frequency prohibited in the bulk. Numerical and experimental studies have shown that such superior guides can be efficiently created, e.g., [**16**, **18**, **23**, **26**, **27**].

In order to create such a guide, one needs to establish several facts. The first, and foremost, is existence of guided waves of frequencies in the band gap. The second, and an easier one, is confinement of these modes. This paper addresses both issues, by finding some sufficient conditions of existence of guided modes and showing their confinement to the guide, in the sense of being evanescent in the bulk.

2010 *Mathematics Subject Classification.* Primary 35P99, 35Q60; Secondary 78A48.

Key words and phrases. Photonic crystal, defect, waveguide, spectrum, Maxwell operator, guided mode.

Work of both authors was partially supported by the NSF Grants DMS 0296150 and 9971674. Work of the first author was also supported by the NSF Grant DMS 0072248 and was also partially supported by the KAUST grant KUS-CI-016-04 through the IAMCS. The authors thank the NSF, IAMCS, and Kaust for the support.

Similar results were previously obtained by the authors in [**22**] for scalar models (i.e., for acoustic analogs of PBG waveguides). In this paper, we will address the above questions for the full Maxwell case.

There is another important question to be resolved. Namely, one needs to show that the impurity spectrum that arises in the spectral gaps due to the presence of a linear defect does not correspond to bound states. This difficult issue is not addressed here (see some relevant remarks and references in Section 5).

In Section 2 we introduce the main model to be investigated. Section 3 contains formulation of the main results, with the proofs provided in Section 4. The paper ends with the sections devoted to final remarks and acknowledgments, as well as the bibliography.

2. Preliminaries

We start by describing the mathematical model studied in this paper. Let $\varepsilon_0(x)$ be a bounded positive measurable functions in $\mathbb{R}^3$ separated from zero:

$$0 < c_0 \leq \varepsilon_0(x) \leq c_1 < \infty. \tag{2.1}$$

It is usually assumed in photonic crystal theory that ε_0 is periodic with respect to a lattice $\Gamma \subset \mathbb{R}^3$, but this is not required for our results.

The function ε_0 represents the dielectric properties of the bulk material. In other words, one can think of the space $\mathbb{R}^3$ filled with a dielectric material with the dielectric function ε_0.

We will use the name "unperturbed Maxwell operator" for the selfadjoint realization of the operator

$$M_0 := \nabla^\times \frac{1}{\varepsilon_0(x)} \nabla^\times \tag{2.2}$$

in $L_2(\mathbb{R}^3;\mathbb{C}^3)$ defined by means of its quadratic form

$$\int \varepsilon_0^{-1} |\nabla^\times u|^2 dx \tag{2.3}$$

with the domain $H^1(\mathbb{R}^3;\mathbb{C}^3)$. We use here the shorthand notation

$$\nabla^\times u = \nabla \times u = \text{curl } u.$$

Usually the name "Maxwell equations" is reserved for the well-know first-order system [**15**]. However, in the monochromatic case it reduces to the spectral problem for the operator (2.2).

A cylindrical domain $\mathcal{S}_l$ (see Figure 1) will represent a linear "defect strip":

$$\mathcal{S}_l := \{x = (x_1, x') \in \mathbb{R}^3 \,|\, x_1 \in \mathbb{R},\ x' \in l\Omega\}.$$

Here the cross-section $l\Omega$ of the strip is a domain Ω in $\mathbb{R}^2$ (e.g., the unit ball centered at the origin), scaled with factor l.

We now introduce the perturbed medium with homogeneous dielectric properties inside the defect strip $\mathcal{S}_l$:

$$\varepsilon(x) = \begin{cases} \varepsilon > 0 & \text{for } x \in \mathcal{S}_l, \\ \varepsilon_0(x) & \text{for } x \notin \mathcal{S}_l, \end{cases} \tag{2.4}$$

where ε is a constant.

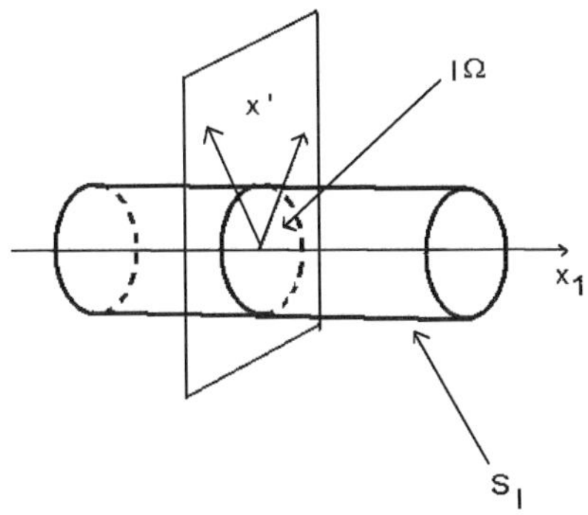

FIGURE 1. The waveguide $\mathcal{S}_l$.

The perturbed Maxwell operator

$$M := \nabla^{\times} \frac{1}{\varepsilon(x)} \nabla^{\times} \tag{2.5}$$

corresponds to the medium with the linear defect. It is defined, analogously to M_0, as a selfadjoint operator in $L_2(\mathbb{R}^3; \mathbb{C}^3)$.

REMARK 2.1. The reader has probably noticed that we disregard the standard restriction of the Maxwell operator to divergence-free fields [**16**]. The difference is essentially in acquiring a huge eigenspace corresponding to the zero frequency, with all other parts of the spectral decomposition staying intact. Since the problem of guided waves concerns the situation inside the spectral gaps of the operator M_0, this difference is irrelevant in this case. On the other hand, abandoning the zero divergence condition will simplify the techniques considerably.

Our goal is the same as in the paper [**22**] devoted to the scalar (acoustic) case, i.e., to show that for any gap (α, β) in the spectrum $\sigma(M_0)$ of the unperturbed medium, under appropriate conditions on the parameters l and ε of the line defect, additional spectrum arises inside the gap, with the corresponding (generalized) eigenmodes being confined to the defect (evanescent in the bulk).

3. Formulation of the results

In order to formulate our first main result, we need to introduce the following quantity:

DEFINITION 3.1. We denote by $\nu > 0$ the lowest eigenvalue of the Laplace operator Δ acting on divergence-free $\mathbb{R}^2$-valued vector fields on Ω with Dirichlet boundary conditions on $\partial\Omega$.

I.e., $\nu > 0$ is the smallest number for which a nontrivial solution of the following problem exists:

$$\begin{cases} -\Delta E = \nu E \text{ in } \Omega, \\ \nabla \cdot E = 0 \text{ in } \Omega, \\ E \mid_{\partial\Omega} = 0, \end{cases} \tag{3.1}$$

where $\frac{\partial}{\partial n}$ is the external normal derivative on $\partial\Omega$.

Our main results are given in the following theorems.

THEOREM 3.2. *Let $G = (\alpha, \beta)$ be a nonempty finite interval, such that $\alpha > 0$ (we will be especially interested in the case when G is a gap in the spectrum of the "background medium" operator M_0). Let the following inequality be satisfied:*

$$l^2(\beta - \alpha)\varepsilon > 2\nu. \tag{3.2}$$

Then the interval G contains at least one point of the spectrum $\sigma(M)$ of the perturbed operator.

This theorem guarantees that when (3.2) is satisfied, eigenmodes of the perturbed medium do arise in the spectral gaps of the background medium. Furthermore, if $\delta > 0$ is such that $l^2\delta\varepsilon > \nu$, the corresponding spectrum forms a δ-net in the gap.

Before one can fully associate these modes with the guided waves, one needs to establish their confinement to the waveguide (i.e., their evanescent nature in the bulk of the material). In order to describe the corresponding result, we need to remind the reader about some notions and results concerning generalized eigenfunction expansions. Here by generalized eigenfunctions one understands solutions of the eigenvalue problem, which do not decay sufficiently fast (or do not decay at all) to belong to the ambient Hilbert space L^2. One can find detailed discussions of generalized eigenfunction expansions, for instance, in [**3**, **4**].

Namely, as is shown in [**19**], for the operator M that we consider, for almost any (with respect to the spectral measure) λ in the spectrum $\sigma(M)$, there is a generalized eigenfunction $u_\lambda(x) \in H^1_{loc}(\mathbb{R}^3, \mathbb{C}^3)$ with the growth estimates

$$(1 + |x|)^{-N}u(x) \in L_2(\mathbb{R}^3; \mathbb{C}^3) \text{ and } (1 + |x|)^{-N}\nabla^\times u(x) \in L_2(\mathbb{R}^3; \mathbb{C}^3) \tag{3.3}$$

for some $N > 0$ (that might depend on the eigenfunction). This system of generalized functions is complete in the whole space (see [**3**,**4**,**19**] for detailed explanations of the meaning of this completeness). For elliptic operators with smooth coefficients, this is a well-known fact [**3**].

DEFINITION 3.3. We refer to the following growth condition as *polynomial boundedness of order N*: for any compact set $K \subset \mathbb{R}^3$ and $x \in \mathbb{R}^3$,

$$\|u\|_{L_2((K+x);\mathbb{C}^3)} + \|\nabla^\times u\|_{L_2((K+x);\mathbb{C}^3)} \leq C_K(1 + |x|)^N. \tag{3.4}$$

In the next result we will use the following notation: for $x = (x_1, x') \in \mathbb{R}^3$, we denote by $\chi_x(y)$ the *characteristic function* of the cube

$$\{y \mid |y_j - x_j| \leq 1 \text{ for } j = 1, 2, 3\}$$

centered at x, i.e., $\chi_x(y)$ is equal to 1 when y is in this cube and 0 otherwise.

THEOREM 3.4. *Let G be a finite spectral gap of M_0 and let u_λ be a polynomially bounded generalized eigenfunction of M corresponding to $\lambda \in G \cap \sigma(M)$. Then there exist positive constants C and $C(\lambda)$ such that*

$$\|\chi_x u_\lambda\| \leq C\,(1 + |x_1|)^N\, e^{-C(\lambda)\text{dist}(x,\, S_l)}, \tag{3.5}$$

where N is the order of polynomial boundedness of u_λ.

When the bulk medium is periodic in the x_1-direction, the polynomial growth in (3.5) disappears:

THEOREM 3.5. *If $\varepsilon_0(x)$ is periodic in the x_1-direction, then one can find a complete family of generalized eigenfunctions that satisfies*

$$\|u_\lambda\|_{L_2((K+x);\mathbb{C}^3)} + \|\nabla^\times u_\lambda\|_{L_2((K+x);\mathbb{C}^3)} \leq C_K(1+|x'|)^N \tag{3.6}$$

for any compact set $K \subset \mathbb{R}^3$, $x \in \mathbb{R}^3$. In this case, for $\lambda \in G \cap \sigma(M)$, one has the estimate

$$\|\chi_x u_\lambda\| \leq Ce^{-C(\lambda)\operatorname{dist}(x,\mathcal{S}_l)}. \tag{3.7}$$

4. Proofs of the results

In what follows, the norm and inner product in $L_2(\mathbb{R}^n;\mathbb{C}^3)$ will be denoted by $\|\cdot\|$ and $\langle\cdot,\cdot\rangle$, respectively.

4.1. Proof of Theorem 3.2. We will show that if $\mu > 0$ and $\delta > 0$ is such that $l^2\delta\epsilon > \nu$, there is some spectrum of the operator M in the δ-vicinity of μ. Then, taking $\mu = (\alpha+\beta)/2$ and $\delta = (\beta-\alpha)/2$, one gets the statement of the theorem.

In order to show this, due to selfadjointness of M, it is sufficient to find a vector function $w \in L_2(\mathbb{R}^3, \mathbb{C}^3)$ of unit norm, such that $w \in \mathcal{D}(M)$ and

$$\|Mw - \mu w\|^2 < \delta^2. \tag{4.1}$$

Let g be a smooth, unit L^2-norm, divergence-free real vector field on $\mathbb{R}^2$ with compact support in Ω and unit L^2-norm, i.e, $g(y,z) = (\phi(y,z), \zeta(y,z))$ where ϕ, $\zeta \in C_0^\infty(\Omega)$ and $\nabla \cdot g = 0$. We define

$$g_l := (\phi_l, \zeta_l) = l^{-1}(\phi(x'/l), \zeta(x'/l)).$$

Then g_l is also a unit L^2-norm divergence-free field. Let $\psi(x_1) \in C_0^\infty(\mathbb{R})$ have unit $L_2(\mathbb{R})$-norm and let $\psi_n(x_1) = n^{-1/2}\psi(x_1/n)$ for $n > 0$. Clearly $\psi_n(x_1)$ also has unit L_2-norm.

Denoting $k = \sqrt{\mu\varepsilon}$, we introduce the following candidate for an approximate eigenfunction:

$$w_{l,n}(x) = \psi_n(x_1)e^{ikx_1}\begin{pmatrix} 0 \\ \phi_l(x') \\ \zeta_l(x') \end{pmatrix}. \tag{4.2}$$

The function $w_{l,n}$ clearly has unit norm in $L_2(\mathbb{R} \times l\Omega)$.

Instead of estimating the left-hand side of inequality (4.1), we will estimate $\|\varepsilon\,(Mw - \mu w)\|^2$. Taking into account that the function w is supported inside the defect, the needed inequality (4.1) can also be rewritten as

$$\|\nabla^\times\nabla^\times w - k^2 w\|^2 < \delta^2\varepsilon^2. \tag{4.3}$$

Using the identity

$$\nabla^\times\nabla^\times w = -\Delta w + \nabla(\nabla \cdot w)$$

and the fact that g_l is divergence-free, we obtain

$$\|\nabla^\times\nabla^\times w - k^2 w\|^2 = \left\| \begin{pmatrix} 0 \\ -(\psi_n'' + 2ik\psi_n')\phi_l - \psi_n\Delta\phi_l \\ -(\psi_n'' + 2ik\psi_n')\zeta_l - \psi_n\Delta\zeta_l \end{pmatrix} \right\|^2$$

where the norms are in $L_2(\mathbb{R}^3;\mathbb{C}^3)$.

Since the functions ϕ, ζ, and ψ are real valued, their assumed normalization shows that the above expression is equal to

$$n^{-4}\|\psi''\|^2_{L_2(\mathbb{R})} + 4k^2 n^{-2}\|\psi'\|^2_{L_2(\mathbb{R})} + l^{-4}\|\Delta g\|^2_{L_2(\Omega,\mathbb{R}^2)} \\ + 2(nl)^{-2}\langle\psi'',\psi\rangle_{L_2(\mathbb{R})}\langle\Delta g, g\rangle_{L_2(\Omega,\mathbb{R}^2)}.$$

Since n can be chosen arbitrarily large, the terms with the factors that are negative powers of n can be made arbitrarily small (uniformly with respect to k on any finite interval). Hence, one needs to control only the remaining terms by an appropriate choice of a divergence-free vector field g. In other words, one is interested in making $l^{-4}\|\Delta g\|^2_{L_2(\Omega,\mathbb{R}^2)}$ smaller than $\delta^2\varepsilon^2$, i.e.,

$$\|\Delta g\|_{L_2(\Omega;\mathbb{R}^2)} < l^2\delta\varepsilon, \tag{4.4}$$

while keeping $\langle\Delta g, g\rangle$ under control.

Since

$$\nu = \inf \|\Delta g\|_{L_2(\Omega;\mathbb{R}^2)},$$

where the *infimum* is taken over real, unit L^2-norm, divergence-free vector fields $g \in C_0^\infty(\Omega;\mathbb{R}^2)$, this condition boils down to

$$l^2\delta\varepsilon > \nu, \tag{4.5}$$

which proves the statement of the theorem. □

4.2. Proof of Theorem 3.4. We assume here that λ belongs to a finite gap G of the spectrum of M_0 and that $u := u_\lambda$ is the corresponding generalized eigenfunction of M satisfying (3.4). Let us also introduce the resolvent $R(\lambda) = (M_0 - \lambda)^{-1}$. We will also use the function χ_x introduced before Theorem 3.4.

We will need the following auxiliary statement concerning the exponential decay of the resolvent, which is a result of [**8**]:

LEMMA 4.1 ([**8**]). *There exists a positive number m_λ that depends only on the distance of the point λ from the gap edges, such that for a positive constant C, the following estimates hold for the local $L_2(\mathbb{R}^3;\mathbb{C}^3)$-norm of the resolvent $R(\lambda)$:*

$$\begin{aligned} \|\chi_u R(\lambda)\chi_v\| &\le Ce^{-m_\lambda|u-v|}, \\ \|\chi_u \nabla^\times R(\lambda)\chi_v\| &\le Ce^{-m_\lambda|u-v|} \end{aligned} \tag{4.6}$$

for any u, $v \in \mathbb{R}^3$. Here the norms in the left-hand side are the operator norms in $L_2(\mathbb{R}^3;\mathbb{C}^3)$.

We now consider the sesquilinear form

$$Q[\varphi, w] := \langle\nabla^\times\varphi, \frac{1}{\varepsilon_0}\nabla^\times w\rangle - \lambda\langle\varphi, w\rangle$$

with the domain $H^1(\mathbb{R}^3;\mathbb{C}^3)$.

Let $\varphi := R(\lambda)\chi_x u$. Note that φ belongs to the domain of the operator M_0.

Let $p = \max(2\text{dist}(x, S_l), 1)$ and let $\xi_x(y)$ be a nonnegative smooth cutoff function that depends on y_1 only, is supported in $(x_1 - (p+1), x_1 + (p+1))$, and such that it is equal to 1 on $[x_1 - p, x_1 + p]$. We assume further that $\xi_x(y) \leq 1$ and $|\nabla \xi_x(y)| \leq C$ for some constant C and all $x, y \in \mathbb{R}^3$. Note that $\xi_x u \in H^1(\mathbb{R}^3; \mathbb{C}^3)$. Using $w = \xi_x u$, one gets

$$Q[\varphi, \xi_x u] = \langle M_0 \varphi, \xi_x u \rangle - \langle \lambda \varphi, \xi_x u \rangle = \langle \chi_x u, \xi_x u \rangle = \|\chi_x u\|^2 .$$

Thus, the goal is to estimate $Q[\varphi, \xi_x u]$ from above. On the other hand, using the equality $Mu = \lambda u$ and integration by parts, one obtains

$$\begin{aligned} Q[\varphi, \xi_x u] &= \langle \nabla^\times \varphi, \varepsilon_0^{-1} \nabla^\times (\xi_x u) \rangle - \langle \varphi, \xi_x \lambda u \rangle \\ &= \langle \nabla^\times \varphi, \tilde{\epsilon} \nabla^\times (\xi_x u) \rangle + \langle \nabla^\times \varphi, \varepsilon^{-1} \nabla^\times (\xi_x u) \rangle - \langle \varphi, \xi_x M u \rangle \\ &= \langle \nabla^\times \varphi, \tilde{\epsilon} \nabla^\times (\xi_x u) \rangle + \langle \nabla^\times \varphi, \varepsilon^{-1} \nabla^\times (\xi_x u) \rangle - \langle \nabla^\times (\xi_x \varphi), \varepsilon^{-1} \nabla^\times u \rangle, \end{aligned} \tag{4.7}$$

where we used the notation

$$\tilde{\epsilon}(x) := \frac{1}{\varepsilon_0(x)} - \frac{1}{\varepsilon(x)}.$$

Notice that $\tilde{\epsilon}$ is supported inside the strip $\mathcal{S}_l$.

Using the identity $\nabla^\times (\xi u) = \xi \nabla^\times u + \nabla \xi \times u$, the first two terms in the last line of (4.7) can be combined to obtain

$$\begin{aligned} &\langle \nabla^\times \varphi, \tilde{\epsilon}(\xi_x \nabla^\times u + \nabla \xi_x \times u) \rangle + \langle \nabla^\times \varphi, \varepsilon^{-1}(\xi_x \nabla^\times u + \nabla \xi_x \times u) \rangle \\ &\quad = \langle \nabla^\times \varphi, \xi_x \tilde{\epsilon} \nabla^\times u \rangle + \langle \nabla^\times \varphi, (\tilde{\epsilon} + \varepsilon^{-1}) \nabla \xi_x \times u \rangle + \langle \nabla^\times \varphi, \varepsilon^{-1} \xi_x \nabla^\times u \rangle. \end{aligned}$$

The term $-\langle \nabla^\times (\xi_x \varphi), \varepsilon^{-1} \nabla^\times u \rangle$ can be expanded to

$$-\langle \xi_x \nabla^\times \varphi, \varepsilon^{-1} \nabla^\times u \rangle - \langle \nabla \xi_x \times \varphi, \varepsilon^{-1} \nabla^\times u \rangle.$$

Combining the last two expressions, we get

$$\langle \nabla^\times \varphi, \xi_x \tilde{\epsilon} \nabla^\times u \rangle + \langle \nabla^\times \varphi, \varepsilon_0^{-1} \nabla \xi_x \times u \rangle - \langle \nabla \xi_x \times \varphi, \varepsilon^{-1} \nabla^\times u \rangle. \tag{4.8}$$

Our last task in proving the theorem is to estimate from above the terms in (4.8). Let $V = [x_1 - p - 1, x_1 + p + 1] \times l\Omega$. This is a compact domain that can be covered by the union of $p + 1$ fixed size domains $V_j = [b_j, b_j + 2] \times l\Omega$ and which contains the support of $(\xi \tilde{\epsilon})$. Also note that $\operatorname{dist}(x, V_j) \geq \operatorname{dist}(x, S_l)$. Using Lemma 4.1 and (3.4), we get for any $0 < \eta < m_\lambda$

$$\begin{aligned} |\langle \nabla^\times \varphi, \xi_x \tilde{\epsilon} \nabla^\times u \rangle| &\leq \|\chi_V \nabla^\times \varphi\| \, \|\xi_x \tilde{\epsilon} \nabla^\times u\| \\ &\leq C \, \|\textstyle\sum_j \chi_{V_j} \nabla^\times R(\lambda) \chi_x u\| \|\textstyle\sum_j \chi_{V_j} \nabla^\times u\| \\ &\leq C p^2 (|x_1| + p + 1)^{2N} e^{-m_\lambda \operatorname{dist}(x, \mathcal{S}_l)} \\ &\leq C (|x_1| + 1)^{2N} e^{-(m_\lambda - \eta) \operatorname{dist}(x, \mathcal{S}_l)}. \end{aligned} \tag{4.9}$$

We used here that $p = \max(2\operatorname{dist}(x, \mathcal{S}_l), 1)$ and denoted different constants by C.

Let us move now to estimating the last term in (4.8). Denote by $a > 0$ a number such that shifts of $l\Omega$ by vectors aj with $j \in \mathbb{Z}^2$ cover the whole space $\mathbb{R}^2$. We denote

$$W_j := ([x_1 - p - 1, x_1 - p] \cup [x_1 + p, x_1 + p + 1]) \times (l\Omega + aj) .$$

Then $W_j = W_0 + (0, aj)$. Notice that $W = \bigcup_j W_j$ covers $\operatorname{supp} \nabla \xi$ and $\operatorname{dist}(x, W_j) \geq C_1(p + |j|) - C_2$.

We are now ready to estimate the last term of (4.8) from above. We proceed as before, using the lemma, the polynomial growth of u, and uniform boundedness of $\nabla\xi_x$.

$$
\begin{aligned}
|\langle\nabla^\times\varphi, \varepsilon_0^{-1}\nabla\xi_x^\times u\rangle| &\leq C\sum_j \|\chi_{W_j}u\|\|\chi_{W_j}\nabla^\times R(z)\chi_x u\| \\
&\leq C\sum_j(|x_1|+p+|j|+1)^{2N}e^{-m_\lambda \mathrm{dist}(x,W_j)} \\
&\leq C(|x_1|+p+1)^{2N}e^{-Cm_\lambda \mathrm{dist}(x,\mathcal{S}_l)}\sum_j(1+|j|)^{2N}e^{-Cm_\lambda|j|} \\
&\leq C(|x_1|+1)^{2N}e^{-Cm_\lambda \mathrm{dist}(x,\mathcal{S}_l)}.
\end{aligned}
\tag{4.10}
$$

The middle term in (4.8) is estimated analogously. Combining these estimates, we get

$$\|\chi_x u\|^2 = Q[\varphi, \xi_x u] \leq C(1+|x_1|)^{2N}e^{-Cm_\lambda \mathrm{dist}(x,\mathcal{S}_l)}.$$

This finishes the proof of the theorem. □

4.3. Proof of Theorem 3.5. In this periodic situation, operator M has a complete family of generalized eigenfunctions that do not grow in the x_1-direction of periodicity. Indeed, according to Bloch-Floquet theory [**20, 25**], a generalized eigenfunction $u = u_\lambda$ of M corresponding to λ can be chosen as $\tilde{u}(x)e^{ikx_1}$, where $\tilde{u}(x)$ is periodic in the x_1-direction with period a and $k = 2\pi/a$. Thus, u satisfies (3.6). Then, repeating the previous proof, one comes up with the estimate (3.7). □

5. Remarks

(1) Theorem 3.2 provides sufficient conditions for the existence of a δ-net of the defect spectrum inside a spectral gap. One wonders how much of the gap the guided mode spectrum can occupy. Can it fill the whole gap? If so, under what conditions? There seem to be no rigorous results available concerning these questions.

(2) Results of [**2**] on improved Combes-Thomas resolvent estimates show that the exponential decay constant m_λ, which clearly depends on the distance of the point λ from the spectrum, behaves as $\sqrt{(\lambda-\alpha)(\beta-\lambda)}$ inside a gap $G = (\alpha, \beta)$.

(3) As has already being mentioned, in order to have the full right to call the discovered modes "guided", one needs to show that they do not correspond to point spectrum (i.e., to bound states). Here the most treatable case should be of a periodic medium with a linear defect aligned along one of the lattice vectors. In this situation one can apply the Floquet-Bloch theory with respect to the axial variable of the waveguide and hope to use standard techniques applied in the case of Schrödinger operators with periodic potential (e.g., [**5, 13, 20, 25**]). This happens to be not an easy task. Even in the case of "hard wall" periodic waveguides, when waves are contained in a periodic waveguide by Dirichlet, Neumann, or more general boundary conditions, this problem is nontrivial. Although it has been considered for a rather long time [**6, 20**], the first real advances are very recent [**11,12,14,28–30**]. The case of photonic crystal waveguides is more complex, due to the absence of complete confinement of the waves,

which exponentially decay into the bulk, but do not vanish completely. Apparently, the most recent work devoted to this issue assumes the bulk material to be homogeneous, with an embedded periodic guide [**10**].

6. Acknowledgements

The authors express their gratitude to H. Ammari, D. Dobson, P. Exner, F. Santosa, R. Shterenberg, A. Sobolev, and T. Suslina for discussions and information. The authors thank the reviewer for useful comments.

References

[1] H. Ammari and F. Santosa, *Guided waves in a photonic bandgap structure with a line defect*, SIAM J. Appl. Math. **64** (2004), no. 6, 2018–2033.
[2] J. M. Barbaroux, J. M. Combes, and P. D. Hislop, *Localization near band edges for random Schrödinger operators*, Helv. Phys. Acta **70** (1997), 16–43.
[3] Ju. M. Berezanskii, *Expansions in Eigenfunctions of Selfadjoint Operators*, AMS, Providence, RI, 1968.
[4] F. A. Berezin and M. Shubin, *The Schrödinger Equation*, Moskov. Gos. Univ., Moscow, 1983 (in Russian), extended English translation: Kluwer Academic Publishers, 1991.
[5] M. Sh. Birman and T. A. Suslina, *Periodic magnetic Hamiltonian with a variable metric. The problem of absolute continuity*, Algebra i Analiz **11** (1999), no. 2; English translation in St. Petersburg Math. J. **11** (2000), no. 2, 203–232.
[6] V. I. Derguzov, *On the discreteness of the spectrum of the periodic boundary-value problem associated with the study of periodic waveguides*, Siberian Math. J. **21** (1980), 664–672.
[7] A. Figotin and A. Klein, *Localization of classical waves I: Acoustic waves*, Commun. Math. Phys. **180** (1996), 439–482.
[8] A. Figotin and A. Klein, *Localization of classical waves II: Electromagnetic waves*, Commun. Math. Phys. **184** (1997), no. 2, 411–441.
[9] A. Figotin and A. Klein, *Localized classical waves created by defects*, J. Stat. Phys. **86** (1997), no. 1-2, 165–177.
[10] N. Filonov and F. Klopp, *Absolute continuity of spectrum of a Schrodinger operator with a potential that is periodic in some direction and decays in others*, Doc. Math. **9** (2004), 107–121.
[11] R. L. Frank, *On the scattering theory of the Laplacian with a periodic boundary condition. I. Existence of wave operators*, Doc. Math. **8** (2003), 547–565.
[12] R. L. Frank and R. G. Shterenberg, *On the scattering theory of the Laplacian with a periodic boundary condition. II. Additional channels of scattering*, Doc. Math. **9** (2004), 57–77.
[13] L. Friedlander, *On the spectrum of a class of second order periodic elliptic differential operators*, Commun. Math. Phys. **229** (2002), 49–55.
[14] L. Friedlander, *Absolute continuity of the spectra of periodic waveguides*, Contemporary Math. **339** (2002), 37–42.
[15] J. D. Jackson, *Classical Electrodynamics*, 3rd ed., Wiley, New York, NY, 1998.
[16] J. D. Joannopoulos, S. G. Johnson, R. D. Meade, and J. N. Winn, *Photonic Crystals, Molding the Flow of Light*, Princeton Univ. Press, 2nd edition, Princeton, NJ, 2008.
[17] S. John, *Strong localization of photons in certain disordered dielectric superlattices*, Phys. Rev. Lett. **58** (1987), 2486–2489.
[18] S. G. Johnson and J. D. Joannopoulos, *Photonic Crystals, the Road from Theory to Practice*, Kluwer Academic Publishers, Boston, 2002.
[19] A. Klein, A. Koines, and M. Seifert, *Generalized eigenfunctions for waves in homogeneous media*, J. Funct. Anal. **190** (2002), no. 1, 255–291.
[20] P. Kuchment, *Floquet Theory for Partial Differential Equations*, Birkhäuser Verlag, Basel, 1993.
[21] P. Kuchment, *The mathematics of photonics crystals*, Chapter 7 in *Mathematical Modeling in Optical Science*, G. Bao, L. Cowsar, and W. Masters (Editors), SIAM, Philadelphia, 2001, 207–272.
[22] P. Kuchment and B.-S. Ong, *On guided waves in photonic crystal waveguides*, in *Waves in Periodic and Random Media*, Contemp. Math. **339**, AMS, Providence, RI, 2002, 105–115.

[23] J. T. Londergan, J. P. Carini, and D. P. Murdock, *Binding and Scattering in Two-Dimensional Systems*, Springer Verlag, Berlin, 1999.

[24] Photonic & Sonic Band-Gap and Metamaterial Bibliography, http://phys.lsu.edu/~jdowling/pbgbib.html.

[25] M. Reed and B. Simon *Methods of Modern Mathematical Physics, IV: Analysis of Operators*, Academic Press, New York, 1978.

[26] K. Sakoda, *Optical Properties of Photonic Crystals*, Springer Verlag, Berlin, 2001.

[27] M. Skorobogatiy and J. Yang, *Fundamentals of Photonic Crystal Guiding*, Cambridge Univ. Press, 2009.

[28] A. V. Sobolev and J. Walthoe, *Absolute continuity in periodic waveguides*, Proc. London Math. Soc. **(3)** (2002), no. 85, 717–741.

[29] R. G. Shterenberg, *Schrödinger operator in a periodic waveguide on the plane and quasi-conformal mappings*, Zap. Nauchn. Sem. POMI **295** (2003), 204–243 (in Russian).

[30] R. G. Shterenberg and T. A. Suslina, *Absolute continuity of the spectrum of a magnetic Schrödinger operator in a two-dimensional periodic waveguide*, Algebra i Analiz **14** (2002), 159–206; English transl. in St. Petersburg Math. J. **14** (2003).

[31] E. Yablonovitch, *Inhibited spontaneous emission in solid-state physics and electronics*, Phys. Rev. Lett. **58** (1987), 2059–2062.

[32] F. Zolla, G. Renversez, A. Nicolet, B. Kuhlmey, S. Guenneau, D. Felbacq, *Foundations Of Photonic Crystal Fibres*, World Sci. Publ., 2005.

Department of Mathematics, Texas A&M University, College Station, TX 77843-3368

E-mail address: kuchment@math.tamu.edu

Veritas, 10300 Town Park Drive, Houston, Texas 77072

E-mail address: beng.ong@cggveritas.com

Amer. Math. Soc. Transl.
Volume **231**, 2010

Spectral Inequalities for a Class of Nonelliptic Operators

Ari Laptev and Fabian Portmann

ABSTRACT. We prove spectral inequalities for the moments of the eigenvalues of the operator $D_x^2 D_y^2 - V(x,y)$, $V \geq 0$, with Dirichlet boundary conditions in $L^2(\mathbb{R}^2_{++})$.

1. Introduction and main result

Well-known Lieb-Thirring inequalities for a Schrödinger operator $-\Delta - V$, $V \in L^{\gamma+d/2}(\mathbb{R}^d)$, state that for the γ-moments of its negative eigenvalues $\{-\lambda_k\}$ the estimate

$$\sum_k \lambda_k^\gamma \leq L_{\gamma,d} \int_{\mathbb{R}^d} V_+^{\gamma+d/2}(x)\,dx \tag{1.1}$$

holds, where $V_+ = (|V| + V)/2$ is the positive part of V. The constants $L_{\gamma,d}$ in this inequality are finite if $\gamma \geq 1/2$ $(d = 1)$, $\gamma > 0$ $(d = 2)$, and $\gamma \geq 0$ $(d \geq 3)$; for details see [**W, LTh, C, L, R**]. If $\gamma = 1$, E. H. Lieb and W. Thirring proved in [**LTh**] that (1.1) is equivalent to a so-called generalized Sobolev inequality for an orthonormal system of functions $\{\varphi_k\}_{k=1}^N$ in $L^2(\mathbb{R}^d)$,

$$\int_{\mathbb{R}^d} [\rho_N(x)]^{(2+d)/d}\,dx \leq C_d \sum_{k=1}^N \int_{\mathbb{R}^d} |\nabla\varphi_k(x)|\,dx, \tag{1.2}$$

where

$$\rho_N(x) = \sum_{k=1}^N |\varphi_k(x)|^2.$$

With the help of the Fourier transform (1.2) can be rewritten as

$$\int_{\mathbb{R}^d} [\rho_N(x)]^{(2+d)/d}\,dx \leq C_d(2\pi)^d \sum_{k=1}^N \int_{\mathbb{R}^d} |\xi|^2|\hat\varphi_k(\xi)|^2\,d\xi, \quad x, \xi \in \mathbb{R}^d. \tag{1.3}$$

Recently, D. S. Barsegyan [**B1**] has obtained Lieb-Thirring type inequalities (1.3) in $\mathbb{R}^2$, where the Laplace operator (whose symbol equals $|\xi|^2$) has been substituted

---

2010 *Mathematics Subject Classification.* Primary 35P15; Secondary 81Q10.

by the product $|D_x D_y|$, $D_x = -i\partial/\partial x$. In this case the latter inequality takes the form

(1.4)

$$\int_{\mathbb{R}^2} [\rho_N(x,y)]^2\,dxdy \le C\,(\log N + 1)\sum_{k=1}^N \int_{\mathbb{R}^2} |\xi\eta||\hat{\varphi}_k(\xi,\eta)|^2\,d\xi d\eta, \quad (x,y),(\xi,\eta)\in\mathbb{R}^2,$$

where the constant C is independent of N. This inequality could be rewritten as an inequality for the negative eigenvalues $\{-\lambda_k\}$ of the operator

$$|D_x D_y| - V \tag{1.5}$$

acting in $L^2(\mathbb{R}^2)$. Let $-\lambda_1 \le -\lambda_2 \le \cdots \le -\lambda_N \le \cdots$ be the sequence of negative eigenvalues. Then (1.4) implies that for any N,

$$\sum_{k=1}^N \lambda_k \le C(\log N + 1)\int_{\mathbb{R}^2} V_+^2(x,y)\,dxdy. \tag{1.6}$$

Indeed, if $\{\varphi_k\}$ is an orthonormal system of eigenfunctions of the operator (1.5), then by (1.4) and the Cauchy-Schwartz inequality we have

$$-\sum_{k=1}^N \lambda_k = \int_{\mathbb{R}^2} |\xi\eta| \sum_{k=1}^N |\hat{\varphi}_k(\xi,\eta)|^2\,d\xi d\eta - \int_{\mathbb{R}^2} V \sum_{k=1}^N |\varphi_k(x,y)|^2\,dxdy$$

$$\ge C(\log N+1)^{-1}\int_{\mathbb{R}^2} [\rho_N(x,y)]^2\,dxdy - \Big(\int_{\mathbb{R}^2} V^2\,dxdy\Big)^{1/2}\Big(\int_{\mathbb{R}^2}[\rho_N(x,y)]^2\,dxdy\Big)^{1/2}.$$

Minimizing the right-hand side with respect to

$$X = \Big(\int_{\mathbb{R}^2} [\rho_N(x,y)]^2\,dxdy\Big)^{1/2}$$

immediately gives (1.6).

Remark 1.1. It easy to prove ([**LTh**]) that (1.6) is equivalent to (1.4).

Remark 1.2. Although it has been proved in [**B1**] that (1.4) is sharp, it does not give a satisfactory inequality for the sum of all negative eigenvalues, because the right-hand side of (1.6) depends on $\log N + 1$. When $d = 2$, estimates for the number of negative eigenvalues even for Schrödinger operators is a delicate problem. Necessary and sufficient conditions for the finiteness of the negative spectrum are so far not known (see [**BL**], [**LN**], [**Sol3**]).

In this paper we consider a related problem and obtain spectral inequalities for the operator

$$D_x^2 D_y^2 u - Vu = -\lambda u, \quad u(x,0) = u(0,y) = 0, \tag{1.7}$$

in $L^2(\mathbb{R}^2_{++})$, where $\mathbb{R}^2_{++} = (0,\infty)^2$. Recall the classical Beta-function

$$\mathcal{B}(a,b) = \int_0^1 t^{a-1}(1-t)_+^{b-1}.$$

The following theorem is our main result.

THEOREM 1.3. *Let $\gamma \geq 1/2$. Then for the negative eigenvalues $\{-\lambda_k\}$ of the operator* (1.7) *we have*

$$\sum_k \lambda_k^\gamma \leq \frac{1}{4^\gamma}(2\pi)^{-2}(R_{\gamma,1})^2\, \mathcal{B}(1/2,\gamma+1) \int_{\mathbb{R}^2_{++}} V_+^{1/2+\gamma} \log(1+4xy\sqrt{V_+})\,dxdy, \tag{1.8}$$

where the constants $L_{\gamma,1}$ are defined in (1.1).

REMARK 1.4. For both (1.5) and (1.7) the phase volume type estimates do not exist, because the classical phase volume is infinite. Differential parts of these operators are highly nonelliptic. Some examples of operators with infinite classical phase volume were previously considered in papers of B. Simon [**S1**], [**S2**], M. Z. Solomyak [**Sol1**], [**Sol2**], and M. Z. Solomyak and I. L. Vulis [**SolV**].

The sharpness of the inequality (1.8) in terms of large potentials is indicated by the following argument. It is known from [**S2**] that for the number $N(\lambda)$ of the eigenvalues $\{\lambda_k\}$ below λ of the operator $D_x^2 + D_y^2 + x^2y^2$ there is the following asymptotic formula:

$$N(\lambda) = \pi^{-1}\lambda^{3/2}\log\lambda + o(\lambda^{3/2}\log\lambda), \quad \lambda \to \infty.$$

This formula ([**AL**]) immediately implies that

$$\sum_k (\lambda - \lambda_k)_+^\gamma = \frac{1}{(\gamma+3/2)\pi}\,\lambda^{\gamma+3/2}\log\lambda + o(\lambda^{\gamma+3/2}\log\lambda), \quad \lambda \to \infty.$$

Using the duality of the Fourier transform, it is equivalent to studying the spectrum below λ of the operator $D_x^2D_y^2 + x^2 + y^2$. We now reduce this problem to studying the negative spectrum of the operator $D_x^2D_y^2 - (\lambda - x^2 - y^2)_-$. By Theorem 1.3 we find that for $\gamma \geq 1/2$

$$\begin{aligned}\sum_k (\lambda - \lambda_k)_+^\gamma &\leq \frac{1}{4^\gamma}(2\pi)^{-2}(R_{\gamma,1})^2\, \mathcal{B}(1/2,\gamma+1) \\ &\quad\times \int_{\mathbb{R}^2_{++}} (\lambda - x^2 - y^2)_+^{1/2+\gamma} \log(1+4xy\sqrt{(\lambda - x^2 - y^2)_+})\,dxdy \\ &\leq C\lambda^{\gamma+3/2}(1+\log(\lambda+1)),\end{aligned}$$

where C is independent of λ. Applying the Legendre transform (see for example [**LW2**]), we can also find that for the sum of eigenvalues of the operator $D_x^2D_y^2 + x^2 + y^2$ we have

$$\sum_{k=1}^N \lambda_k \geq C\frac{N^{5/3}}{1+\log^{2/3}N}.$$

The latter formula has also been obtained by D. S. Barsegyan in [**B2**].

2. Proof of Theorem 1.3

In order to prove Theorem 1.3 we need a version of the operator-valued spectral inequality which has been proved in [**LW1**], [**HLW**], and [**DLL**] (see also [**BL**]).

LEMMA 2.1. *Let $\gamma \geq 0$ and let $A(x)$, $x \in \mathbb{R}$, be a self-adjoint operator-valued function in a Hilbert space $\mathcal{H}$. Assume that*

$$\int_{\mathbb{R}} \|A_+(x)\|_{\mathcal{S}_{\gamma+1/2}}^{\gamma+1/2} < \infty,$$

where $\mathcal{S}_p$ is the Schatten class of compact operators. Then for the negative eigenvalues $\{-\lambda_k\}$ of the operator $b(-d^2/dx^2) - A$, $b > 0$, we have

$$\sum_k \lambda_k^\gamma \leq (2\pi)^{-1} R_{\gamma,1} \int_{\mathbb{R}^2} \operatorname{Tr}(b\,\xi^2 - A(x))_-^\gamma \, dx d\xi$$
$$= L_{\gamma,1}\, b^{-1/2} \int_{\mathbb{R}} \operatorname{Tr} A_+^{\gamma+1/2}(x)\, dx,$$

where

$$L_{\gamma,1} = 4\pi R_{\gamma,1} \int_0^1 (1-\xi^2)^\gamma \, d\xi.$$

REMARK 2.2. It is known that $R_{1/2,1} = 2$ and $R_{\gamma,1} \leq 2$ if $1/2 \leq \gamma < 1$ ([**HLW**]), $R_{\gamma,1} \leq 1.8138...$ if $1 \leq \gamma < 3/2$ ([**DLL**]), and $R_{\gamma,1} = 1$ if $3/2 \leq \gamma < \infty$ ([**LW1**]). Note that in the scalar case the sharp value of the constant $L_{1/2,1} = 1/2$ has first been obtained in the paper by D. Hundertmark, E. H. Lieb, and L. Thomas [**HLT**], and $L_{3/2,1} = 3/16$ in [**LTh**].

PROOF OF THEOREM 1.3. By using Hardy's inequality $(D_x^2)^{\mathcal{D}} \geq 1/4x^2$ on $\mathbb{R}_+$ and applying the Fourier transform with respect to the y-variable, we obtain

$$\operatorname{Tr}((D_x^2 D_y^2)^{\mathcal{D}} - V)_-^\gamma \leq \operatorname{Tr}\left(\left(\frac{1}{2}(D_x^2)^{\mathcal{D}} + \frac{1}{8x^2}\right)(D_y^2)^{\mathcal{D}} - V\right)_-^\gamma$$
$$= \operatorname{Tr}\left(\left(\frac{1}{2}(D_x^2)^{\mathcal{D}} + \frac{1}{8x^2}\right)\eta^2 - \hat{\mathbf{V}}\right)_-^\gamma,$$

where $\hat{\mathbf{V}}$ is the integral operator defined by

$$\hat{\mathbf{V}}u(x,\eta) = \int_{\mathbb{R}} \hat{V}(x, \eta - \theta) u(x,\theta)\, d\theta.$$

If $\gamma \geq 1/2$, then Lemma 2.1 with the operator-valued potential

$$\frac{1}{8x^2}\eta^2 - \hat{\mathbf{V}}$$

yields

$$\operatorname{Tr}((D_x^2 D_y^2)^{\mathcal{D}} - V)_-^\gamma \leq (2\pi)^{-1} R_{\gamma,1} \int_{\mathbb{R}} \int_0^\infty \operatorname{Tr}\left(\left(\frac{1}{2}\xi^2 + \frac{1}{8x^2}\right)\eta^2 - \hat{\mathbf{V}}\right)_-^\gamma dx d\xi$$
$$= (2\pi)^{-1} R_{\gamma,1} \int_{\mathbb{R}} \int_0^\infty \operatorname{Tr}\left(\left(\frac{1}{2}\xi^2 + \frac{1}{8x^2}\right)(D_y^2)^{\mathcal{D}} - V\right)_-^\gamma dx d\xi.$$

Using the same argument with respect to the y-variable, we find

$$\operatorname{Tr}((D_x^2 D_y^2)^{\mathcal{D}} - V)_-^\gamma$$
$$\leq (2\pi)^{-1} R_{\gamma,1} \int_{\mathbb{R}} \int_0^\infty \operatorname{Tr}\left(\left(\frac{1}{2}\xi^2 + \frac{1}{8x^2}\right)\left(\frac{1}{2}(D_y^2)^{\mathcal{D}} + \frac{1}{8y^2}\right) - V\right)_-^\gamma dx d\xi$$
$$\leq \frac{1}{4^\gamma}(2\pi)^{-2}(R_{\gamma,1})^2 \int_{\mathbb{R}^2} \int_{\mathbb{R}^2_{++}} \left(\left(\xi^2 + \frac{1}{4x^2}\right)\left(\eta^2 + \frac{1}{4y^2}\right) - V\right)_-^\gamma dx dy d\xi d\eta.$$

The rest of the proof consists of computing the latter integral. With a change of variables

$$t^2 = \xi^2 \left(\eta^2 + \frac{1}{4y^2}\right)\left(\frac{1}{4x^2}\left(\eta^2 + \frac{1}{4y^2}\right) - V\right)_-^{-1}$$

we arrive at

$$\int_{\mathbb{R}^2}\int_{\mathbb{R}^2_{++}} \Big(\Big(\xi^2+\frac{1}{4x^2}\Big)\Big(\eta^2+\frac{1}{4y^2}\Big)-V\Big)_-^\gamma\,dxdyd\xi d\eta$$
$$=\int_{\mathbb{R}^2}\int_{\mathbb{R}^2_{++}} \Big(\frac{1}{4x^2}\Big(\eta^2+\frac{1}{4y^2}\Big)-V\Big)_-^{1/2+\gamma}\Big(\eta^2+\frac{1}{4y^2}\Big)^{-1/2}(t^2-1)_-^\gamma\,dxdydtd\eta$$
$$=\mathcal{B}(1/2,\gamma{+}1)\,\frac{1}{2}\int_{\mathbb{R}}\int_{\mathbb{R}^2_{++}}(16x^2y^2)^{-1/2+\gamma}(s^2{+}1{-}16\,x^2y^2V)_-^{1/2+\gamma}(s^2{+}1)^{-1/2}\,dxdyds,$$

where $s^2=\eta^2\,4y^2$. Applying $(s+1)^2/2\le s^2+1$, dropping the term s^2+1 in $(s^2+1-16\,x^2y^2V)_-$, and also keeping in mind that $|s|\le 4xy\sqrt{V_+}$, we finally obtain

$$\operatorname{Tr}((D_x^2D_y^2)^{\mathcal{D}}-V)_-^\gamma$$
$$\le\frac{1}{4^\gamma}(2\pi)^{-2}(R_{\gamma,1})^2\,\mathcal{B}(1/2,\gamma+1)\int_0^{4xy\sqrt{V_+}}\int_{\mathbb{R}^2_{++}}V_+^{1/2+\gamma}(s+1)^{-1}\,dxdyds$$
$$=\frac{1}{4^\gamma}\,(2\pi)^{-2}(R_{\gamma,1})^2\,\mathcal{B}(1/2,\gamma+1)\int_{\mathbb{R}^2_{++}}V_+^{1/2+\gamma}\log(1+4xy\sqrt{V}_+)\,dxdy.$$

The proof is complete. □

References

[AL] M. Aizenman and E. H. Lieb, *On semi-classical bounds for eigenvalues of Schrödinger operators.* Phys. Lett. **66A** (1978), 427–429.

[B1] D. S. Barsegyan, *On the possibility of strengthening the Lieb-Thirring inequality.* Mat. Zametki **86:6** (2009), 803-818.

[B2] D. S. Barsegyan, private communication.

[BL] R. Benguria and M. Loss, *A simple proof of a theorem by Laptev and Weidl.* Math. Res. Lett. **7** (2000), 195–203.

[BA] M. Birman and A. Laptev, *The negative discrete spectrum of a two-dimensional Schrödinger operators.* Comm. Pure Appl. Math. **XLIX** (1996), 967–997.

[C] M. Cwikel, *Weak type estimates for singular values and the number of bound states of Schrödinger operators.* Trans. AMS **224** (1977), 93–100.

[DLL] J. Dolbeault, A. Laptev, and M.Loss, *Lieb-Thirring inequalities with improved constants.* JEMS **10** (2008), No. 4, 1121–1126.

[HLW] D. Hundertmark, A. Laptev, and T. Weidl, *New bounds on the Lieb-Thirring constants.* Inv. Math. **140** (2000), 693–704.

[HLT] D. Hundertmark, E. H. Lieb, and L. E. Thomas, *A sharp bound for an eigenvalue moment of the one-dimensional Schrödinger operator.* Adv. Theor. Math. Phys. **2** (1998), 719–731.

[LN] A. Laptev and Yu. Netrusov, *On the negative eigenvalues of a class of Schrödinger operators.* Amer. Math. Soc., Providence, Differential operators and spectral theory, Amer. Math. Soc. Transl. Ser. 2 **189** (1999), 173–186.

[LW1] A. Laptev and T. Weidl, *Sharp Lieb-Thirring inequalities in high dimensions.* Acta Math. **184** (2000), 87–111.

[LW2] A. Laptev and T. Weidl, *Recent results on Lieb-Thirring inequalities*, Journées "Équations aux Dérivées Partielles" (La Chapelle sur Erdre, 2000), Exp. No. XX, Univ. Nantes, Nantes (2000).

[L] E. H. Lieb, *Bounds on the eigenvalues of the Laplace and Schrödinger operators.* Bull. Amer. Math. Soc. **82** (1976), 751–753.

[LTh] E.H. Lieb and W. Thirring, *Inequalities for the moments of the eigenvalues of the Schrödinger Hamiltonian and their relation to Sobolev inequalities.* Studies in Math. Phys., Essays in Honor of Valentine Bargmann, Princeton (1976), 269–303.

[R] G. V. Rozenblum, *Distribution of the discrete spectrum of singular differential operators.* Dokl. AN SSSR **202** (1972), 1012–1015.

[S1] B. Simon, *Some quantum mechanical operators with discrete spectrum but classically continuous spectrum.* Ann. Phys. **146** (1983), 209–220.

[S2] B. Simon, *Nonclassical eigenvalue asymptotics.* J. Func. Anal. **53** (1983), 84–98.

[Sol1] M. Z. Solomyak, *On the asymptotics of the spectrum of the Schrödinger operator with a nonregular operator with a nonregular homogeneous potential.* (English. Russian original) Sov. Math. Dokl. **30** (1984), 379–383.

[Sol2] M. Z. Solomyak, *Asymptotics of the spectrum of the Schrödinger operator with a nonregular homogeneous potential.* (Russian) Mat. Sb., N. Ser. **127(169)** (1985), No.1 (5), 21–39.

[Sol3] M. Z. Solomyak, *Piecewise-polynomial approximation of functions from* $H^l((0,1)^d)$, $2l = d$ *and applications to the spectral theory of Schrödinger operator.* Israel J. Math. **86** (1994), 253–276.

[SolV] M. Z. Solomyak and I. L. Vulis, *Spectral asymptotics of second-order degenerate elliptic operators.* (English) Math. USSR Izv. **8** (1974), 1343–1371.

[W] T. Weidl, *On the Lieb-Thirring constants* $L_{\gamma,1}$ *for* $\gamma \geq 1/2$. Comm. in Math. Phys. **178** (1996), 135-146.

Department of Mathematics, Imperial College London, London SW7 2AZ, United Kingdom and Department of Mathematics, Royal Institute of Technology, 100 44 Stockholm, Sweden

E-mail address: a.laptev@imperial.ac.uk; laptev@math.kth.se

Department of Mathematics, Royal Institute of Technology, 100 44 Stockholm, Sweden

E-mail address: fabianpo@kth.se

Amer. Math. Soc. Transl.
Volume **231**, 2010

On the Near Periodicity of Eigenvalues of Toeplitz Matrices

Michael Levitin, Alexander V. Sobolev, and Daphne Sobolev

1. Introduction

In order to compute the spectrum of a selfadjoint operator A in a Hilbert space H one can approximate A by a sequence of finite-dimensional operators $A_n = P_nAP_n$, $n = 1, 2, \dots$, where P_n are orthogonal projections on finite-dimensional subspaces of H with the property $P_n \to I$ in the strong sense as $n \to \infty$. However, it is well known that the operators A_n may have eigenvalues which in the limit do not converge to spectral points of A. Such eigenvalues are termed *spurious eigenvalues* and their presence is often described as *spectral pollution* (see, e.g., [**3**], [**5**]). The spectral pollution usually happens in spectral gaps of the operator A. This phenomenon has been extensively studied both in the abstract setting (see, e.g., [**7**], [**4**]) and in various special cases (see, e.g., [**2**], [**8**], [**6**]). In particular, it was shown in [**7**], [**4**], [**9**], and [**5**] that the spectral pollution may occur at any point in a gap of the essential spectrum of A.

Perhaps the simplest example illustrating the spectral pollution is the classical Toeplitz matrix. Let a be a real-valued piecewise continuous function on the interval $[-\pi, \pi)$, and let P_n be the orthogonal projection in $L^2(-\pi, \pi)$ on the subspace spanned by the exponentials $(2\pi)^{-1/2}e^{-ikx}$, $k = 1, 2, \dots, n$. We define the Toeplitz operator with the symbol a as

$$T_n = T_n[a] = P_nAP_n,$$

where A is the operator of multiplication by a. If the range of the function a is a disconnected set, then the spurious eigenvalues, in the limit $n \to \infty$, fill in the gaps separating the components of the range. More precisely, it can be inferred from [**1**] that an open interval I strictly inside the gap contains $W \log n + O(1)$ eigenvalues of T_n, where $W = W(a, I)$ is an explicitly computable constant. The objective of this note is to study in more detail the spectrum of the Toeplitz matrix for the piecewise constant symbol of the form

$$a(x) = \begin{cases} 0, & x \in [-\pi, L), \\ 1, & x \in [L, \pi), \end{cases} \tag{1.1}$$

where $L \in [0, \pi)$. In [**5**], it was shown numerically that if $Lp = \pi q$ with some (co-prime) integer numbers p, q, then the spurious eigenvalues of the operator $T_n[a]$

2010 *Mathematics Subject Classification.* Primary 47A75, 47B35.

are "nearly periodic" in n with a "period" $\omega = \omega(p, q)$; see formula (2.1) below. Specifically, let $\lambda^{(k)} \in (0, 1)$ be an eigenvalue of the matrix $T_k[a]$. Then for each of the operators $T_{k+\omega l}[a]$, $l = 1, 2, \ldots$, there exists an eigenvalue $\lambda^{(k+\omega l)}$ such that the difference $\lambda^{(k+\omega(l+1))} - \lambda^{(k+\omega l)}$ tends to zero as $l \to \infty$. Although the convergence rate of the latter was not estimated in [**5**], graphs of the eigenvalues qualitatively showed a rate which is faster than the logarithmic filling rate of the gap. As an example, Figure 1 (see [**5**]) gives a diagram of the numerically computed eigenvalues of $P_n A P_n$ vs. n, with $L = \frac{\pi}{2}$. The symbols "o" and "×" mark the eigenvalues for even and odd n, respectively. The diagram clearly suggests periodicity with period 4.

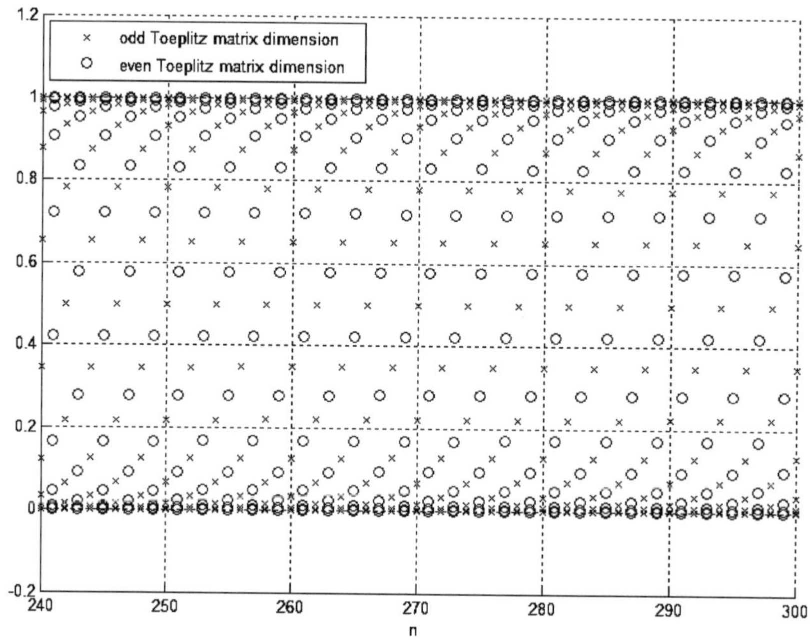

FIGURE 1. Eigenvalues of $T_n[a]$, with $L = \frac{\pi}{2}$, plotted vs. n.

The paper [**5**] (see also [**10**] in this volume) offered no quantitative measurement of the properties of the periodic behavior. Moreover no rigorous proof has been given; see [**10**] for some intuitive discussion of the phenomenon.

In this article we address the question of the near periodicity of eigenvalues, but instead of the symbol (1.1) it is more convenient to take a of the form (2.1). We have not been able to find a proof of this effect for the operator $T_n[a]$ itself, but we can show its presence for the squared Toeplitz operator, i.e., for $M_n[a] = (T_n[a])^2$; see Theorem 2.1. Numerical examples and more detailed conjectures with regard to the periodicity will be presented in a further publication.

2. The main result

We are concerned with the spectrum of the squared Toeplitz operator

$$M_n = M_n[a] = (T_n[a])^2.$$

First, we introduce some consistent notation for eigenvalues of various operators which appear later in the paper. For any selfadjoint matrix operator $S = S_n$ of size n, we shall denote by $\lambda_j(S)$, $j = 1, \dots, n$, its eigenvalues labeled in the descending order, and by $\mu_k(S)$, $k = 1, \dots, n$, its eigenvalues labeled in the ascending order, so that

$$\mu_j(S) = \lambda_{n+1-j}(S) .$$

Also, for brevity, in the particular case of operators M_n, we shall write

$$\mu_j^{(n)} = \mu_j(M_n) .$$

The next theorem is the main result of the paper:

THEOREM 2.1. *Let $a(x)$ be the function*

$$a(x) = \begin{cases} -1, & x \in [-\pi, L), \\ 1, & x \in [L, \pi), \end{cases} \tag{2.1}$$

with some $L \in [0, \pi)$. Suppose that

$$pL = \pi q \tag{2.2}$$

for some co-prime $p \in \mathbb{N}$ and $q \in \mathbb{Z}$. Define

$$\omega = \omega(p, q) = \begin{cases} 2, & q = 0, \\ p, & p \text{ and } q \text{ are odd}, \\ 2p, & \text{either } p \text{ or } q \text{ is even}. \end{cases}$$

Let $\epsilon \in (0, 1)$ be a fixed number, and let j be such that $\mu_j^{(n)} < 1 - \epsilon$. Then for a sufficiently large number $K > 0$ and $n \geq K\omega\epsilon^{-1}$,

$$|\mu_j^{(n)} - \mu_j^{(n+\omega)}| \leq \frac{C\omega(1 + \log^2 n)}{\epsilon n}, \tag{2.3}$$

with a constant $C > 0$ independent of p, q, n, j, ϵ.

In general, by C and c we denote various positive constants independent of n, p, q, and ϵ, whose precise value is of no importance.

Note that under the condition (2.2) we have

$$\omega(\pi + L) \equiv 0 \mod 2\pi. \tag{2.4}$$

REMARK 2.2. Theorem 2.1 immediately implies that under the condition $\mu_j^{(n)} < 1 - \epsilon$, $\epsilon > 0$, for any fixed $N = 1, 2, \dots$ and all sufficiently large n,

$$|\mu_j^{(n+\omega(m+1))} - \mu_j^{(n+\omega m)}| \leq \frac{C\omega(1 + \log^2 n)}{\epsilon n}, \tag{2.5}$$

for all $m = 0, 1, \dots, N$, uniformly in j. This means that for any N the spectra of $M_n[a]$ have strings of length N of "nearly equal" eigenvalues.

Throughout the proof of Theorem 2.1 we omit the subscript n for brevity whenever possible. Denote $Q = Q_n = I - P_n$ and compute, remembering that $A^2 = I$:

$$M_n[a] = (PAP)^2 = PAPAP = PA^2P - PAQAP = P - PAQAP.$$

This means that

$$\mu_j^{(n)} = 1 - \lambda_j(B_n) ,$$

where $B_n = B_n[a] = PAQAP$. Therefore it suffices to prove the inequality (2.3) for the eigenvalues $\lambda_j(B_n)$ which satisfy $\lambda_j(B_n) > \epsilon$, instead.

The entries of the matrix B_n are easy to find:

$$b_{rl} = b_{rl}^{(n)} = \frac{1}{2\pi} \sum_{\substack{m\le 0,\\ m\ge n+1}} a_{r-m}a_{m-l}, \ r,l = 1,2,\dots,n, \tag{2.6}$$

where the Fourier coefficients

$$a_k = \frac{1}{\sqrt{2\pi}} \int_{-\pi}^{\pi} a(x)e^{ikx}dx$$

are given by

$$a_k = \begin{cases} -\sqrt{\dfrac{2}{\pi}}L, & k = 0, \\ \sqrt{\dfrac{2}{\pi}}\dfrac{(-1)^k}{ik}\left[1 - e^{ik(L+\pi)}\right], & k \neq 0. \end{cases} \tag{2.7}$$

The crucial point of our argument is that due to (2.4) the exponential on the right-hand side of (2.7) is ω-periodic as a function of k. This fact is used only once, in the proof of Lemma 2.6.

We need to establish the following theorem.

THEOREM 2.3. *Let $a(x)$ be as defined in (2.1), and suppose that the condition (2.2) is satisfied. Let $\epsilon \in (0,1)$ be a fixed number, and let j be such that $\lambda_j(B_n) > \epsilon$. Then for all sufficiently large $n \ge 1$,*

$$|\lambda_j(B_n) - \lambda_j(B_{n+\omega})| \le \frac{C\omega(1+\log^2 n)}{\epsilon n}, \tag{2.8}$$

uniformly in j.

First we estimate the entries $b_{rl}^{(n)}$.

LEMMA 2.4. *Let $a(x)$ be defined by (2.1), and let $b_{rl}^{(n)}$ be defined by (2.6). Then for all $1 \le l \le r \le n$ we have*

$$|b_{rl}^{(n)}| \le \frac{16}{\pi^2}\frac{1+\log n}{|l-r|}, \ l \neq r, \tag{2.9}$$

and

$$|b_{rl}^{(n)}| \le \frac{8}{\pi^2}\left(\frac{1}{n+1-r} + \frac{1}{l}\right). \tag{2.10}$$

PROOF. Substituting (2.7) in (2.6), we get

$$\begin{aligned} b_{rl} = -\frac{(-1)^{r-l}}{\pi^2} &\sum_{\substack{m\le 0,\\ m\ge n+1}} \frac{1}{(r-m)(m-l)} \\ &\times \left(1 + e^{i(r-l)(L+\pi)} - e^{i(r-m)(L+\pi)} - e^{i(m-l)(L+\pi)}\right). \end{aligned} \tag{2.11}$$

Therefore

$$|b_{rl}| \le \frac{4}{\pi^2} \sum_{\substack{m \le 0,\\ m \ge n+1}} \left| \frac{1}{(r-m)(m-l)} \right|$$

$$= \frac{4}{\pi^2} \left(\frac{1}{|r-(n+1)|\,|(n+1)-l|} + \frac{1}{rl} \right) + \frac{4}{\pi^2} \sum_{\substack{m \le -1,\\ m \ge n+2}} \frac{1}{|r-m|\,|m-l|}.$$

The second term can be easily estimated by an appropriate integral. If $r > l$, then

$$\sum_{\substack{m \le -1,\\ m \ge n+2}} \frac{1}{|r-m|\,|m-l|} \le \int_{-\infty}^{0} \frac{1}{(r-x)(l-x)} dx + \int_{n+1}^{\infty} \frac{1}{(x-r)(x-l)} dx$$

$$= \frac{1}{(r-l)} \left(\log\left(\frac{n+1-l}{n+1-r} \right) + \log\left(\frac{r}{l} \right) \right). \tag{2.12}$$

The bound (2.9) follows immediately.

Let us derive (2.10). In view of the straightforward bounds

$$\log\left(\frac{n+1-l}{n+1-r} \right) \le \frac{r-l}{n+1-r}, \qquad \log\left(\frac{r}{l}\right) \le \frac{r-l}{l},$$

the right-hand side of (2.12) does not exceed

$$\frac{1}{n+1-r} + \frac{1}{l}.$$

For the case $r = l$, a similar estimate can be obtained:

$$\sum_{\substack{m \le -1,\\ m \ge n+2}} \frac{1}{|r-m|^2} \le \int_{n+1}^{\infty} \frac{1}{(x-r)^2} dx + \int_{-\infty}^{0} \frac{1}{(r-x)^2} dx$$

$$= \frac{1}{n+1-r} + \frac{1}{r}.$$

Hence for all $r \ge l$ we have

$$|b_{rl}| \le \frac{8}{\pi^2} \left(\frac{1}{n+1-r} + \frac{1}{l} \right),$$

which coincides with (2.10). □

Let $\omega \in \mathbb{N}$, $1 < \omega < n$, and let

$$k = k_n = \left\lceil \frac{n}{2} \right\rceil = \begin{cases} \dfrac{n}{2}, & n \text{ even},\\ \dfrac{n+1}{2}, & n \text{ odd}. \end{cases}$$

Let us construct, using matrix $B_{n+\omega}$, two new auxiliary matrices. The $(n+\omega) \times (n+\omega)$ matrix $D_{n+\omega}$ has the entries

$$d_{rl}^{(n+\omega)} = \begin{cases} b_{rl}^{(n+\omega)}, & k \le r < k+\omega \text{ or } k \le l < k+\omega,\\ 0, & \text{otherwise}. \end{cases} \tag{2.13}$$

The $n \times n$ matrix F_n has the entries

$$
f_{rl}^{(n)} = \begin{cases} b_{r,l}^{(n+\omega)}, & 1 \le r \le k-1,\ 1 \le l \le k-1, \\ b_{r+\omega,l+\omega}^{(n+\omega)}, & k \le r \le n,\ k \le l \le n, \\ b_{r,l+\omega}^{(n+\omega)}, & 1 \le r \le k-1,\ k \le l \le n, \\ b_{r+\omega,l}^{(n+\omega)}, & k \le r \le n,\ 1 \le l \le k-1. \end{cases} \tag{2.14}
$$

The method of constructing the matrices $D_{n+\omega}$ and F_n from $B_{n+\omega}$ is illustrated by Figure 2. First, we shade the central "cross" of ω rows and ω columns in $B_{n+\omega}$, starting with row and column number k. The matrix $D_{n+\omega}$ is constructed by replacing everything outside the cross by zeros, and the matrix F_n by "removing" the cross and pulling the remaining four blocks together.

$$
\begin{pmatrix}
b_{1,1} & \cdots & b_{1,k-1} & b_{1,k} & \cdots & b_{1,k+\omega-1} & b_{1,k+\omega} & \cdots & b_{1,n+\omega} \\
\vdots & \ddots & \vdots & \vdots & \vdots & \vdots & \vdots & \vdots & \vdots \\
b_{k-1,1} & \cdots & b_{k-1,k-1} & b_{k-1,k} & \cdots & b_{k-1,k+\omega-1} & b_{k-1,k+\omega} & \cdots & b_{k-1,n+\omega} \\
b_{k,1} & \cdots & b_{k,k-1} & b_{k,k} & \cdots & b_{k,k+\omega-1} & b_{k,k+\omega} & \cdots & b_{k,n+\omega} \\
\vdots & \vdots & \vdots & \vdots & \ddots & \vdots & \vdots & \vdots & \vdots \\
b_{k+\omega-1,1} & \cdots & b_{k+\omega-1,k-1} & b_{k+\omega-1,k} & \cdots & b_{k+\omega-1,k+\omega-1} & b_{k+\omega-1,k+\omega} & \cdots & b_{k+\omega-1,n+\omega} \\
b_{k+\omega,1} & \cdots & b_{k+\omega,k-1} & b_{k+\omega,k} & \cdots & b_{k+\omega,k+\omega-1} & b_{k+\omega,k+\omega} & \cdots & b_{k+\omega,n+\omega} \\
\vdots & \vdots & \vdots & \vdots & \vdots & \vdots & \vdots & \ddots & \vdots \\
b_{n+\omega,1} & \cdots & b_{n+\omega,k-1} & b_{n+\omega,k} & \cdots & b_{n+\omega,k+\omega-1} & b_{n+\omega,k+\omega} & \cdots & b_{n+\omega,n+\omega}
\end{pmatrix}
$$

FIGURE 2. Matrix $B_{n+\omega}$ and the central "cross" used in the construction of matrices $D_{n+\omega}$ and F_n.

Introduce the projections $\Pi_{n,\omega} = P_{k+\omega-1} - P_{k-1}$ and $\Xi_{n,\omega} = P_{n+\omega}(I - \Pi_{n,\omega})$. In terms of these projections, the operator $D^{(n+\omega)}$ can be represented as follows:

$$
D_{n+\omega} = \Pi_{n,\omega} B_{n+\omega} \Pi_{n,\omega} + \Pi_{n,\omega} B_{n+\omega} \Xi_{n,\omega} + \Xi_{n,\omega} B_{n+\omega} \Pi_{n,\omega},
$$

so that

$$
B_{n+\omega} = \Xi_{n,\omega} B_{n+\omega} \Xi_{n,\omega} + D_{n+\omega}.
$$

The next lemma estimates the matrix $D_{n+\omega}$ in terms of these projections.

LEMMA 2.5. *Let $b_{rl}^{(n)}$ be defined as in (2.6), and $d_{rl}^{(n+\omega)}$ as in (2.13). Then for any $\epsilon > 0$ and all $n \ge \omega$ we have*

$$
|(D_{n+\omega} u, u)| \le (\epsilon + C\omega n^{-1}) \|\Pi_{n,\omega} u\|^2 + \frac{C\omega}{\epsilon} \frac{1 + \log^2 n}{n} \|\Xi_{n,\omega} u\|^2, \tag{2.15}
$$

for all $u \in L^2(-\pi, \pi)$.

PROOF. Write the straightforward estimate

$$|(D_{n+\omega}u, u)| \le |(\Pi_{n,\omega}B_{n+\omega}\Pi_{n,\omega}u, \Pi_{n,\omega}u)| + 2|(\Pi_{n,\omega}B_{n+\omega}\Xi_{n,\omega}u, \Pi_{n,\omega}u)|$$

$$\le \|\Pi_{n,\omega}B_{n+\omega}\Pi_{n,\omega}\| \; \|\Pi_{n,\omega}u\|^2 + 2\|\Pi_{n,\omega}B_{n+\omega}\Xi_{n,\omega}u\| \; \|\Pi_{n,\omega}u\|$$

$$\le (\|\Pi_{n,\omega}B_{n+\omega}\Pi_{n,\omega}\| + \epsilon)\|\Pi_{n,\omega}u\|^2 + \epsilon^{-1}\|\Pi_{n,\omega}B_{n+\omega}\Xi_{n,\omega}\|^2\|\Xi_{n,\omega}u\|^2. \tag{2.16}$$

Here we have used the elementary estimate $2ab \le \epsilon a^2 + \epsilon^{-1}b^2$, $\epsilon, a, b > 0$.

Let us estimate the matrix norms entering the above inequality. For the first term we estimate the Hilbert-Schmidt norm using the bound (2.10):

$$\|\Pi_{n,\omega}B_{n+\omega}\Pi_{n,\omega}\|^2_{\mathfrak{S}_2} \le 2\frac{16^2}{\pi^2}\sum_{r=k}^{k+\omega-1}\sum_{l=k}^{r}\left(\frac{1}{n+\omega+1-r}+\frac{1}{l}\right)^2 \le C\frac{\omega^2}{k^2}. \tag{2.17}$$

For the matrix $\|\Pi_{n,\omega}B_{n+\omega}\Xi_{n,\omega}\|$ we also estimate its Hilbert-Schmidt norm, but now we need both (2.9) and (2.10):

$$\|\Pi_{n,\omega}B_{n+\omega}\Xi_{n,\omega}\|^2_{\mathfrak{S}_2} \le \left(\sum_{r=k}^{k+\omega-1}\sum_{l=1}^{k-1} + \sum_{r=k+\omega}^{n+\omega}\sum_{l=k}^{k+\omega-1}\right)|b_{rl}^{(n+\omega)}|^2. \tag{2.18}$$

Let us estimate the first sum, which we denote S_1. Split it into two parts: $1 \le l \le k/2$ and $k/2 < l \le k-1$. For the first part we use the estimate (2.9), so that

$$|b_{rl}^{(n+\omega)}| \le C\frac{1+\log n}{n}$$

and

$$\sum_{r=k}^{k+\omega-1}\sum_{l\le k/2}|b_{rl}^{(n+\omega)}|^2 \le C\frac{(1+\log n)^2}{n^2}\sum_{r=k}^{k+\omega-1}\sum_{l=1}^{k/2}1 \le C\omega\frac{(1+\log n)^2}{n}.$$

The part with $k/2 < l \le k-1$ is estimated with the help of (2.10), so that

$$|b_{rl}^{(n+\omega)}| \le \frac{C}{n}$$

and

$$\sum_{r=k}^{k+\omega-1}\sum_{k/2<l\le k-1}|b_{rl}^{(n+\omega)}|^2 \le C\frac{\omega}{n}.$$

Thus

$$S_1 \le C\omega\frac{(1+\log n)^2}{n}.$$

The same bound holds for the second sum on the right-hand side of (2.18). Together with (2.17) and (2.16) these bounds lead to the claimed estimate (2.15). □

LEMMA 2.6. *Suppose that (2.2) is satisfied. Then for all $n \ge 1$ we have*

$$\|F_n - B_n\| \le \frac{C\omega(1+\log n)}{n}. \tag{2.19}$$

PROOF. First we estimate the difference $f_{rl}^{(n)} - b_{rl}^{(n)}$. For convenience we rewrite the formula (2.6) for $b_{rl}^{(n+\omega)}$:

$$b_{rl}^{(n+\omega)} = \frac{1}{2\pi} \sum_{\substack{m\le 0,\\ m\ge n+\omega+1}} a_{r-m}a_{m-l}$$

$$= \frac{1}{2\pi}\sum_{m\le 0} a_{r-m}a_{m-l} + \frac{1}{2\pi}\sum_{m\ge n+1} a_{r-m-\omega}a_{m+\omega-l}. \tag{2.20}$$

Our plan is to estimate the Hilbert-Schmidt norm of $F^{(n)} - B^{(n)}$. So, we estimate carefully the sum of squares $|f_{rl} - b_{rl}|^2$ over the four ranges of r and l, specified in (2.14).

Case 1. $1 \le r \le k-1$, $1 \le l \le k-1$ **(upper-left block in Figure 2).** According to (2.20) we have

$$f_{rl}^{(n)} - b_{rl}^{(n)} = \frac{1}{2\pi}\sum_{m\ge n+1}\big(a_{r-m-\omega}a_{m+\omega-l} - a_{r-m}a_{m-l}\big).$$

Using (2.7) and (2.4), we get the bound

$$\begin{aligned}
&|a_{r-m-\omega}\, a_{m+\omega-l} - a_{r-m}a_{m-l}| \\
&\quad= \frac{2}{\pi}|(1-e^{i(r-m)(L+\pi)})(1-e^{i(m-l)(L+\pi)})| \\
&\qquad\times\left|\frac{1}{(r-m-\omega)(m+\omega-l)} - \frac{1}{(r-m)(m-l)}\right| \\
&\quad\le \frac{8\omega}{\pi}\left[\frac{1}{(r-m)^2(m-l)} + \frac{1}{(m-r)(m-l)^2}\right].
\end{aligned}$$

Therefore, we can estimate

$$\begin{aligned}
|f_{rl}^{(n)} - b_{lr}^{(n)}| &\le \frac{4\omega}{\pi^2}\sum_{m\ge n+1}\left[\frac{1}{(r-m)^2(m-l)} + \frac{1}{(m-r)(m-l)^2}\right] \\
&\le \frac{C\omega}{(n+1-r)(n+1-l)} \le \frac{C\omega}{n^2},\quad 1\le r,\ l\le k-1,
\end{aligned}$$

and hence

$$\sum_{l,r=1}^{k-1} |f_{rl}^{(n)} - b_{rl}^{(n)}|^2 \le \frac{C\omega^2}{n^2}. \tag{2.21}$$

Case 2. $k\le r\le n$, $k\le l\le n$ **(lower-right block in Figure 2).** Using again (2.20), from (2.14) we get

$$f_{rl}^{(n)} - b_{rl}^{(n)} = \frac{1}{2\pi}\sum_{m\le 0}\big(a_{r-m+\omega}a_{m-l-\omega} - a_{r-m}a_{m-l}\big).$$

Arguing as in Case 1, we obtain the same bound:

$$\sum_{l,r=k}^{n} |f_{rl}^{(n)} - b_{rl}^{(n)}|^2 \le \frac{C\omega^2}{n^2}. \tag{2.22}$$

The remaining two cases are trickier:

Case 3. $1 \le r \le k-1$, $k \le l \le n$ **(upper-right block in Figure 2)**. By (2.14),

$$f_{rl}^{(n)} - b_{lr}^{(n)} = b_{r,l+\omega}^{(n+\omega)} - b_{lr}^{(n)}$$
$$= \frac{1}{2\pi} \sum_{m\le 0} \big(a_{r-m}a_{m-l-\omega} - a_{r-m}a_{m-l}\big)$$
$$+ \frac{1}{2\pi} \sum_{m\ge n+1} \big(a_{r-m-\omega}a_{m-l} - a_{r-m}a_{m-l}\big).$$

Arguing as in the previous case, we obtain

$$|a_{r-m}a_{m-l-\omega} - a_{r-m}a_{m-l}| \le \frac{8\omega}{\pi}\frac{1}{(r-m)(m-l)^2}, \quad m \le 0,$$

$$|a_{r-m-\omega}a_{m-l} - a_{r-m}a_{m-l}| \le \frac{8\omega}{\pi}\frac{1}{(r-m)^2(m-l)}, \quad m \ge n+1.$$

We estimate

$$s_{rl} := \sum_{m\le 0} \frac{1}{(r-m)(m-l)^2} \le C\frac{1+\log n}{n^2},$$

and

$$q_{rl} := \sum_{m\ge n+1} \frac{1}{(r-m)^2(m-l)} \le C\frac{1+\log n}{n^2}.$$

Therefore

$$\sum_{r=1}^{k-1}\sum_{l=k}^{n} s_{rl}^2 + \sum_{r=1}^{k-1}\sum_{l=k}^{n} q_{rl}^2 \le \frac{C(1+\log^2 n)}{n^2}.$$

Combining these bounds, we arrive at the estimate

$$\sum_{r=1}^{k-1}\sum_{l=k}^{n} |f_{rl}^{(n)} - b_{rl}^{(n)}|^2 \le \frac{C\omega^2(1+\log^2 n)}{n^2}. \tag{2.23}$$

Case 4. $k \le r \le n$, $1 \le l \le k-1$ **(lower-left block in Figure 2)**. Since the matrices $F^{(n)}$ and $B^{(n)}$ are Hermitian, we can use the estimate obtained in Case 3, and hence

$$\sum_{r=k}^{n}\sum_{l=1}^{k-1} |f_{rl}^{(n)} - b_{rl}^{(n)}|^2 \le \frac{C\omega^2(1+\log^2 n)}{n^2}. \tag{2.24}$$

End of proof. We estimate the norm by the Hilbert-Schmidt norm:

$$\|F_n - B_n\| \le \|F_n - B_n\|_{\mathfrak{S}_2}.$$

In view of (2.21), (2.22), (2.23), (2.24),

$$\|F_n - B_n\|_{\mathfrak{S}_2} \le C\omega\frac{1+\log n}{n},$$

which coincides with the proclaimed estimate. □

Our next step is to estimate the difference between eigenvalues of $B_{n+\omega}$ and F_n. Instead of the eigenvalues themselves, it is more convenient to work with their counting function. Denote by $n_+(\lambda, S)$, $\lambda > 0$, the number of eigenvalues of a Hermitian matrix S strictly above λ.

LEMMA 2.7. *Suppose that $\lambda > 0$ and $n \geq K\omega\lambda^{-1}$ with a sufficiently large constant $K > 0$. Then*

$$n_+\left(\lambda + \frac{C\omega}{\lambda}\frac{1+\log^2 n}{n}, F_n\right) \leq n_+(\lambda, B_{n+\omega}) \leq n_+\left(\lambda - \frac{C\omega}{\lambda}\frac{1+\log^2 n}{n}, F_n\right).$$

PROOF. We use the projections $\Xi_{n,\omega}, \Pi_{n,\omega}$ introduced earlier. Recall that

$$B_{n+\omega} = \Xi_{n,\omega} B_{n+\omega} \Xi_{n,\omega} + D_{n+\omega}.$$

By Lemma 2.5 this implies that

$$B_{n+\omega} \leq \Xi_{n,\omega} B_{n+\omega} \Xi_{n,\omega} + \frac{C\omega}{\delta}\frac{1+\log^2 n}{n}\Xi_{n,\omega} + (\delta + C\omega n^{-1})\Pi_{n,\omega},$$

$$B_{n+\omega} \geq \Xi_{n,\omega} B_{n+\omega} \Xi_{n,\omega} - \frac{C\omega}{\delta}\frac{1+\log^2 n}{n}\Xi_{n,\omega} - (\delta + C\omega n^{-1})\Pi_{n,\omega},$$

for any $\delta > 0$. Thus the counting function $n_+(\lambda, B_{n+\omega})$ satisfies the following estimates:

$$\begin{aligned} n_+(\lambda, B_{n+\omega}) \leq n_+\left(\lambda, \Xi_{n,\omega} B_{n+\omega} \Xi_{n,\omega} + \frac{C\omega}{\delta}\frac{1+\log^2 n}{n}\Xi_{n,\omega}\right) \\ + n_+(\lambda, (\delta + C\omega n^{-1})\Pi_{n,\omega}), \end{aligned} \tag{2.25}$$

$$n_+(\lambda, B_{n+\omega}) \geq n_+\left(\lambda, \Xi_{n,\omega} B_{n+\omega} \Xi_{n,\omega} - \frac{C\omega}{\delta}\frac{1+\log^2 n}{n}\Xi_{n,\omega}\right), \tag{2.26}$$

with an arbitrary $\delta > 0$. Now take $\delta = \lambda/2$, so that for sufficiently large K under the condition $n \geq K\omega\lambda^{-1}$ we have $\delta + C\omega n^{-1} < \lambda$. Therefore the second term on the right-hand side of (2.25) equals zero. The matrix F_n is obviously similar to $\Xi_{n,\omega} B_{n+\omega} \Xi_{n,\omega}$, so that their positive eigenvalues coincide. Thus (2.25) and (2.26) lead to the required inequalities. □

PROOF OF THEOREM 2.3. By Lemma 2.6, the elementary perturbation theory yields

$$n_+\left(\lambda + \frac{C\omega(1+\log n)}{n}, B_n\right) \leq n_+(\lambda, F_n) \leq n_+\left(\lambda - \frac{C\omega(1+\log n)}{n}, B_n\right),$$

for any $\lambda \in (0, 1]$. Using this bound in combination with Lemma 2.7, we get

$$n_+\left(\lambda + \frac{C\omega}{\epsilon}\frac{1+\log^2 n}{n}, B_{n+\omega}\right) \leq n_+(\lambda, B_n) \leq n_+\left(\lambda - \frac{C\omega}{\epsilon}\frac{1+\log^2 n}{n}, B_{n+\omega}\right),$$

for all $\lambda \in [\epsilon, 1]$ and $n \geq K\omega\epsilon^{-1}$, with a sufficiently large constant K. This means that if $\lambda_j(B_n) > \epsilon$, then

$$|\lambda_j(B_n) - \lambda_j(B_{n+\omega})| \leq \frac{C\omega}{\epsilon}\frac{1+\log^2 n}{n},$$

for all $n \geq K\omega\epsilon^{-1}$. □

As we have already pointed out, Theorem 2.1 follows from Theorem 2.3 due to the equality $\mu_j^{(n)} = 1 - \lambda_j(B_n)$.

References

[1] E. L. Basor, *Trace formulas for Toeplitz matrices with piecewise continuous symbols*, J. Math. Anal. Appl. **120** (1986), no. 1, 25–38.

[2] M. Dauge and M. Suri, *Numerical approximation of the spectra of non-compact operators arising in buckling problems*, J. of Num. Math. **10** (2002), 193–219.

[3] E. B. Davies and M. Plum, *Spectral pollution*, IMA J. Numer. Anal. **24** (2004), no. 3, 417–438.

[4] J. Descloux, *Essential numerical range of an operator with respect to a coercive form and the approximation of its spectrum by the Galerkin Method*, SIAM J. Numer. Anal. **18**(6) (1981), 1128–1133.

[5] M. Levitin and E. Shargorodsky, *Spectral pollution and second order relative spectra for self-adjoint operators*, IMA J. Numer. Anal. **24** (2004), no. 3, 393–416.

[6] M. Lewin, E. Séré, *Spectral pollution and how to avoid it (with applications to Dirac and periodic Schrodinger operators)*, Proc. London Math. Soc. (3) **100** (2010), 864–900.

[7] A. Pokrzywa, *Method of orthogonal projections and approximation of the spectrum of a bounded operator*, Studia Math. **65**(1) (1979), 21–29.

[8] J. Rappaz, J. Sanchez Hubert, E. Sanchez Palencia, and D. Vassiliev, *On spectral pollution in the finite element approximation of thin elastic "membrane" shells*, Numer. Math. **75** (1997), 473–500.

[9] E. Shargorodsky, *Geometry of higher order relative spectra and projection methods*, J. Operator Theory **44** (2000), 43-62.

[10] E. Shargorodsky, *On some open problems in spectral theory*, in this volume.

DEPARTMENT OF MATHEMATICS AND STATISTICS, UNIVERSITY OF READING, WHITEKNIGHTS, PO BOX 220, READING RG6 6AX, UNITED KINGDOM

E-mail address: M.Levitin@reading.ac.uk

DEPARTMENT OF MATHEMATICS, UNIVERSITY COLLEGE LONDON, GOWER STREET, LONDON WC1E 6BT, UNITED KINGDOM

E-mail address: asobolev@math.ucl.ac.uk

MATHEMATICS AND COMPUTING FACULTY, THE OPEN UNIVERSITY IN LONDON, 1-11 HAWLEY CRESCENT, LONDON NW1 8NP, UNITED KINGDOM

E-mail address: ds8788@tutor.open.ac.uk

Amer. Math. Soc. Transl.
Volume **231**, 2010

Green's Kernels for Transmission Problems in Bodies with Small Inclusions

V. Maz′ya, A. Movchan, and M. Nieves

In memory of V. B. Lidskii

ABSTRACT. The uniform asymptotic approximation of Green's kernel for the transmission problem of antiplane shear is obtained for domains with small inclusions. The remainder estimates are provided. Numerical simulations are presented to illustrate the effectiveness of the approach.

1. Introduction

Our goal is to obtain a uniform asymptotic approximation of Green's function for a transmission problem of antiplane shear in a domain with small inclusions.

Exact solutions to singularly perturbed problems corresponding to bodies with defects are often unavailable. For complicated geometries involved in problems of this kind, i.e., domains with multiple small perforations, numerical algorithms may become incapable of reaching the required accuracy. Also, when the right-hand sides of such boundary value problems are singular, numerical procedures can suffer from the same deficiencies. In this case, asymptotic solutions to these problems are desirable.

The approximation of Green's kernels for regularly perturbed problems, for the Laplace operator and the biharmonic operator, was first studied by Hadamard in [**3**]. More recently, the question of uniformity for the approximations of Green's kernels for boundary value problems in domains with singularly and regularly perturbed boundaries was addressed in [**8**]. The uniform approximations in [**8**] were derived using the method of compound asymptotic expansions.

The paper [**6**] contains the rigorous proofs and remainder estimates for uniform approximations of Green's kernels, given in [**8**], for $-\Delta$ in an n-dimensional domain ($n \geq 2$), with a single small rigid inclusion. Uniform asymptotic formulae of Green's functions for mixed problems of antiplane shear in domains with a small hole or a crack are discussed in detail in [**7**]. In [**9**], an extension of the theory developed in [**6**] is made to the case of Green's tensors of vector elasticity, for an elastic body

2010 *Mathematics Subject Classification.* Primary 35J08, 35Q70; Secondary 41A60, 80M35.
The authors would like to acknowledge the financial support of the UK Engineering and Physical Sciences Research Council through the research grant EP/F005563/1.

with a small inclusion. This was followed by [**10**], where uniform asymptotics of Green's kernels for planar and three-dimensional elasticity in bodies with multiple rigid inclusions are given. The paper [**10**] also includes analysis of Green's kernels and numerical simulations for antiplane shear and plane strain.

In the present paper, the new feature of the problem tackled is that on the small inclusions we prescribe transmission conditions (the continuity of tractions and displacements). The inclusions are assumed to be occupied by materials which are different from that of the ambient medium. Compared to previous expositions into the uniform approximation of Green's kernels in [**6, 7, 9**], where the kernels are approximated in the bodies containing small holes, we also must approximate the Green's kernel inside the inclusions. The analysis also brings additional boundary layers when the point force is placed inside the inclusion.

Below, we illustrate one of the main results in this article, for the case when the domain has a single inclusion. Let ω_ε be a small planar inclusion, occupied by a material of shear modulus μ_I, containing the origin $\mathbf{O}$ and which is perfectly bonded to the rest of the matrix $\Omega_\varepsilon \subset \mathbb{R}^2$ whose shear modulus is μ_O. Here, ε is a small positive parameter characterizing the normalized size of the inclusion. Consider the antiplane shear Green's function N_ε for the transmission problem inside the perturbed domain $\Omega_\varepsilon \cup \omega_\varepsilon$. We also use $N^{(\Omega)}$, as the Neumann function in the unperturbed domain (without the inclusion) and $R^{(\Omega)}$ as its regular part. We denote by $\mathcal{N}$ the Green's function for the transmission problem, in the unbounded domain with the scaled inclusion containing the origin. Also let us define the vector function $\mathcal{D} = (\mathcal{D}_1, \mathcal{D}_2)^T$, where the components $\mathcal{D}_j$, $j = 1, 2$, are the dipole fields for the scaled inclusion in the unbounded domain.

As one of the results we state

Theorem. *The approximation of Green's function for the transmission problem of antiplane shear in* $\Omega_\varepsilon \cup \omega_\varepsilon \subset \mathbb{R}^2$ *is given by*

$$\boxed{\begin{aligned} N_\varepsilon(\mathbf{x},\mathbf{y}) &= N^{(\Omega)}(\mathbf{x},\mathbf{y}) + \mathcal{N}(\varepsilon^{-1}\mathbf{x},\varepsilon^{-1}\mathbf{y}) + (2\pi\mu_O)^{-1}\log(\varepsilon^{-1}|\mathbf{x}-\mathbf{y}|) \\ &+\varepsilon\mathcal{D}(\varepsilon^{-1}\mathbf{x})\cdot\nabla_{\mathbf{x}}R^{(\Omega)}(\mathbf{O},\mathbf{y}) + \varepsilon\mathcal{D}(\varepsilon^{-1}\mathbf{y})\cdot\nabla_{\mathbf{y}}R^{(\Omega)}(\mathbf{x},\mathbf{O}) + O(\varepsilon^2)\end{aligned}} \tag{1.1}$$

uniformly for $\mathbf{x}, \mathbf{y} \in \Omega_\varepsilon \cup \omega_\varepsilon$.

The structure of the article is as follows. In Section 2, we introduce the main notation that will be adopted throughout the text and define Green's function for the domain containing several inclusions. Section 3 contains the description of model fields used to construct the uniform asymptotic approximations of the Green's function. In Section 4, we state and prove an estimate for solutions to model transmission problems in an unbounded domain with an inclusion. Solutions to transmission problems in a domain with several small inclusions are studied in Section 5. Results of these sections will aid us in deriving the remainder estimate present in the generalization of approximation (1.1) to the case of multiple inclusions. Asymptotic properties of the boundary layer fields, involving the regular part of Green's function for the transmission problem, in the unbounded domain with an inclusion are investigated in Sections 6. Then we consider the uniform approximation of Green's function for the transmission problem in the domain with small inclusions in Section 7 and give the formal algorithm together with the remainder estimates. Finally, in Section 8 we demonstrate the effectiveness of our approach and present the numerical simulations comparing the asymptotic formula in Section 7 with the finite element calculations in COMSOL.

The asymptotic formulae obtained in the sequel are readily applicable to numerical simulations. As an example, Figure 1 shows the comparison between an asymptotic approximation and COMSOL computation for the modulus of the gradient of the regular part of Green's function for the transmission problem in a domain with a circular inclusion, for the case when the point force is applied outside the inclusion in a planar body. This figure represents the regular part of the displacement field produced by a point force inside a cast iron disk containing an aluminum inclusion: Figure 1(a) shows the computations obtained through the formula (1.1) when $\mathbf{y} \in \Omega_\varepsilon$, while Figure 1(b) corresponds to the numerical finite element solution produced in COMSOL. The surface plots shown are very similar.

2. Main notation

Let Ω be a subset of $\mathbb{R}^2$, with smooth boundary $\partial\Omega$ and compact closure $\bar{\Omega}$. Also let $\omega^{(j)} \subset \mathbb{R}^2$ have smooth boundary $\partial\omega^{(j)}$ and compact closure $\bar{\omega}^{(j)}$, whose complement in the infinite plane is $C\bar{\omega}^{(j)} = \mathbb{R}^2 \backslash \bar{\omega}^{(j)}$, $j = 1, \dots, M$. The sets $\omega^{(j)}$, $j = 1, \dots, M$, are assumed to contain the origin $\mathbf{O}$, and the maximum distance between $\mathbf{O}$ and $\partial\omega^{(j)}$ is 1. Let $\omega_\varepsilon^{(j)}$ be a subset of Ω, with center $\mathbf{O}^{(j)}$, $1 \le j \le M$. We assume that the minimum distance between $\mathbf{O}^{(j)}$ and $\mathbf{O}^{(k)}$, $k \neq j$, $k = 1, \dots, M$, and the minimum distance between $\mathbf{O}^{(j)}$ and $\partial\Omega$ is 1. We relate the domain $\omega_\varepsilon^{(j)}$ to $\omega^{(j)}$ via $\omega_\varepsilon^{(j)} = \{\mathbf{x} : \varepsilon^{-1}(\mathbf{x} - \mathbf{O}^{(j)}) \in \omega^{(j)}\}$, $j = 1, \dots, M$. The perturbed geometry is defined by $\Omega_\varepsilon = \Omega \backslash \overline{\bigcup_j \omega_\varepsilon^{(j)}}$, and we say that this domain is occupied by a material with shear modulus μ_O and the domain $\omega_\varepsilon^{(j)}$ is occupied by a material with shear modulus μ_{I_j}, where $\mu_O, \mu_{I_j} > 0$, $1 \le j \le M$. In the subsequent sections, along with $\mathbf{x}$ and $\mathbf{y}$ we will also use the scaled variables $\boldsymbol{\xi}_j = \varepsilon^{-1}(\mathbf{x} - \mathbf{O}^{(j)})$, $\boldsymbol{\eta}_j = \varepsilon^{-1}(\mathbf{y} - \mathbf{O}^{(j)})$.

By χ_T we mean the characteristic function of the set T, that is,

$$\chi_T(\mathbf{x}) = \begin{cases} 1, & \text{if } \mathbf{x} \in T, \\ 0, & \text{otherwise}. \end{cases}$$

Our goal is to obtain uniform asymptotics for the Green's function N_ε, of the transmission problem in $\bigcup_j \omega_\varepsilon^{(j)} \cup \Omega_\varepsilon$, which is a solution of

$$\mu_O \Delta_{\mathbf{x}} N_\varepsilon(\mathbf{x}, \mathbf{y}) + \delta(\mathbf{x} - \mathbf{y}) = 0, \quad \mathbf{x} \in \Omega_\varepsilon, \ \mathbf{y} \in \bigcup_l \omega_\varepsilon^{(l)} \cup \Omega_\varepsilon, \tag{2.1}$$

$$\mu_{I_j} \Delta_{\mathbf{x}} N_\varepsilon(\mathbf{x}, \mathbf{y}) + \delta(\mathbf{x} - \mathbf{y}) = 0, \quad \mathbf{x} \in \omega_\varepsilon^{(j)}, \ j = 1, \dots, M, \ \mathbf{y} \in \bigcup_l \omega_\varepsilon^{(l)} \cup \Omega_\varepsilon. \tag{2.2}$$

The normal derivative of N_ε on the exterior boundary satisfies

$$\mu_O \frac{\partial N_\varepsilon}{\partial n_{\mathbf{x}}}(\mathbf{x}, \mathbf{y}) = -\frac{1}{|\partial\Omega|}, \quad \mathbf{x} \in \partial\Omega, \ \mathbf{y} \in \bigcup_l \omega_\varepsilon^{(l)} \cup \Omega_\varepsilon, \tag{2.3}$$

where $|\partial\Omega|$ is the one-dimensional measure of the set $\partial\Omega$; $\partial/\partial n_{\mathbf{x}} = \boldsymbol{n} \cdot \nabla_{\mathbf{x}}$ is the normal derivative, where $\boldsymbol{n}$ is the unit outward normal.

Assuming that the small inclusions $\omega_\varepsilon^{(j)}$, $j = 1, \dots, M$, are perfectly bonded to the matrix Ω_ε, we write transmission conditions across the interface $\partial\omega_\varepsilon^{(j)}$ in the

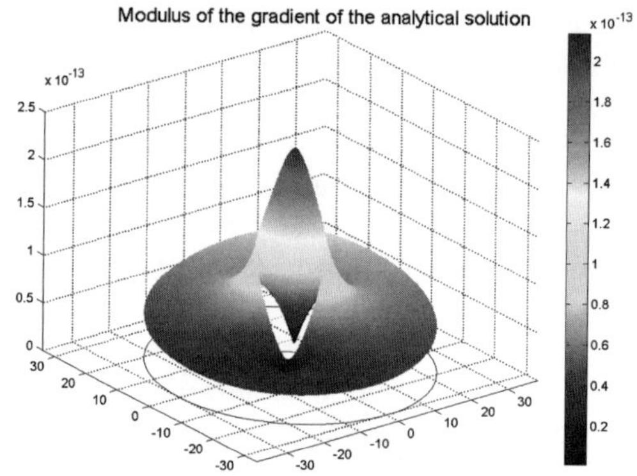

(a)

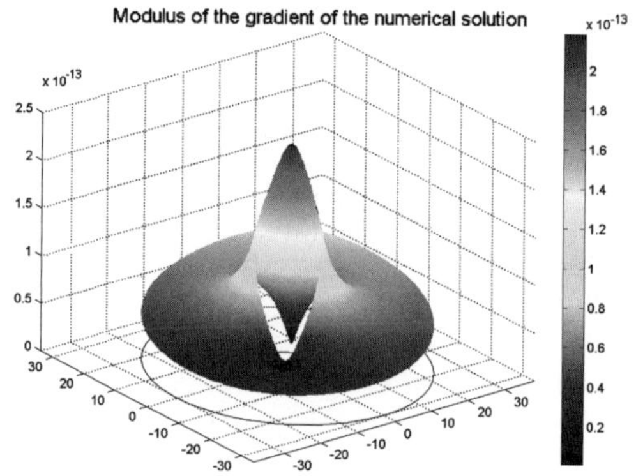

(b)

FIGURE 1. (a) Modulus of the gradient of the regular part of the Green's function, computed with the aid of the asymptotic formula (1.1) when $\mathbf{y} \in \Omega_\varepsilon$; (b) a finite element computation (produced in COMSOL) for the modulus of the gradient of the regular part of the Green's function for the transmission problem in a domain with an inclusion. Here we consider a circular cylinder of radius 30m containing a circular inclusion of radius 7m; the shear modulus of the inclusion is $\mu_I = 2.6316 \times 10^{10}\text{Nm}^{-2}$ (an aluminum inclusion), where the shear modulus for the rest of the matrix is $\mu_O = 5.6 \times 10^{10}\text{Nm}^{-2}$. The position of the unit point force is $\mathbf{y} = (10\text{m}, 10\text{m})$, which is quite close to the boundary of the inclusion.

form

$$\mu_O \frac{\partial N_\varepsilon}{\partial n_{\mathbf{x}}}(\mathbf{x}, \mathbf{y})\big|_{\mathbf{x} \in \partial\omega_\varepsilon^{(j)+}} = \mu_{I_j} \frac{\partial N_\varepsilon}{\partial n_{\mathbf{x}}}(\mathbf{x}, \mathbf{y})\big|_{\mathbf{x} \in \partial\omega_\varepsilon^{(j)-}} , \tag{2.4}$$

$$N_\varepsilon(\mathbf{x}, \mathbf{y})\big|_{\mathbf{x} \in \partial\omega_\varepsilon^{(j)+}} = N_\varepsilon(\mathbf{x}, \mathbf{y})\big|_{\mathbf{x} \in \partial\omega_\varepsilon^{(j)-}} ,$$

for $j = 1, \ldots, M$ and $\mathbf{y} \in \bigcup_l \omega_\varepsilon^{(l)} \cup \Omega_\varepsilon$; the notation $\partial\omega_\varepsilon^{(j)\pm}$ indicates the exterior or interior boundary of the set $\omega_\varepsilon^{(j)}$.

The symmetry of N_ε, i.e.,

$$N_\varepsilon(\mathbf{x}, \mathbf{y}) = N_\varepsilon(\mathbf{y}, \mathbf{x}) ,$$

is guaranteed by the condition

$$\int_{\partial\Omega} N_\varepsilon(\mathbf{x},\mathbf{y})\, dS_{\mathbf{x}} = 0\,. \tag{2.5}$$

The proof of symmetry of N_ε, for this problem, is addressed below.

The symmetry of the Green's function N_ε.

PROPOSITION 1. *Green's function N_ε for the transmission problem in $\bigcup_{l=1}^{M}\omega_\varepsilon^{(l)} \cup \Omega_\varepsilon$ satisfies*

$$N_\varepsilon(\mathbf{x},\mathbf{y}) = N_\varepsilon(\mathbf{y},\mathbf{x})\,. \tag{2.6}$$

PROOF. Let

$$\begin{aligned} N_\varepsilon^{(O)}(\mathbf{x},\mathbf{y}) &= N_\varepsilon(\mathbf{x},\mathbf{y})\,, \qquad && \text{for } \mathbf{x} \subset \Omega_\varepsilon,\ \mathbf{y} \in \textstyle\bigcup_{l=1}^{M}\omega_\varepsilon^{(l)} \cup \Omega_\varepsilon\,,\\ N_\varepsilon^{(I_j)}(\mathbf{x},\mathbf{y}) &= N_\varepsilon(\mathbf{x},\mathbf{y})\,, && \text{for } \mathbf{x} \in \omega_\varepsilon^{(j)},\ \mathbf{y} \in \textstyle\bigcup_{l=1}^{M}\omega_\varepsilon^{(l)} \cup \Omega_\varepsilon\,. \end{aligned} \tag{2.7}$$

We apply Green's formula for the functions $N_\varepsilon^{(O)}$ and $N_\varepsilon^{(I_j)}$, $j = 1,\dots,M$:

$$\begin{aligned} &\mu_O \int_{\Omega_\varepsilon} \Big\{N_\varepsilon^{(O)}(\mathbf{z},\mathbf{x})\Delta_{\mathbf{z}} N_\varepsilon^{(O)}(\mathbf{z},\mathbf{y}) - N_\varepsilon^{(O)}(\mathbf{z},\mathbf{y})\Delta_{\mathbf{z}} N_\varepsilon^{(O)}(\mathbf{z},\mathbf{x})\Big\}\, d\mathbf{z}\\ &\quad + \sum_{l=1}^{M} \mu_{I_l} \int_{\omega_\varepsilon^{(l)}} \Big\{N_\varepsilon^{(I_l)}(\mathbf{z},\mathbf{x})\Delta_{\mathbf{z}} N_\varepsilon^{(I_l)}(\mathbf{z},\mathbf{y}) - N_\varepsilon^{(I_l)}(\mathbf{z},\mathbf{y})\Delta_{\mathbf{z}} N_\varepsilon^{(I_l)}(\mathbf{z},\mathbf{x})\Big\}\, d\mathbf{z}\\ &= \mu_O \int_{\partial\Omega} \Big\{N_\varepsilon^{(O)}(\mathbf{z},\mathbf{x}) \frac{\partial N_\varepsilon^{(O)}}{\partial n_{\mathbf{z}}}(\mathbf{z},\mathbf{y}) - N_\varepsilon^{(O)}(\mathbf{z},\mathbf{y}) \frac{\partial N_\varepsilon^{(O)}}{\partial n_{\mathbf{z}}}(\mathbf{z},\mathbf{x})\Big\}\, dS_{\mathbf{z}}\\ &\quad + \sum_{l=1}^{M} \int_{\partial\omega_\varepsilon^{(l)}} \Bigg\{ N_\varepsilon^{(O)}(\mathbf{z},\mathbf{x}) \left[\mu_{I_l} \frac{\partial N_\varepsilon^{(I_l)}}{\partial n_{\mathbf{z}}}(\mathbf{z},\mathbf{y}) - \mu_O \frac{\partial N_\varepsilon^{(O)}}{\partial n_{\mathbf{z}}}(\mathbf{z},\mathbf{y})\right]\\ &\qquad - N_\varepsilon^{(I_l)}(\mathbf{z},\mathbf{y}) \left[\mu_{I_l} \frac{\partial N_\varepsilon^{(I_l)}}{\partial n_{\mathbf{z}}}(\mathbf{z},\mathbf{x}) - \mu_O \frac{\partial N_\varepsilon^{(O)}}{\partial n_{\mathbf{z}}}(\mathbf{z},\mathbf{x})\right]\Bigg\}\, dS_{\mathbf{z}}\,, \end{aligned}$$

where the right-hand side is zero as a result of the transmission conditions (2.4), the exterior boundary condition (2.3), and the normalization (2.5). Thus

$$\begin{aligned} 0 \;=\; &\mu_O \int_{\Omega_\varepsilon} \Big\{N_\varepsilon^{(O)}(\mathbf{z},\mathbf{x})\Delta_{\mathbf{z}} N_\varepsilon^{(O)}(\mathbf{z},\mathbf{y}) - N_\varepsilon^{(O)}(\mathbf{z},\mathbf{y})\Delta_{\mathbf{z}} N_\varepsilon^{(O)}(\mathbf{z},\mathbf{x})\Big\}\, d\mathbf{z}\\ &+ \sum_{l=1}^{M} \mu_{I_l} \int_{\omega_\varepsilon^{(l)}} \Big\{N_\varepsilon^{(I_l)}(\mathbf{z},\mathbf{x})\Delta_{\mathbf{z}} N_\varepsilon^{(I_l)}(\mathbf{z},\mathbf{y}) - N_\varepsilon^{(I_l)}(\mathbf{z},\mathbf{y})\Delta_{\mathbf{z}} N_\varepsilon^{(I_l)}(\mathbf{z},\mathbf{x})\Big\}\, d\mathbf{z}\,. \end{aligned} \tag{2.8}$$

The next step involves using the governing equations (2.1) and (2.2) along with the above definitions of $N_\varepsilon^{(O)}$ and $N_\varepsilon^{(I_j)}$, $j = 1,\dots,M$. When $\mathbf{x},\mathbf{y} \in \Omega_\varepsilon$, (2.8) gives

$$N_\varepsilon^{(O)}(\mathbf{x},\mathbf{y}) = N_\varepsilon^{(O)}(\mathbf{y},\mathbf{x})$$

whereas if $\mathbf{x} \in \Omega_\varepsilon$, $\mathbf{y} \in \omega_\varepsilon^{(j)}$,

$$N_\varepsilon^{(O)}(\mathbf{x},\mathbf{y}) = N_\varepsilon^{(I_j)}(\mathbf{y},\mathbf{x})\,, \quad j = 1,\dots,M\,.$$

Similarly, for $\mathbf{x} \in \omega_\varepsilon^{(j)}$, $\mathbf{y} \in \Omega_\varepsilon$, we deduce

$$N_\varepsilon^{(I_j)}(\mathbf{x},\mathbf{y}) = N_\varepsilon^{(O)}(\mathbf{y},\mathbf{x})\,, \quad j = 1,\dots,M\,,$$

and if $\mathbf{x}, \mathbf{y} \in \omega_\varepsilon^{(j)}$,

$$N_\varepsilon^{(I_j)}(\mathbf{x}, \mathbf{y}) = N_\varepsilon^{(I_j)}(\mathbf{y}, \mathbf{x}) \,, \quad j = 1, \dots, M \,.$$

Finally, with $\mathbf{x} \in \omega_\varepsilon^{(j)}$, $\mathbf{y} \in \omega_\varepsilon^{(k)}$, $k \neq j$, we have

$$N_\varepsilon^{(I_j)}(\mathbf{x}, \mathbf{y}) = N_\varepsilon^{(I_k)}(\mathbf{y}, \mathbf{x}), \quad 1 \le j, k \le M, \quad k \neq j \,.$$

The above relations together with (2.7) lead to (2.6). □

3. Special solutions in model domains

The asymptotic algorithm uses special fields defined in model domains including the unperturbed set and the exterior of a scaled inclusion.

1. The regular part $R^{(\Omega)}$ of the Neumann function $N^{(\Omega)}$ in Ω is defined as a solution of

$$\mu_O \Delta_{\mathbf{x}} R^{(\Omega)}(\mathbf{x}, \mathbf{y}) = 0 \,, \quad \mathbf{x} \in \Omega \,,$$

$$\mu_O \frac{\partial R^{(\Omega)}}{\partial n_{\mathbf{x}}}(\mathbf{x}, \mathbf{y}) = -\frac{\partial}{\partial n_{\mathbf{x}}}((2\pi)^{-1} \log |\mathbf{x} - \mathbf{y}|) + \frac{1}{|\partial\Omega|} \,, \quad \mathbf{x} \in \partial\Omega \,, \tag{3.1}$$

where $\mathbf{y} \in \Omega$.

To guarantee the symmetry of $R^{(\Omega)}$, we impose the orthogonality condition

$$\int_{\partial\Omega} R^{(\Omega)}(\mathbf{x}, \mathbf{y}) \, dS_{\mathbf{x}} = -\frac{1}{2\pi\mu_O} \int_{\partial\Omega} \log |\mathbf{x} - \mathbf{y}| \, dS_{\mathbf{x}} \,.$$

The function $N^{(\Omega)}$ is related to $R^{(\Omega)}$ by

$$N^{(\Omega)}(\mathbf{x}, \mathbf{y}) = -(2\pi\mu_O)^{-1} \log |\mathbf{x} - \mathbf{y}| - R^{(\Omega)}(\mathbf{x}, \mathbf{y}) \,. \tag{3.2}$$

2. The next model field is Green's function for the transmission problem in the domain $C\bar{\omega}^{(j)} \cup \omega^{(j)}$, $j = 1, \dots, M$. This function is denoted by $\mathcal{N}^{(j)}$ and, for $\boldsymbol{\eta} \in C\bar{\omega}^{(j)} \cup \omega^{(j)}$, is subject to

$$\mu_O \Delta_{\boldsymbol{\xi}} \mathcal{N}^{(j)}(\boldsymbol{\xi}, \boldsymbol{\eta}) + \delta(\boldsymbol{\xi} - \boldsymbol{\eta}) = 0 \,, \quad \boldsymbol{\xi} \in C\bar{\omega}^{(j)} \,,$$

$$\mu_{I_j} \Delta_{\boldsymbol{\xi}} \mathcal{N}^{(j)}(\boldsymbol{\xi}, \boldsymbol{\eta}) + \delta(\boldsymbol{\xi} - \boldsymbol{\eta}) = 0 \,, \quad \boldsymbol{\xi} \in \omega^{(j)} \,,$$

where the transmission conditions across the interface of the inclusion are given as

$$\begin{aligned} \mu_O \frac{\partial \mathcal{N}^{(j)}}{\partial n_{\boldsymbol{\xi}}}(\boldsymbol{\xi}, \boldsymbol{\eta})\big|_{\boldsymbol{\xi} \in \partial\omega^{(j)}+} &= \mu_{I_j} \frac{\partial \mathcal{N}^{(j)}}{\partial n_{\boldsymbol{\xi}}}(\boldsymbol{\xi}, \boldsymbol{\eta})\big|_{\boldsymbol{\xi} \in \partial\omega^{(j)}-} \,, \\ \mathcal{N}^{(j)}(\boldsymbol{\xi}, \boldsymbol{\eta})\big|_{\boldsymbol{\xi} \in \partial\omega^{(j)}+} &= \mathcal{N}^{(j)}(\boldsymbol{\xi}, \boldsymbol{\eta})\big|_{\boldsymbol{\xi} \in \partial\omega^{(j)}-} \,, \end{aligned} \tag{3.3}$$

and at infinity we will prescribe the condition

$$\mathcal{N}^{(j)}(\boldsymbol{\xi}, \boldsymbol{\eta}) = -(2\pi\mu_O)^{-1} \log |\boldsymbol{\xi}| + c(\boldsymbol{\eta}) + O(|\boldsymbol{\xi}|^{-1}), \quad \text{as} \quad |\boldsymbol{\xi}| \to \infty \,. \tag{3.4}$$

Symmetry of $\mathcal{N}^{(j)}$. We choose

$$c(\boldsymbol{\eta}) \equiv 0 \,, \tag{3.5}$$

and then it can be shown that

$$\mathcal{N}^{(j)}(\boldsymbol{\xi}, \boldsymbol{\eta}) = \mathcal{N}^{(j)}(\boldsymbol{\eta}, \boldsymbol{\xi}) \,,$$

i.e., $\mathcal{N}^{(j)}$ is symmetric.

The proof is analogous to the one of Proposition 1.

We set

$$\mathcal{N}^{(j,O)}(\boldsymbol{\xi},\boldsymbol{\eta}) = \mathcal{N}^{(j)}(\boldsymbol{\xi},\boldsymbol{\eta}) \quad \text{for } \boldsymbol{\eta} \in C\bar{\omega}^{(j)},\ \boldsymbol{\xi} \in C\bar{\omega}^{(j)} \cup \omega^{(j)}\ ,$$
$$\mathcal{N}^{(j,I)}(\boldsymbol{\xi},\boldsymbol{\eta}) = \mathcal{N}^{(j)}(\boldsymbol{\xi},\boldsymbol{\eta}) \quad \text{for } \boldsymbol{\eta} \in \omega^{(j)},\ \boldsymbol{\xi} \in C\bar{\omega}^{(j)} \cup \omega^{(j)}\ ,$$

and

$$c^{(O)}(\boldsymbol{\eta}) = c(\boldsymbol{\eta}) \text{ for } \boldsymbol{\eta} \in C\bar{\omega}^{(j)}\ ,$$
$$c^{(I)}(\boldsymbol{\eta}) = c(\boldsymbol{\eta}) \text{ otherwise}\ .$$

Assume $\boldsymbol{\eta} \in C\bar{\omega}^{(j)}$, and let $B_R = \{\boldsymbol{\xi} : |\boldsymbol{\xi}| < R\}$ be a disk, with sufficiently large radius R, so that $\boldsymbol{\eta} \in B_R\backslash\bar{\omega}^{(j)}$. By applying Green's formula to $\mathcal{N}^{(j)}(\mathbf{z},\boldsymbol{\xi})$ and $\mathcal{N}^{(j)}(\mathbf{z},\boldsymbol{\eta})$ in $B_R\backslash\bar{\omega}^{(j)}$ and $\omega^{(j)}$, we obtain

$$\mu_O \int_{B_R\backslash\bar{\omega}^{(j)}} \Big\{\mathcal{N}^{(j)}(\mathbf{z},\boldsymbol{\xi})\Delta_{\mathbf{z}}\mathcal{N}^{(j,O)}(\mathbf{z},\boldsymbol{\eta}) - \mathcal{N}^{(j,O)}(\mathbf{z},\boldsymbol{\eta})\Delta_{\mathbf{z}}\mathcal{N}^{(j)}(\mathbf{z},\boldsymbol{\xi})\Big\}\, d\mathbf{z}$$
$$+\mu_{I_j} \int_{\omega^{(j)}} \Big\{\mathcal{N}^{(j)}(\mathbf{z},\boldsymbol{\xi})\Delta_{\mathbf{z}}\mathcal{N}^{(j,O)}(\mathbf{z},\boldsymbol{\eta}) - \mathcal{N}^{(j)}(\mathbf{z},\boldsymbol{\eta})\Delta_{\mathbf{z}}\mathcal{N}^{(j)}(\mathbf{z},\boldsymbol{\xi})\Big\}\, d\mathbf{z}$$
$$= \mu_O \int_{\partial B_R} \Big\{\mathcal{N}^{(j)}(\mathbf{z},\boldsymbol{\xi})\frac{\partial \mathcal{N}^{(j,O)}}{\partial n_{\mathbf{z}}}(\mathbf{z},\boldsymbol{\eta}) - \mathcal{N}^{(j,O)}(\mathbf{z},\boldsymbol{\eta})\frac{\partial \mathcal{N}^{(j)}}{\partial n_{\mathbf{z}}}(\mathbf{z},\boldsymbol{\xi})\Big\}\, dS_{\mathbf{z}}\ .$$

The transmission conditions (3.3) for $\mathcal{N}^{(j)}$ imply that the integral over $\partial\omega^{(j)}$ is zero. When $\boldsymbol{\xi} \in B_R\backslash\bar{\omega}^{(j)}$,

$$\mathcal{N}^{(j)}(\boldsymbol{\xi},\boldsymbol{\eta}) - \mathcal{N}^{(j)}(\boldsymbol{\eta},\boldsymbol{\xi})$$
$$= \mu_O \int_{\partial B_R} \Big\{\mathcal{N}^{(j,O)}(\mathbf{z},\boldsymbol{\xi})\frac{\partial \mathcal{N}^{(j,O)}}{\partial n_{\mathbf{z}}}(\mathbf{z},\boldsymbol{\eta}) - \mathcal{N}^{(j,O)}(\mathbf{z},\boldsymbol{\eta})\frac{\partial \mathcal{N}^{(j,O)}}{\partial n_{\mathbf{z}}}(\mathbf{z},\boldsymbol{\xi})\Big\}\, dS_{\mathbf{z}}\ .$$

Taking the limit as $R \to \infty$ and employing (3.4), we deduce

$$\mathcal{N}^{(j)}(\boldsymbol{\xi},\boldsymbol{\eta}) - \mathcal{N}^{(j)}(\boldsymbol{\eta},\boldsymbol{\xi})$$
$$= \lim_{R\to\infty} \mu_O \int_{\partial B_R} \Big\{\Big(\frac{1}{2\pi\mu_O}\log|\mathbf{z}|^{-1} + c^{(O)}(\boldsymbol{\xi}) + O\Big(\frac{1}{R}\Big)\Big)$$
$$\times\Big(\frac{1}{2\pi\mu_O}\frac{\partial \log|\mathbf{z}|^{-1}}{\partial n_{\mathbf{z}}} + O\Big(\frac{1}{R^2}\Big)\Big)$$
$$-\Big(\frac{1}{2\pi\mu_O}\log|\mathbf{z}|^{-1} + c^{(O)}(\boldsymbol{\eta}) + O\Big(\frac{1}{R}\Big)\Big)\Big(\frac{1}{2\pi\mu_O}\frac{\partial \log|\mathbf{z}|^{-1}}{\partial n_{\mathbf{z}}} + O\Big(\frac{1}{R^2}\Big)\Big)\Big\}dS_{\mathbf{z}}\ ,$$

which is equivalent to

$$\begin{aligned}\mathcal{N}^{(j)}(\boldsymbol{\xi},\boldsymbol{\eta}) - \mathcal{N}^{(j)}(\boldsymbol{\eta},\boldsymbol{\xi}) &= (c^{(O)}(\boldsymbol{\eta}) - c^{(O)}(\boldsymbol{\xi}))\lim_{R\to\infty}\int_{\partial B_R}\frac{\partial}{\partial n_{\mathbf{z}}}((2\pi)^{-1}\log|\mathbf{z}|)dS_{\mathbf{z}}\\ &= c^{(O)}(\boldsymbol{\eta}) - c^{(O)}(\boldsymbol{\xi}) = 0\ .\end{aligned}$$

Hence $\mathcal{N}^{(j)}(\boldsymbol{\xi},\boldsymbol{\eta})$ is symmetric for $\boldsymbol{\xi},\boldsymbol{\eta} \in C\bar{\omega}^{(j)}$. In a similar way, it can be shown that

$$\mathcal{N}^{(j)}(\boldsymbol{\xi},\boldsymbol{\eta}) - \mathcal{N}^{(j)}(\boldsymbol{\eta},\boldsymbol{\xi}) = c^{(I)}(\boldsymbol{\eta}) - c^{(O)}(\boldsymbol{\xi}) \text{ for } \boldsymbol{\xi} \in C\bar{\omega}^{(j)},\ \boldsymbol{\eta} \in \omega^{(j)}\ ,$$
$$\mathcal{N}^{(j)}(\boldsymbol{\xi},\boldsymbol{\eta}) - \mathcal{N}^{(j)}(\boldsymbol{\eta},\boldsymbol{\xi}) = c^{(O)}(\boldsymbol{\eta}) - c^{(I)}(\boldsymbol{\xi}) \text{ for } \boldsymbol{\eta} \in C\bar{\omega}^{(j)},\ \boldsymbol{\xi} \in \omega^{(j)}\ ,$$
$$\mathcal{N}^{(j)}(\boldsymbol{\xi},\boldsymbol{\eta}) - \mathcal{N}^{(j)}(\boldsymbol{\eta},\boldsymbol{\xi}) = c^{(I)}(\boldsymbol{\eta}) - c^{(I)}(\boldsymbol{\xi}) \text{ for } \boldsymbol{\xi} \in \omega^{(j)},\ \boldsymbol{\eta} \in \omega^{(j)}\ ,$$

and the condition (3.5) implies that the above right-hand sides are zero. Thus $\mathcal{N}^{(j)}$ is symmetric.

Regular part of $\mathcal{N}^{(j)}$. Let the function $\mathcal{N}^{(j)}$ have the form

$$\mathcal{N}^{(j)}(\boldsymbol{\xi},\boldsymbol{\eta}) = \chi_{C\bar{\omega}^{(j)}}(\boldsymbol{\eta})\mathcal{N}^{(j,O)}(\boldsymbol{\xi},\boldsymbol{\eta}) + \chi_{\omega^{(j)}}(\boldsymbol{\eta})\mathcal{N}^{(j,I)}(\boldsymbol{\xi},\boldsymbol{\eta}) ,$$

where

$$\begin{aligned} \mathcal{N}^{(j,O)}(\boldsymbol{\xi},\boldsymbol{\eta}) &= -(2\pi\mu_O)^{-1}\log|\boldsymbol{\xi}-\boldsymbol{\eta}| - h_N^{(j,O)}(\boldsymbol{\xi},\boldsymbol{\eta}) , \\ \mathcal{N}^{(j,I)}(\boldsymbol{\xi},\boldsymbol{\eta}) &= -(2\pi\mu_{I_j})^{-1}\log|\boldsymbol{\xi}-\boldsymbol{\eta}| - h_N^{(j,I)}(\boldsymbol{\xi},\boldsymbol{\eta}) , \end{aligned} \tag{3.6}$$

and here $h_N^{(j,O)}$ and $h_N^{(j,I)}$ are the regular parts of $\mathcal{N}^{(j,O)}$ and $\mathcal{N}^{(j,I)}$, $j = 1,\dots,M$, respectively. Moreover, we also set

$$\begin{aligned} h_N^{(j,O,O)}(\boldsymbol{\xi},\boldsymbol{\eta}) &= h_N^{(j,O)}(\boldsymbol{\xi},\boldsymbol{\eta}) , && \text{for } \boldsymbol{\xi},\boldsymbol{\eta} \in C\bar{\omega}^{(j)} , \\ h_N^{(j,I,O)}(\boldsymbol{\xi},\boldsymbol{\eta}) &= h_N^{(j,O)}(\boldsymbol{\xi},\boldsymbol{\eta}) , && \text{for } \boldsymbol{\xi} \in \omega^{(j)}, \boldsymbol{\eta} \in C\bar{\omega}^{(j)} , \text{ and} \\ h_N^{(j,O,I)}(\boldsymbol{\xi},\boldsymbol{\eta}) &= h_N^{(j,I)}(\boldsymbol{\xi},\boldsymbol{\eta}) , && \text{for } \boldsymbol{\xi} \in C\bar{\omega}^{(j)}, \boldsymbol{\eta} \in \omega^{(j)} , \\ h_N^{(j,I,I)}(\boldsymbol{\xi},\boldsymbol{\eta}) &= h_N^{(j,I)}(\boldsymbol{\xi},\boldsymbol{\eta}) , && \text{for } \boldsymbol{\xi},\boldsymbol{\eta} \in \omega^{(j)} . \end{aligned} \tag{3.7}$$

Then the above definitions for $h_N^{(j,O)}$ and $h_N^{(j,I)}$ lead to this representation for $\mathcal{N}^{(j)}$:

$$\begin{aligned} \mathcal{N}^{(j)}(\boldsymbol{\xi},\boldsymbol{\eta}) \;=\; & \chi_{C\bar{\omega}^{(j)}}(\boldsymbol{\xi})\chi_{C\bar{\omega}^{(j)}}(\boldsymbol{\eta})\left\{-\frac{1}{2\pi\mu_O}\log|\boldsymbol{\xi}-\boldsymbol{\eta}| - h_N^{(j,O,O)}(\boldsymbol{\xi},\boldsymbol{\eta})\right\} \\ & +\chi_{\omega^{(j)}}(\boldsymbol{\xi})\chi_{C\bar{\omega}^{(j)}}(\boldsymbol{\eta})\left\{-\frac{1}{2\pi\mu_O}\log|\boldsymbol{\xi}-\boldsymbol{\eta}| - h_N^{(j,I,O)}(\boldsymbol{\xi},\boldsymbol{\eta})\right\} \\ & +\chi_{C\bar{\omega}^{(j)}}(\boldsymbol{\xi})\chi_{\omega^{(j)}}(\boldsymbol{\eta})\left\{-\frac{1}{2\pi\mu_{I_j}}\log|\boldsymbol{\xi}-\boldsymbol{\eta}| - h_N^{(j,O,I)}(\boldsymbol{\xi},\boldsymbol{\eta})\right\} \\ & +\chi_{\omega^{(j)}}(\boldsymbol{\xi})\chi_{\omega^{(j)}}(\boldsymbol{\eta})\left\{-\frac{1}{2\pi\mu_{I_j}}\log|\boldsymbol{\xi}-\boldsymbol{\eta}| - h_N^{(j,I,I)}(\boldsymbol{\xi},\boldsymbol{\eta})\right\} . \end{aligned}$$

The symmetry of $\mathcal{N}^{(j)}$ then implies the conditions

$$\left.\begin{aligned} h_N^{(j,O,O)}(\boldsymbol{\xi},\boldsymbol{\eta}) = h_N^{(j,O,O)}(\boldsymbol{\eta},\boldsymbol{\xi}) \quad & \text{for } \boldsymbol{\xi},\boldsymbol{\eta} \in C\bar{\omega}^{(j)} , \\ h_N^{(j,I,O)}(\boldsymbol{\xi},\boldsymbol{\eta}) = h_N^{(j,O,I)}(\boldsymbol{\eta},\boldsymbol{\xi}) + \frac{1}{2\pi}\left\{\frac{1}{\mu_{I_j}} - \frac{1}{\mu_O}\right\}\log|\boldsymbol{\xi}-\boldsymbol{\eta}| \quad & \text{for } \boldsymbol{\eta} \in C\bar{\omega}^{(j)},\ \boldsymbol{\xi} \in \omega^{(j)}, \\ \text{and} \quad h_N^{(j,I,I)}(\boldsymbol{\xi},\boldsymbol{\eta}) = h_N^{(j,I,I)}(\boldsymbol{\eta},\boldsymbol{\xi}) \quad & \text{for } \boldsymbol{\xi},\boldsymbol{\eta} \in \omega^{(j)} . \end{aligned}\right\} \tag{3.8}$$

3. We also make use of model solutions known as the dipole fields $\mathcal{D}_k^{(j)}$, $k = 1, 2$, $j = 1,\dots,M$, which play the role of the boundary layers in the asymptotic algorithm. Let $\mathcal{D}^{(j)} = (\mathcal{D}_1^{(j)}, \mathcal{D}_2^{(j)})^T$, where

$$\mathcal{D}^{(j)}(\boldsymbol{\xi}) = \chi_{C\bar{\omega}^{(j)}}(\boldsymbol{\xi})\mathcal{D}^{(j,O)}(\boldsymbol{\xi}) + \chi_{\omega^{(j)}}(\boldsymbol{\xi})\mathcal{D}^{(j,I)}(\boldsymbol{\xi})$$

and the vector functions $\mathcal{D}^{(j,O)}$, $\mathcal{D}^{(j,I)}$ solve the problem

$$\mu_O\Delta_{\boldsymbol{\xi}}\mathcal{D}^{(j,O)}(\boldsymbol{\xi}) = \mathbf{O} , \quad \boldsymbol{\xi} \in C\bar{\omega}^{(j)} ,$$

$$\mu_{I_j}\Delta_{\boldsymbol{\xi}}\mathcal{D}^{(j,I)}(\boldsymbol{\xi}) = \mathbf{O} , \quad \boldsymbol{\xi} \in \omega^{(j)} .$$

The transmission conditions on the boundary of the inclusion $\omega^{(j)}$ are

$$\mu_{I_j}\frac{\partial \mathcal{D}^{(j,I)}}{\partial n_{\boldsymbol{\xi}}}(\boldsymbol{\xi}) - \mu_O\frac{\partial \mathcal{D}^{(j,O)}}{\partial n_{\boldsymbol{\xi}}}(\boldsymbol{\xi}) = (\mu_{I_j} - \mu_O)\boldsymbol{n}^{(j)}\,, \text{ on } \partial\omega^{(j)}\,, \tag{3.9}$$

$$\mathcal{D}^{(j,O)}(\boldsymbol{\xi}) = \mathcal{D}^{(j,I)}(\boldsymbol{\xi})\,, \text{ on } \partial\omega^{(j)}\,,$$

where in (3.9), $\boldsymbol{n}^{(j)}$ is the unit normal to $\omega^{(j)}$. At infinity the vector function $\mathcal{D}^{(j,O)}$ satisfies

$$\mathcal{D}^{(j,O)}(\boldsymbol{\xi}) = O(|\boldsymbol{\xi}|^{-1}), \quad \text{as} \quad |\boldsymbol{\xi}| \to \infty\,. \tag{3.10}$$

4. We introduce the function $\zeta^{(j)}$ which is a solution of

$$\Delta_{\boldsymbol{\xi}}\zeta^{(j)}(\boldsymbol{\xi}) = 0\,, \quad \boldsymbol{\xi} \in C\bar{\omega}^{(j)}\,,$$

$$\zeta^{(j)}(\boldsymbol{\xi}) = 0\,, \quad \boldsymbol{\xi} \in \partial\omega^{(j)}\,,$$

$$\zeta^{(j)}(\boldsymbol{\xi}) = (2\pi\mu_O)^{-1}\log|\boldsymbol{\xi}| + \zeta_\infty^{(j)} + O(|\boldsymbol{\xi}|^{-1})\,, \quad \text{as} \quad |\boldsymbol{\xi}| \to \infty\,,$$

where $\zeta_\infty^{(j)}$ is a constant.

4. An estimate for solutions to transmission problems for antiplane shear in unbounded domains

The next result plays an important role in the asymptotic algorithm. It will allow us to obtain estimates for the boundary layer fields and derive the estimate for a solution of the transmission problem in a domain with multiple inclusions.

LEMMA 4.1. *Let $U^{(j)}$ be a solution of the transmission problem*

$$\mu_O\Delta U^{(j)}(\boldsymbol{\xi}) = 0\,, \quad \boldsymbol{\xi} \in C\bar{\omega}^{(j)}\,,$$

$$\mu_{I_j}\Delta U^{(j)}(\boldsymbol{\xi}) = 0\,, \quad \boldsymbol{\xi} \in \omega^{(j)}\,,$$

$$U^{(j)}(\boldsymbol{\xi})\big|_{\boldsymbol{\xi}\in\partial\omega^{(j)}+} = U^{(j)}(\boldsymbol{\xi})\big|_{\boldsymbol{\xi}\in\partial\omega^{(j)}-}\,,$$

$$\mu_O\frac{\partial U^{(j)}}{\partial n}(\boldsymbol{\xi})\big|_{\boldsymbol{\xi}\in\partial\omega^{(j)}+} - \mu_{I_j}\frac{\partial U^{(j)}}{\partial n}(\boldsymbol{\xi})\big|_{\boldsymbol{\xi}\in\partial\omega^{(j)}-} = \varphi^{(j)}(\boldsymbol{\xi})\,,$$

$$U^{(j)}(\boldsymbol{\xi}) \to 0 \quad \textit{as} \quad |\boldsymbol{\xi}| \to \infty\,,$$

where $\varphi^{(j)} \in L_\infty(\partial\omega^{(j)})$, $\partial/\partial n$ is the normal derivative on the smooth boundary $\partial\omega^{(j)}$, outward with respect to $\omega^{(j)}$, and

$$\int_{\partial\omega^{(j)}} \varphi^{(j)}(\boldsymbol{\xi})\, dS_{\boldsymbol{\xi}} = 0\,. \tag{4.1}$$

We also assume that

$$\int_{\partial\omega^{(j)}} U^{(j)}(\boldsymbol{\xi})\frac{\partial\zeta^{(j)}}{\partial n}(\boldsymbol{\xi})\big|_{\boldsymbol{\xi}\in\partial\omega^{(j)}+} dS_{\boldsymbol{\xi}} = 0\,,$$

where $\zeta^{(j)}$ is given as solution of problem 4 of Section 3. Then

$$\sup_{\boldsymbol{\xi}\in C\bar{\omega}^{(j)}\cup\omega^{(j)}} \{(|\boldsymbol{\xi}|+1)|U^{(j)}(\boldsymbol{\xi})|\} \le \text{ const } \|\varphi^{(j)}\|_{L_\infty(\partial\omega^{(j)})}\,, \tag{4.2}$$

where the constant depends on μ_O, μ_{I_j}, and $\partial\omega^{(j)}$, for $j = 1, \ldots, M$.

PROOF. We note that transmission problems are studied in detail in [**1**] and [**2**], in the context of boundary integral equations and their solvability.

Let us first represent the solution $U^{(j)}$ by two functions $U^{(j,O)}$ and $U^{(j,I)}$, harmonic in the domains $C\bar{\omega}^{(j)}$ and $\omega^{(j)}$, respectively. These functions satisfy

$$U^{(j,O)}(\boldsymbol{\xi}) = U^{(j,I)}(\boldsymbol{\xi}) , \quad \boldsymbol{\xi} \in \partial\omega^{(j)} ,$$
$$\mu_O \frac{\partial U^{(j,O)}}{\partial n}(\boldsymbol{\xi}) - \mu_{I_j} \frac{\partial U^{(j,I)}}{\partial n}(\boldsymbol{\xi}) = \varphi^{(j)}(\boldsymbol{\xi}) , \quad \boldsymbol{\xi} \in \partial\omega^{(j)} .$$

For the function $U^{(j,O)}$, the condition

$$U^{(j,O)}(\boldsymbol{\xi}) \to 0 \quad \text{as} \quad |\boldsymbol{\xi}| \to \infty \tag{4.3}$$

also holds.

Applying Green's formula to the functions $\mathcal{N}^{(j)}$ (see problem 2, Section 3), $U^{(j,O)}$ and $U^{(j,I)}$, one obtains

$$U^{(j,O)}(\boldsymbol{\xi}) = -\int_{\partial\omega^{(j)}} \mathcal{N}^{(j)}(\boldsymbol{\xi},\boldsymbol{\eta})\varphi^{(j)}(\boldsymbol{\eta})\, dS_{\boldsymbol{\eta}} , \tag{4.4}$$

for $\boldsymbol{\xi} \in C\bar{\omega}^{(j)}$, and

$$U^{(j,I)}(\boldsymbol{\xi}) = -\int_{\partial\omega^{(j)}} \mathcal{N}^{(j)}(\boldsymbol{\xi},\boldsymbol{\eta})\varphi^{(j)}(\boldsymbol{\eta})\, dS_{\boldsymbol{\eta}} , \tag{4.5}$$

for $\boldsymbol{\xi} \in \omega^{(j)}$.

First, let $|\boldsymbol{\xi}| \ge 2$. Then using the asymptotics for the function $\mathcal{N}^{(j)}$ at infinity and the condition (4.1), we deduce

$$\begin{aligned}(1+|\boldsymbol{\xi}|)|U^{(j,O)}(\boldsymbol{\xi})| &\le \text{const}\Big((1+|\boldsymbol{\xi}|)|\log|\boldsymbol{\xi}||\Big|\int_{\partial\omega^{(j)}}\varphi^{(j)}(\boldsymbol{\eta})dS_{\boldsymbol{\eta}}\Big| \\ &\qquad + \|\varphi^{(j)}\|_{L_1(\partial\omega^{(j)})}\Big) \\ &\le \text{const}\|\varphi^{(j)}\|_{L_\infty(\partial\omega^{(j)})} .\end{aligned} \tag{4.6}$$

Also by the Cauchy-Schwarz inequality and (4.4)

$$|U^{(j,O)}(\boldsymbol{\xi})| \le \text{const}\ \|\varphi^{(j)}\|_{L_2(\partial\omega^{(j)})} \le \text{const}\|\varphi^{(j)}\|_{L_\infty(\partial\omega^{(j)})} , \quad \text{for} \quad \boldsymbol{\xi} \in B_3\backslash\bar{\omega}^{(j)} , \tag{4.7}$$

where $B_3 = \{\boldsymbol{\xi} : |\boldsymbol{\xi}| < 3\}$.

Similarly the integral representation (4.5) gives

$$|U^{(j,I)}(\boldsymbol{\xi})| \le \text{const}\ \|\varphi\|_{L_2(\partial\omega^{(j)})} \le \text{const}\|\varphi^{(j)}\|_{L_\infty(\partial\omega^{(j)})} , \quad \text{for} \quad \boldsymbol{\xi} \in \omega^{(j)} . \tag{4.8}$$

The combination of (4.6), (4.7), and (4.8) leads to (4.2). □

As an immediate corollary of Lemma 4.1, we have an estimate for the dipole fields associated with the scaled inclusion $\omega^{(j)}$, $j = 1, \dots, M$:

LEMMA 4.2. *For the dipole fields* $\mathcal{D}_j^{(i)}$, $j = 1, 2$, $i = 1, \dots, M$,

$$\sup_{\boldsymbol{\xi} \in C\bar{\omega}^{(i)} \cup \omega^{(i)}} \{(|\boldsymbol{\xi}|+1)|\mathcal{D}_j^{(i)}(\boldsymbol{\xi})|\} \le \text{const}$$

holds, where the constant on the right-hand side can depend on μ_O, μ_{I_i}, *and* $\omega^{(i)}$.

5. An estimate for the maximum modulus of solutions to transmission problems for antiplane shear in a domain with several small inclusions

Here we obtain an estimate for solutions to transmission problems for antiplane shear in domains with small inclusions. The next lemma will be used in Section 7 incorporating the remainder estimates produced by the approximations of Green's function in a perforated domain.

LEMMA 5.1. *Let u be a function in $L_\infty(\bigcup_{l=1}^M \omega_\varepsilon^{(l)} \cup \Omega_\varepsilon)$ such that ∇u is square integrable in a neighborhood of $\partial\omega_\varepsilon^{(i)}$, $i = 1, \dots, M$. Also, let u be a solution of the transmission problem*

$$
(5.1) \qquad \left.
\begin{aligned}
&\mu_O \Delta u(\mathbf{x}) = 0\,, \quad \mathbf{x} \in \Omega_\varepsilon\,,\\
&\mu_{I_i} \Delta u(\mathbf{x}) = 0\,, \quad \mathbf{x} \in \omega_\varepsilon^{(i)},\ i = 1, \dots, M\,,\\
&\mu_O \frac{\partial u}{\partial n}(\mathbf{x}) = \psi(\mathbf{x})\,, \quad \mathbf{x} \in \partial\Omega\,,\\
&\mu_O \frac{\partial u}{\partial n}(\mathbf{x})\Big|_{\mathbf{x}\in\partial\omega_\varepsilon^{(i)}+} - \mu_{I_i} \frac{\partial u}{\partial n}(\mathbf{x})\Big|_{\mathbf{x}\in\partial\omega_\varepsilon^{(i)}-} = \varphi_\varepsilon^{(i)}(\mathbf{x})\,,\\
&u(\mathbf{x})\Big|_{\mathbf{x}\in\partial\omega_\varepsilon^{(i)}+} = u(\mathbf{x})\Big|_{\mathbf{x}\in\partial\omega_\varepsilon^{(i)}-}
\end{aligned}
\right\}
$$

where $\psi \in L_\infty(\partial\Omega)$, $\varphi_\varepsilon^{(i)} \in L_\infty(\partial\omega_\varepsilon^{(i)})$, for $1 \le i \le M$,

$$\int_{\partial\Omega} \psi(\mathbf{x})\, dS_{\mathbf{x}} = 0\,, \qquad \int_{\partial\omega_\varepsilon^{(i)}} \varphi_\varepsilon^{(i)}(\mathbf{x})\, dS_{\mathbf{x}} = 0\,,$$

and $\varphi_\varepsilon^{(i)}(\mathbf{x}) = \varepsilon^{-1}\varphi^{(i)}(\varepsilon^{-1}(\mathbf{x} - \mathbf{O}^{(i)}))$, $i = 1, \dots, M$. To provide uniqueness we also assume

$$\int_{\partial\Omega} u(\mathbf{x}) dS_{\mathbf{x}} = 0\,.$$

Then there exists a positive constant A, independent of ε and such that

$$
(5.2) \qquad \|u\|_{L_\infty(\bigcup_{l=1}^M \omega_\varepsilon^{(l)} \cup \Omega_\varepsilon)} \le A \Big\{ \|\psi\|_{L_\infty(\partial\Omega)} + \varepsilon \max_{1\le k\le M} \|\varphi_\varepsilon^{(k)}\|_{L_\infty(\partial\omega_\varepsilon^{(k)})} \Big\}\,.
$$

PROOF. (a) *The inverse operators to model problems in Ω and $C\bar{\omega}^{(j)}$, $j = 1, \dots, M$.* Let us introduce the operators

$$
(5.3) \qquad \mathbf{N} : \psi \to w \quad \text{and} \quad \mathfrak{N}^{(j)} : \varphi^{(j)} \to v^{(j)}\,,
$$

which are the inverse operators of the problems

$$
(5.4) \qquad \left.
\begin{aligned}
&\mu_O \Delta w(\mathbf{x}) = 0\,, \quad \mathbf{x} \in \Omega\,,\\
&\mu_O \frac{\partial w}{\partial n}(\mathbf{x}) = \psi(\mathbf{x})\,, \quad \mathbf{x} \in \partial\Omega\,,\\
&\int_{\partial\Omega} w(\mathbf{x})\, dS_{\mathbf{x}} = 0\,,
\end{aligned}
\right\}
$$

and

$$\left.\begin{aligned}
&\mu_O \Delta v^{(j)}(\boldsymbol{\xi}) = 0\,, \quad \boldsymbol{\xi} \in C\bar{\omega}^{(j)}\,,\\
&\mu_{I_j} \Delta v^{(j)}(\boldsymbol{\xi}) = 0\,, \quad \boldsymbol{\xi} \in \omega^{(j)}\,,\\
&\mu_O \frac{\partial v^{(j)}}{\partial n}(\boldsymbol{\xi})\big|_{\boldsymbol{\xi}\in\partial\omega^{(j)}+} - \mu_{I_j}\frac{\partial v^{(j)}}{\partial n}(\boldsymbol{\xi})\big|_{\boldsymbol{\xi}\in\partial\omega^{(j)}-} = \varphi^{(j)}(\boldsymbol{\xi})\,,\\
&v^{(j)}(\boldsymbol{\xi})\big|_{\boldsymbol{\xi}\in\partial\omega^{(j)}+} = v^{(j)}(\boldsymbol{\xi})\big|_{\boldsymbol{\xi}\in\partial\omega^{(j)}-}\,,\\
&v^{(j)}(\boldsymbol{\xi}) \to 0 \quad \text{as} \quad |\boldsymbol{\xi}| \to \infty\,,
\end{aligned}\right\} \tag{5.5}$$

where $\psi \in L_\infty(\partial\Omega)$, $\varphi \in L_\infty(\partial\omega^{(j)})$, $j = 1, \dots, M$, and also

$$\int_{\partial\omega^{(j)}} \varphi^{(j)}(\boldsymbol{\xi})\, dS_{\boldsymbol{\xi}} = 0 \quad \text{and} \quad \int_{\partial\Omega} \psi(\mathbf{x})\, dS_{\mathbf{x}} = 0\,.$$

In scaled coordinates $\boldsymbol{\xi}_j = \varepsilon^{-1}(\mathbf{x} - \mathbf{O}^{(j)})$, $j = 1, \dots, M$, the operator $\mathfrak{N}^{(j)}_\varepsilon$ is defined by

$$(\mathfrak{N}^{(j)}_\varepsilon \varphi^{(j)}_\varepsilon)(\mathbf{x}) = (\mathfrak{N}^{(j)}\varphi^{(j)})(\boldsymbol{\xi}_j)\,,$$

where $\varphi^{(j)}_\varepsilon(\mathbf{x}) = \varepsilon^{-1}\varphi^{(j)}(\varepsilon^{-1}(\mathbf{x} - \mathbf{O}^{(j)}))$.

(b) *An estimate for solutions to the model Neumann problem in Ω.* Let $N^{(\Omega)}$ denote the Neumann function (3.2) in Ω.

Then an application of Green's formula to $N^{(\Omega)}(\mathbf{x},\mathbf{y})$ and $w(\mathbf{y})$ yields the representation for w

$$w(\mathbf{x}) = \int_{\partial\Omega} N^{(\Omega)}(\mathbf{y},\mathbf{x})\psi(\mathbf{y})\, dS_{\mathbf{y}} + \frac{1}{|\partial\Omega|}\int_{\partial\Omega} w(\mathbf{y})\, dS_{\mathbf{y}}\,. \tag{5.6}$$

The solution w of the Neumann problem in Ω is subject to the orthogonality condition in problem (5.4), and hence the last term on the right-hand side of (5.6) is zero, so that

$$w(\mathbf{x}) = \int_{\partial\Omega} N^{(\Omega)}(\mathbf{y},\mathbf{x})\psi(\mathbf{y})\, dS_{\mathbf{y}}\,.$$

From this we obtain the estimate

$$\sup_{\Omega} |w| \le \text{const} \sup_{\partial\Omega} |\psi|\,. \tag{5.7}$$

(c) *The case of the homogeneous boundary condition on $\partial\Omega$.* When the right-hand side of the Neumann condition on $\partial\Omega$ in (5.1) is zero, we look for a solution of the form

$$u_1 = \sum_{j=1}^{M} \mathfrak{N}^{(j)}_\varepsilon g^{(j)}_\varepsilon - \mathbf{N}\Big(\mathrm{Tr}_{\partial\Omega}\, \mu_O \sum_{j=1}^{M} \frac{\partial}{\partial n}(\mathfrak{N}^{(j)}_\varepsilon g^{(j)}_\varepsilon)\Big)\,, \tag{5.8}$$

where $g^{(j)}_\varepsilon$ is an unknown function defined on $\partial\omega^{(j)}_\varepsilon$ such that

$$\int_{\partial\omega^{(j)}} g^{(j)}(\boldsymbol{\xi}_j)\, dS_{\boldsymbol{\xi}_j} = 0$$

and $g^{(j)}(\boldsymbol{\xi}_j) = \varepsilon g_\varepsilon^{(j)}(\mathbf{x})$. The function u_1 is harmonic inside $\bigcup_{l=1}^M \omega_\varepsilon^{(l)} \cup \Omega_\varepsilon$ and is continuous across the boundaries of the small inclusions. On $\partial\Omega$, u_1 satisfies

$$\mu_O \frac{\partial u_1}{\partial n}(\mathbf{x}) = 0 \, .$$

Computing the jump in tractions of u_1 on $\partial\omega_\varepsilon^{(m)}$, we obtain

$$\varphi_\varepsilon^{(m)}(\mathbf{x}) = \mu_O \frac{\partial u_1}{\partial n}(\mathbf{x})\big|_{\mathbf{x}\in\partial\omega_\varepsilon^{(m)+}} - \mu_{I_m} \frac{\partial u_1}{\partial n}(\mathbf{x})\big|_{\mathbf{x}\in\partial\omega_\varepsilon^{(m)-}} = g_\varepsilon^{(m)} + (S_\varepsilon^{(m)} g_\varepsilon)(\mathbf{x})$$

where $g_\varepsilon(\mathbf{x}) = (g^{(1)}(\mathbf{x}), \dots, g^{(M)}(\mathbf{x}))^T$ and

$$(5.9)\quad \begin{aligned}(S_\varepsilon^{(m)} g_\varepsilon)(\mathbf{x}) = (\mu_O - \mu_{I_m})\Big[\frac{\partial}{\partial n}\Big\{ \sum_{\substack{j\neq m\\ 1\le j\le M}} \mathfrak{N}_\varepsilon^{(j)} g_\varepsilon^{(j)}\Big\}\big|_{\mathbf{x}\in\partial\omega_\varepsilon^{(m)}} \\ - \frac{\partial}{\partial n}\Big\{\mathbf{N}\Big(\mathrm{Tr}_{\partial\Omega}\, \mu_O \sum_{j=1}^M \frac{\partial}{\partial n}(\mathfrak{N}_\varepsilon^{(j)} g_\varepsilon^{(j)})\Big)\Big\}\big|_{\mathbf{x}\in\partial\omega_\varepsilon^{(m)}}\Big].\end{aligned}$$

Let $B^{(m)}$ denote a sufficiently small disk centered at $\mathbf{O}^{(m)}$ and containing $\omega_\varepsilon^{(m)}$. Using a local estimate for solutions of Laplace's equation, along with Lemma 4.1 and the definition of $g_\varepsilon^{(j)}$ above, we obtain

$$(5.10)\quad \begin{aligned}\Big\|\frac{\partial}{\partial n}\Big\{ \sum_{\substack{j\neq m\\ 1\le j\le M}} \mathfrak{N}_\varepsilon^{(j)} g_\varepsilon^{(j)}\Big\}\Big\|_{L_\infty(\partial\omega_\varepsilon^{(m)})} &\le \text{const}\, \Big\| \sum_{\substack{j\neq m\\ 1\le j\le M}} \mathfrak{N}_\varepsilon^{(j)} g_\varepsilon^{(j)}\Big\|_{L_\infty(B^{(m)})} \\ &\le \text{const} \sum_{\substack{j\neq m\\ 1\le j\le M}} \varepsilon^2 \|g_\varepsilon^{(j)}\|_{L_\infty(\partial\omega_\varepsilon^{(j)})} \, .\end{aligned}$$

Then, from the local estimate for harmonic functions, we can also assert the inequality

$$\begin{aligned}&\Big\|\frac{\partial}{\partial n}\Big\{\mathbf{N}\Big(\mathrm{Tr}_{\partial\Omega}\, \mu_O \sum_{j=1}^M \frac{\partial}{\partial n}(\mathfrak{N}_\varepsilon^{(j)} g_\varepsilon^{(j)})\Big)\Big\}\Big\|_{L_\infty(\partial\omega_\varepsilon^{(m)})} \\ &\quad\le \text{const}\, \Big\|\mathbf{N}\Big(\mathrm{Tr}_{\partial\Omega}\, \mu_O \sum_{j=1}^M \frac{\partial}{\partial n}(\mathfrak{N}_\varepsilon^{(j)} g_\varepsilon^{(j)})\Big)\Big\|_{L_\infty(B^{(m)})} \\ &\quad\le \text{const} \sum_{j=1}^M \Big\|\frac{\partial}{\partial n}(\mathfrak{N}_\varepsilon^{(j)} g_\varepsilon^{(j)})\Big\|_{L_\infty(\partial\Omega)} \le \text{const} \sum_{j=1}^M \varepsilon^2 \|g_\varepsilon^{(j)}\|_{L_\infty(\partial\omega_\varepsilon^{(j)})} \, ,\end{aligned}$$

where in moving to the second inequality we used estimate (5.7), and then Lemma 4.1 brings us to the last inequality.

Then the preceding estimate and (5.9), (5.10) lead to

$$\|S_\varepsilon^{(m)} g_\varepsilon\|_{L_\infty(\partial\omega_\varepsilon^{(m)})} \le \text{const}\, \varepsilon^2 \sum_{j=1}^M \|g_\varepsilon^{(j)}\|_{L_\infty(\partial\omega_\varepsilon^{(j)})} \, .$$

Hence, from the smallness of $S_\varepsilon^{(m)}$, we can write

$$g_\varepsilon = (\mathbf{I} + \mathbf{S}_\varepsilon)^{-1} \varphi_\varepsilon \, ,$$

where $g_\varepsilon(\mathbf{x}) = (g_\varepsilon^{(1)}(\mathbf{x}), \dots, g_\varepsilon^{(M)}(\mathbf{x}))^T$, $\varphi_\varepsilon(\mathbf{x}) = (\varphi_\varepsilon^{(1)}(\mathbf{x}), \dots, \varphi_\varepsilon^{(M)}(\mathbf{x}))^T$, $\mathbf{S}_\varepsilon$ is a matrix whose rows are $S_\varepsilon^{(1)}, \dots, S_\varepsilon^{(M)}$, and

$$\|g_\varepsilon^{(j)}\|_{L_\infty(\partial\omega_\varepsilon^{(j)})} \leq \text{const} \max_{1\leq k\leq M} \|\varphi_\varepsilon^{(k)}\|_{L_\infty(\partial\omega_\varepsilon^{(k)})} \,. \tag{5.11}$$

From (5.8), together with (5.7) and Lemma 4.1, we obtain

$$\|u_1\|_{L_\infty(\bigcup_{l=1}^M \omega_\varepsilon^{(l)} \cup \Omega_\varepsilon)} \leq \text{const} \sum_{j=1}^M \varepsilon \|g_\varepsilon^{(j)}\|_{L_\infty(\partial\omega_\varepsilon^{(j)})} \,.$$

Now this and (5.11) give

$$\|u_1\|_{L_\infty(\bigcup_{l=1}^M \omega_\varepsilon^{(l)} \cup \Omega_\varepsilon)} \leq \text{const}\ \varepsilon \max_{1\leq k\leq M} \|\varphi_\varepsilon^{(k)}\|_{L_\infty(\partial\omega_\varepsilon^{(k)})} \,. \tag{5.12}$$

(d) *The case of continuous tractions on* $\partial\omega_\varepsilon^{(j)}$, $j = 1, \dots, M$. In this situation we look for the solution u_2 in the form

$$u_2 = \mathbf{N}\psi + v \,. \tag{5.13}$$

Then v is a solution of (5.1) with the boundary conditions

$$\mu_O \frac{\partial v}{\partial n}(\mathbf{x}) = 0 \,, \text{ on } \partial\Omega \,,$$

and

$$v(\mathbf{x})\big|_{\mathbf{x}\in\partial\omega_\varepsilon^{(j)+}} = v(\mathbf{x})\big|_{\mathbf{x}\in\partial\omega_\varepsilon^{(j)-}}$$

$$\mu_O \frac{\partial v}{\partial n}(\mathbf{x})\big|_{\mathbf{x}\in\partial\omega_\varepsilon^{(j)+}} - \mu_{I_j}\frac{\partial v}{\partial n}(\mathbf{x})\big|_{\mathbf{x}\in\partial\omega_\varepsilon^{(j)-}} = (\mu_{I_j} - \mu_O)\frac{\partial \mathbf{N}\psi}{\partial n}(\mathbf{x})\big|_{\mathbf{x}\in\partial\omega_\varepsilon^{(j)}}$$

for $1 \leq j \leq M$, where the right-hand sides of the above traction conditions on $\partial\omega_\varepsilon^{(j)}$, $j = 1, \dots, M$, are self-balanced, so by part (c) of the present proof

$$\begin{aligned}\|v\|_{L_\infty(\bigcup_{l=1}^M \omega_\varepsilon^{(l)} \cup \Omega_\varepsilon)} &\leq \text{const}\ \varepsilon \max_{1\leq k\leq M} \Big\|\frac{\partial \mathbf{N}\psi}{\partial n}(\mathbf{x})\Big\|_{L_\infty(\partial\omega_\varepsilon^{(k)})} \\ &\leq \text{const}\ \varepsilon \max_{1\leq k\leq M} \|\mathbf{N}\psi\|_{L_\infty(B^{(k)})} \\ &\leq \text{const}\ \varepsilon \|\psi\|_{L_\infty(\partial\Omega)} \,.\end{aligned}$$

This inequality, (5.7), and (5.13) give

$$\|u_2\|_{L_\infty(\bigcup_{l=1}^M \omega_\varepsilon^{(l)} \cup \Omega_\varepsilon)} \leq \text{const}\ \|\psi\|_{L_\infty(\partial\Omega)} \,. \tag{5.14}$$

Finally, we obtain (5.2) through the combination of (5.12) and (5.14). □

6. An asymptotic approximation for the regular part of the Green's function $\mathcal{N}^{(j)}$ at infinity

In this section, we shall prove a result which concerns the asymptotics of $h_N^{(j,O)}$ and $h_N^{(j,I)}$, $j = 1, \dots, M$ (see (3.6)), at infinity.

LEMMA 6.1. *Let* $|\boldsymbol{\xi}| > 2$. *Then*
(a)

$$h_N^{(j,O)}(\boldsymbol{\xi}, \boldsymbol{\eta}) = -\mathcal{D}^{(j,O)}(\boldsymbol{\eta}) \cdot \nabla_{\boldsymbol{\xi}}((2\pi\mu_O)^{-1} \log |\boldsymbol{\xi}|^{-1}) + O(|\boldsymbol{\xi}|^{-2}(|\boldsymbol{\eta}| + 1)^{-1}) \,,$$

for $\boldsymbol{\eta} \in C\bar{\omega}^{(j)}$,

(b)

$$h_N^{(j,I)}(\boldsymbol{\xi},\boldsymbol{\eta}) = -(\mu_{I_j}^{-1}-\mu_O^{-1})\{(2\pi)^{-1}\log|\boldsymbol{\xi}| - \boldsymbol{\eta}\cdot\nabla_{\boldsymbol{\xi}}((2\pi)^{-1}\log|\boldsymbol{\xi}|)\} -\mathcal{D}^{(j,I)}(\boldsymbol{\eta})\cdot\nabla_{\boldsymbol{\xi}}((2\pi\mu_O)^{-1}\log|\boldsymbol{\xi}|^{-1}) + O(|\boldsymbol{\xi}|^{-2})$$

for $\boldsymbol{\eta}\in\omega^{(j)}$.

PROOF. First we study the asymptotic behavior of the functions $h_N^{(j,O,O)}$, $h_N^{(j,I,O)}$, $h_N^{(j,O,I)}$ introduced in problem 2 of Section 3, in the neighborhood of infinity. We show that for $|\boldsymbol{\xi}|>2$,

(a)

$$h_N^{(j,O,O)}(\boldsymbol{\xi},\boldsymbol{\eta}) = -\mathcal{D}^{(j,O)}(\boldsymbol{\eta})\cdot\nabla_{\boldsymbol{\xi}}((2\pi\mu_O)^{-1}\log|\boldsymbol{\xi}|^{-1}) + O(|\boldsymbol{\xi}|^{-2}(|\boldsymbol{\eta}|+1)^{-1})\,, \tag{6.1}$$

for $\boldsymbol{\eta}\in C\bar{\omega}^{(j)}$,

(b)

$$h_N^{(j,I,O)}(\boldsymbol{\eta},\boldsymbol{\xi}) = -\mathcal{D}^{(j,I)}(\boldsymbol{\eta})\cdot\nabla_{\boldsymbol{\xi}}((2\pi\mu_O)^{-1}\log|\boldsymbol{\xi}|^{-1}) + O(|\boldsymbol{\xi}|^{-2})\,, \tag{6.2}$$

for $\boldsymbol{\eta}\in\omega^{(j)}$,

(c)

$$\begin{aligned} h_N^{(j,O,I)}(\boldsymbol{\xi},\boldsymbol{\eta}) &= -(\mu_{I_j}^{-1}-\mu_O^{-1})\{(2\pi)^{-1}\log|\boldsymbol{\xi}| - \boldsymbol{\eta}\cdot\nabla_{\boldsymbol{\xi}}((2\pi)^{-1}\log|\boldsymbol{\xi}|)\} \\ &\quad -\mathcal{D}^{(j,I)}(\boldsymbol{\eta})\cdot\nabla_{\boldsymbol{\xi}}((2\pi\mu_O)^{-1}\log|\boldsymbol{\xi}|^{-1}) + O(|\boldsymbol{\xi}|^{-2})\,, \end{aligned} \tag{6.3}$$

for $\boldsymbol{\eta}\in\omega^{(j)}$.

(i) *Auxiliary functions* $h_N^{(j)}$ *and* $\Upsilon^{(j)}$. Let us introduce the function $h_N^{(j)}$ as

$$h_N^{(j)}(\boldsymbol{\xi},\boldsymbol{\eta}) = \chi_{C\bar{\omega}^{(j)}}(\boldsymbol{\eta})h_N^{(j,O)}(\boldsymbol{\xi},\boldsymbol{\eta}) + \chi_{\omega^{(j)}}(\boldsymbol{\eta})h_N^{(j,I)}(\boldsymbol{\xi},\boldsymbol{\eta})\,. \tag{6.4}$$

From problem 2 of Section 3, when $\boldsymbol{\eta}\in C\bar{\omega}^{(j)}\cup\omega^{(j)}$, $h_N^{(j)}$ is a solution of the transmission problem

$$\mu_O\Delta_{\boldsymbol{\xi}}h_N^{(j)}(\boldsymbol{\xi},\boldsymbol{\eta}) = 0\,,\quad \boldsymbol{\xi}\in C\bar{\omega}^{(j)}\,,$$

$$\mu_{I_j}\Delta_{\boldsymbol{\xi}}h_N^{(j)}(\boldsymbol{\xi},\boldsymbol{\eta}) = 0\,,\quad \boldsymbol{\xi}\in \omega^{(j)}\,,$$

$$\begin{aligned} &\mu_{I_j}\frac{\partial h_N^{(j)}}{\partial n_{\boldsymbol{\xi}}}(\boldsymbol{\xi},\boldsymbol{\eta})\big|_{\boldsymbol{\xi}\in\partial\omega^{(j)}-} - \mu_O\frac{\partial h_N^{(j)}}{\partial n_{\boldsymbol{\xi}}}(\boldsymbol{\xi},\boldsymbol{\eta})\big|_{\boldsymbol{\xi}\in\partial\omega^{(j)}+} \\ &\quad = -\frac{(\mu_{I_j}-\mu_O)}{2\pi}\left(\frac{\chi_{C\bar{\omega}^{(j)}}(\boldsymbol{\eta})}{\mu_O}+\frac{\chi_{\omega^{(j)}}(\boldsymbol{\eta})}{\mu_{I_j}}\right)\frac{\partial}{\partial n_{\boldsymbol{\xi}}}(\log|\boldsymbol{\xi}-\boldsymbol{\eta}|)\,, \end{aligned} \tag{6.5}$$

$$h_N^{(j)}(\boldsymbol{\xi},\boldsymbol{\eta})\big|_{\boldsymbol{\xi}\in\partial\omega^{(j)}+} = h_N^{(j)}(\boldsymbol{\xi},\boldsymbol{\eta})\big|_{\boldsymbol{\xi}\in\partial\omega^{(j)}-}\,, \tag{6.6}$$

$$h_N^{(j)}(\boldsymbol{\xi},\boldsymbol{\eta}) = -(\mu_{I_j}^{-1}-\mu_O^{-1})\chi_{\omega^{(j)}}(\boldsymbol{\eta})(2\pi)^{-1}\log|\boldsymbol{\xi}| + O(|\boldsymbol{\xi}|^{-1}),\quad \text{as}\quad |\boldsymbol{\xi}|\to\infty\,.$$

We introduce one more auxiliary vector function $\Upsilon^{(j)}(\boldsymbol{\xi}) = \{\Upsilon_i^{(j)}(\boldsymbol{\xi})\}_{i=1}^2$, defined by

$$\Upsilon^{(j)}(\boldsymbol{\xi}) = \boldsymbol{\xi} - \mathcal{D}^{(j)}(\boldsymbol{\xi})\,. \tag{6.7}$$

It solves the transmission problem

$$
(6.8)\qquad \left.\begin{aligned}
&\mu_O \Delta_{\boldsymbol{\xi}} \Upsilon^{(j)}(\boldsymbol{\xi}) = \mathbf{O}\,, \quad \boldsymbol{\xi} \in C\bar{\omega}^{(j)}\,,\\
&\mu_{I_j} \Delta_{\boldsymbol{\xi}} \Upsilon^{(j)}(\boldsymbol{\xi}) = \mathbf{O}\,, \quad \boldsymbol{\xi} \in \omega^{(j)}\,,\\
&\mu_{I_j} \frac{\partial \Upsilon^{(j)}}{\partial n_{\boldsymbol{\xi}}}(\boldsymbol{\xi})\big|_{\boldsymbol{\xi}\in\partial\omega^{(j)-}} = \mu_O \frac{\partial \Upsilon^{(j)}}{\partial n_{\boldsymbol{\xi}}}(\boldsymbol{\xi})\big|_{\boldsymbol{\xi}\in\partial\omega^{(j)+}}\,,\\
&\Upsilon^{(j)}(\boldsymbol{\xi})\big|_{\boldsymbol{\xi}\in\partial\omega^{(j)+}} = \Upsilon^{(j)}(\boldsymbol{\xi})\big|_{\boldsymbol{\xi}\in\partial\omega^{(j)-}}\,,\\
&\Upsilon^{(j)}(\boldsymbol{\xi}) = \boldsymbol{\xi} + O(|\boldsymbol{\xi}|^{-1}), \quad \text{as} \quad |\boldsymbol{\xi}| \to \infty\,,
\end{aligned}\right\}
$$

which is consistent with problem 3 of Section 3. One can also represent $\Upsilon^{(j)}$ as

$$
\Upsilon^{(j)}(\boldsymbol{\xi}) = \chi_{C\bar{\omega}^{(j)}}(\boldsymbol{\xi})\Upsilon^{(j,O)}(\boldsymbol{\xi}) + \chi_{\omega^{(j)}}(\boldsymbol{\xi})\Upsilon^{(j,I)}(\boldsymbol{\xi})\,.
$$

(ii) *The asymptotics of $h_N^{(j,O,O)}$ at infinity.* For $|\boldsymbol{\xi}| > 2$, $\boldsymbol{\eta} \in C\bar{\omega}^{(j)}$, we have from Lemma 2 in [**5**] that the function $h_N^{(j,O,O)}$ defined in (6.4) has the asymptotic representation

$$
(6.9)\qquad h_N^{(j,O,O)}(\boldsymbol{\xi}, \boldsymbol{\eta}) = \mathcal{C}^{(j,O)}(\boldsymbol{\eta}) \cdot \nabla_{\boldsymbol{\xi}}((2\pi)^{-1} \log |\boldsymbol{\xi}|^{-1}) + r_N^{(j,O,O)}(\boldsymbol{\xi}, \boldsymbol{\eta})\,,
$$

where the remainder $r_N^{(j,O,O)}(\boldsymbol{\xi}, \boldsymbol{\eta})$ satisfies

$$
|r_N^{(j,O,O)}(\boldsymbol{\xi}, \boldsymbol{\eta})| \le \text{const}\; (1 + |\boldsymbol{\eta}|)^{-1} |\boldsymbol{\xi}|^{-2} \quad \text{for } |\boldsymbol{\xi}| > 2, \boldsymbol{\eta} \in C\bar{\omega}^{(j)}\,.
$$

The vector function $\mathcal{C}^{(j,O)}(\boldsymbol{\eta}) = \{\mathcal{C}_i^{(j,O)}(\boldsymbol{\eta})\}_{i=1}^2$ is evaluated below.

Evaluation of $\mathcal{C}^{(j,O)}(\boldsymbol{\eta})$. Let B_R be a disk centered at the origin with a sufficiently large radius R. We apply Green's formula to $h_N^{(j)}$ and $\Upsilon_i^{(j)}$ in the domain $B_R \backslash \bar{\omega}^{(j)} \cup \omega^{(j)}$:

$$
\begin{aligned}
0 \;=\;& \mu_O \int_{B_R \backslash \bar{\omega}^{(j)}} \left\{ \Upsilon_i^{(j)}(\boldsymbol{\xi}) \Delta_{\boldsymbol{\xi}} h_N^{(j)}(\boldsymbol{\xi}, \boldsymbol{\eta}) - h_N^{(j)}(\boldsymbol{\xi}, \boldsymbol{\eta}) \Delta_{\boldsymbol{\xi}} \Upsilon_i^{(j)}(\boldsymbol{\xi}) \right\} d\boldsymbol{\xi}\\
&+ \mu_{I_j} \int_{\omega^{(j)}} \left\{ \Upsilon_i^{(j)}(\boldsymbol{\xi}) \Delta_{\boldsymbol{\xi}} h_N^{(j)}(\boldsymbol{\xi}, \boldsymbol{\eta}) - h_N^{(j)}(\boldsymbol{\xi}, \boldsymbol{\eta}) \Delta_{\boldsymbol{\xi}} \Upsilon_i^{(j)}(\boldsymbol{\xi}) \right\} d\boldsymbol{\xi}\\
=\;& \mu_O \int_{\partial B_R} \left\{ \Upsilon_i^{(j)}(\boldsymbol{\xi}) \frac{\partial h_N^{(j)}}{\partial n_{\boldsymbol{\xi}}}(\boldsymbol{\xi}, \boldsymbol{\eta}) - h_N^{(j)}(\boldsymbol{\xi}, \boldsymbol{\eta}) \frac{\partial \Upsilon_i^{(j)}}{\partial n_{\boldsymbol{\xi}}}(\boldsymbol{\xi}) \right\} dS_{\boldsymbol{\xi}}\\
&+ \int_{\partial\omega^{(j)}} \Bigg[\Upsilon_i^{(j)}(\boldsymbol{\xi}) \left\{ \mu_{I_j} \frac{\partial h_N^{(j)}}{\partial n_{\boldsymbol{\xi}}}(\boldsymbol{\xi}, \boldsymbol{\eta})\big|_{\partial\omega^{(j)-}} - \mu_O \frac{\partial h_N^{(j)}}{\partial n_{\boldsymbol{\xi}}}(\boldsymbol{\xi}, \boldsymbol{\eta})\big|_{\partial\omega^{(j)+}} \right\}\\
&- h_N^{(j)}(\boldsymbol{\xi}, \boldsymbol{\eta}) \left\{ \mu_{I_j} \frac{\partial \Upsilon_i^{(j)}}{\partial n_{\boldsymbol{\xi}}}(\boldsymbol{\xi})\big|_{\partial\omega^{(j)-}} - \mu_O \frac{\partial \Upsilon_i^{(j)}}{\partial n_{\boldsymbol{\xi}}}(\boldsymbol{\xi})\big|_{\partial\omega^{(j)+}} \right\} \Bigg] dS_{\boldsymbol{\xi}}\,,
\end{aligned} \qquad (6.10)
$$

where while combining the integrals over $\partial\omega^{(j)+}$ and $\partial\omega^{(j)-}$ we have used the continuity conditions for the functions $h_N^{(j)}$ and $\Upsilon_i^{(j)}$, respectively (see (6.6) and problem (6.8)). With the use of the transmission conditions for $\Upsilon_i^{(j)}$ and $h_N^{(j)}$, when $\boldsymbol{\eta} \in C\bar{\omega}^{(j)}$, this equality becomes

$$0 = \mu_O \int_{\partial B_R} \left\{ \Upsilon_i^{(j)}(\boldsymbol{\xi}) \frac{\partial h_N^{(j)}}{\partial n_{\boldsymbol{\xi}}}(\boldsymbol{\xi},\boldsymbol{\eta}) - h_N^{(j)}(\boldsymbol{\xi},\boldsymbol{\eta}) \frac{\partial \Upsilon_i^{(j)}}{\partial n_{\boldsymbol{\xi}}}(\boldsymbol{\xi}) \right\} dS_{\boldsymbol{\xi}}$$

$$- \frac{(\mu_{I_j} - \mu_O)}{\mu_O} \int_{\partial \omega^{(j)}} \Upsilon_i^{(j)}(\boldsymbol{\xi}) \frac{\partial}{\partial n_{\boldsymbol{\xi}}} ((2\pi)^{-1} \log|\boldsymbol{\xi} - \boldsymbol{\eta}|)\, dS_{\boldsymbol{\xi}}\,, \tag{6.11}$$

for $\boldsymbol{\eta} \in C\bar{\omega}^{(j)}$. The last integral, owing to (6.7), is equal to

$$\int_{\partial \omega^{(j)}} \Upsilon_i^{(j)}(\boldsymbol{\xi}) \frac{\partial}{\partial n_{\boldsymbol{\xi}}} ((2\pi)^{-1} \log|\boldsymbol{\xi} - \boldsymbol{\eta}|)\, dS_{\boldsymbol{\xi}}$$

$$= \int_{\partial \omega^{(j)}} \left\{ \xi_i \frac{\partial}{\partial n_{\boldsymbol{\xi}}} ((2\pi)^{-1} \log|\boldsymbol{\xi} - \boldsymbol{\eta}|) - \mathcal{D}_i^{(j)}(\boldsymbol{\xi}) \frac{\partial}{\partial n_{\boldsymbol{\xi}}} ((2\pi)^{-1} \log|\boldsymbol{\xi} - \boldsymbol{\eta}|) \right\} dS_{\boldsymbol{\xi}}$$

$$= \int_{\partial \omega^{(j)}} \left\{ (2\pi)^{-1} \log|\boldsymbol{\xi} - \boldsymbol{\eta}| \frac{\partial \xi_i}{\partial n_{\boldsymbol{\xi}}} - \mathcal{D}_i^{(j)}(\boldsymbol{\xi}) \frac{\partial}{\partial n_{\boldsymbol{\xi}}} ((2\pi)^{-1} \log|\boldsymbol{\xi} - \boldsymbol{\eta}|) \right\} dS_{\boldsymbol{\xi}}\,, \tag{6.12}$$

where $\boldsymbol{\eta} \in C\bar{\omega}^{(j)}$. Now, using the jump in tractions for the dipole fields, stated in problem 3 of Section 3, we can write

$$\frac{(\mu_{I_j} - \mu_O)}{\mu_O} \int_{\partial \omega^{(j)}} \Upsilon_i^{(j)}(\boldsymbol{\xi}) \frac{\partial}{\partial n_{\boldsymbol{\xi}}} ((2\pi)^{-1} \log|\boldsymbol{\xi} - \boldsymbol{\eta}|^{-1})\, dS_{\boldsymbol{\xi}}$$

$$= \frac{1}{\mu_O} \int_{\partial \omega^{(j)}} \left\{ (2\pi)^{-1} \log|\boldsymbol{\xi} - \boldsymbol{\eta}|^{-1} \left[\mu_{I_j} \frac{\partial \mathcal{D}_i^{(j,I)}}{\partial n_{\boldsymbol{\xi}}}(\boldsymbol{\xi}) - \mu_O \frac{\partial \mathcal{D}_i^{(j,O)}}{\partial n_{\boldsymbol{\xi}}}(\boldsymbol{\xi}) \right] \right.$$

$$\left. -(\mu_{I_j} - \mu_O) \mathcal{D}_i^{(j)}(\boldsymbol{\xi}) \frac{\partial}{\partial n_{\boldsymbol{\xi}}} ((2\pi)^{-1} \log|\boldsymbol{\xi} - \boldsymbol{\eta}|^{-1}) \right\} dS_{\boldsymbol{\xi}}\,.$$

Using the continuity of the displacements for the dipole fields (see problem 3, Section 3), we write the above right-hand side as

$$\frac{\mu_{I_j}}{\mu_O} \int_{\partial \omega^{(j)}} \left\{ (2\pi)^{-1} \log|\boldsymbol{\xi} - \boldsymbol{\eta}|^{-1} \frac{\partial \mathcal{D}_i^{(j,I)}}{\partial n_{\boldsymbol{\xi}}}(\boldsymbol{\xi}) - \mathcal{D}_i^{(j,I)}(\boldsymbol{\xi}) \frac{\partial}{\partial n_{\boldsymbol{\xi}}} ((2\pi)^{-1} \log|\boldsymbol{\xi} - \boldsymbol{\eta}|^{-1}) \right\} dS_{\boldsymbol{\xi}}$$

$$- \int_{\partial \omega^{(j)}} \left\{ (2\pi)^{-1} \log|\boldsymbol{\xi} - \boldsymbol{\eta}|^{-1} \frac{\partial \mathcal{D}_i^{(j,O)}}{\partial n_{\boldsymbol{\xi}}}(\boldsymbol{\xi}) - \mathcal{D}_i^{(j,O)}(\boldsymbol{\xi}) \frac{\partial}{\partial n_{\boldsymbol{\xi}}} ((2\pi)^{-1} \log|\boldsymbol{\xi} - \boldsymbol{\eta}|^{-1}) \right\} dS_{\boldsymbol{\xi}}\,.$$

Then, upon applying Green's formula to $\mathcal{D}_i^{(j,O)}$, $\mathcal{D}_i^{(j,I)}$ in $B_R \backslash \bar{\omega}^{(j)}$, $\omega^{(j)}$, respectively, with the fundamental solution of $-\Delta$, and using the definition of $\Upsilon^{(j)}$, we obtain the above in the form

$$\mathcal{D}_i^{(j,O)}(\boldsymbol{\eta}) - \int_{\partial B_R} \left\{ (2\pi)^{-1} \log|\boldsymbol{\xi} - \boldsymbol{\eta}|^{-1} \frac{\partial \mathcal{D}_i^{(j,O)}}{\partial n_{\boldsymbol{\xi}}}(\boldsymbol{\xi}) \right.$$

$$\left. -\mathcal{D}_i^{(j,O)}(\boldsymbol{\xi}) \frac{\partial}{\partial n_{\boldsymbol{\xi}}} ((2\pi)^{-1} \log|\boldsymbol{\xi} - \boldsymbol{\eta}|^{-1}) \right\} dS_{\boldsymbol{\xi}}$$

where $\boldsymbol{\eta} \in C\bar{\omega}^{(j)}$ and by (3.10) the integral over ∂B_R decays as $R \to \infty$ like $O(|\log R|/R)$. Therefore, we have

$$\frac{(\mu_{I_j} - \mu_O)}{\mu_O} \int_{\partial \omega^{(j)}} \Upsilon_i^{(j)}(\boldsymbol{\xi}) \frac{\partial}{\partial n_{\boldsymbol{\xi}}} ((2\pi)^{-1} \log|\boldsymbol{\xi} - \boldsymbol{\eta}|^{-1})\, dS_{\boldsymbol{\xi}} = \mathcal{D}_i^{(j,O)}(\boldsymbol{\eta})$$

for $\boldsymbol{\eta} \in C\bar{\omega}^{(j)}$. Equation (6.11), with substitution of the preceding equality, leads to

$$(6.13) \quad -\mathcal{D}_i^{(j,O)}(\boldsymbol{\eta}) = \mu_O \int_{\partial B_R} \left\{ \Upsilon_i^{(j)}(\boldsymbol{\xi}) \frac{\partial h_N^{(j)}}{\partial n_{\boldsymbol{\xi}}}(\boldsymbol{\xi}, \boldsymbol{\eta}) - h_N^{(j)}(\boldsymbol{\xi}, \boldsymbol{\eta}) \frac{\partial \Upsilon_i^{(j)}}{\partial n_{\boldsymbol{\xi}}}(\boldsymbol{\xi}) \right\} dS_{\boldsymbol{\xi}} \, .$$

We now aim to determine the vector function $\mathcal{C}^{(j,O)}(\boldsymbol{\eta}) = \{\mathcal{C}_i^{(j,O)}(\boldsymbol{\eta})\}_{i=1}^2$ in (6.9). Taking the limit $R \to \infty$ in (6.13) and using (6.9) (where $h_N^{(j,O,O)}(\boldsymbol{\xi}, \boldsymbol{\eta}) = h_N^{(j)}(\boldsymbol{\xi}, \boldsymbol{\eta})$ for $\boldsymbol{\xi}, \boldsymbol{\eta} \in C\bar{\omega}^{(j)}$), we have

$$\begin{aligned} -\mathcal{D}_i^{(j,O)}(\boldsymbol{\eta}) &= \lim_{R\to\infty} \mu_O \mathcal{C}^{(j,O)}(\boldsymbol{\eta}) \cdot \int_{\partial B_R} \Big\{ \xi_i \frac{\partial}{\partial n_{\boldsymbol{\xi}}} (\nabla_{\boldsymbol{\xi}}((2\pi)^{-1} \log |\boldsymbol{\xi}|^{-1})) \\ &\qquad\qquad - \nabla_{\boldsymbol{\xi}}((2\pi)^{-1} \log |\boldsymbol{\xi}|^{-1}) \frac{\partial \xi_i}{\partial n_{\boldsymbol{\xi}}} \Big\} dS_{\boldsymbol{\xi}} \\ &= \mu_O \mathcal{C}_i^{(j,O)}(\boldsymbol{\eta}) \, . \end{aligned}$$

Thus, we have derived that

$$\mathcal{C}^{(j,O)}(\boldsymbol{\eta}) = -\frac{\mathcal{D}^{(j,O)}(\boldsymbol{\eta})}{\mu_O} \, ,$$

which corresponds to the leading order term of the function $h_N^{(j,O,O)}$, stated in the current lemma in (6.1).

(iii) *The asymptotics of $h^{(j,I,O)}$ at infinity.* We have already shown that for $|\boldsymbol{\xi}| \geq 2$, $\boldsymbol{\eta} \in C\bar{\omega}^{(j)}$

$$(6.14) \quad h_N^{(j,O,O)}(\boldsymbol{\xi}, \boldsymbol{\eta}) = -\mathcal{D}^{(j,O)}(\boldsymbol{\eta}) \cdot \nabla_{\boldsymbol{\xi}}((2\pi\mu_O)^{-1} \log |\boldsymbol{\xi}|^{-1}) + r_N^{(j,O,O)}(\boldsymbol{\xi}, \boldsymbol{\eta}) \, .$$

In order to deduce the leading order term of $h^{(j,I,O)}$ at infinity, we recall the relations (3.8). First, since $h_N^{(j,O,O)}$ is symmetric for $\boldsymbol{\xi}, \boldsymbol{\eta} \in C\bar{\omega}^{(j)}$, we have that the above asymptotic representation also holds for $h_N^{(j,O,O)}(\boldsymbol{\eta}, \boldsymbol{\xi})$. Next we allow $\boldsymbol{\eta}$ to approach the boundary of the inclusion $\omega^{(j)}$. For $\boldsymbol{\eta} \in \partial\omega^{(j)}$ we have

$$\mathcal{D}^{(j,O)}(\boldsymbol{\eta}) = \mathcal{D}^{(j,I)}(\boldsymbol{\eta}) \, ,$$

$$h_N^{(j,O,O)}(\boldsymbol{\eta}, \boldsymbol{\xi}) = h_N^{(j,I,O)}(\boldsymbol{\eta}, \boldsymbol{\xi}).$$

Therefore, allowing $\boldsymbol{\eta} \in \omega^{(j)}$, for $|\boldsymbol{\xi}| > 2$, we arrive at

$$(6.15) \quad h_N^{(j,I,O)}(\boldsymbol{\eta}, \boldsymbol{\xi}) = -\mathcal{D}^{(j,I)}(\boldsymbol{\eta}) \cdot \nabla_{\boldsymbol{\xi}}((2\pi\mu_O)^{-1} \log |\boldsymbol{\xi}|^{-1}) + r_N^{(j,O,I)}(\boldsymbol{\xi}, \boldsymbol{\eta}) \, ,$$

where $r_N^{(j,O,I)}$ is the remainder term and subject to its smallness the leading order part of (6.2) has been formally deduced.

Remainder estimate. Consider the function

$$r_N^{(j)}(\boldsymbol{\xi}, \boldsymbol{\eta}) = \chi_{C\bar{\omega}^{(j)}}(\boldsymbol{\eta}) r_N^{(j,O,O)}(\boldsymbol{\xi}, \boldsymbol{\eta}) + \chi_{\omega^{(j)}}(\boldsymbol{\eta}) r_N^{(j,O,I)}(\boldsymbol{\xi}, \boldsymbol{\eta})$$

which by (6.14) and (6.15) is

$$\begin{aligned} r_N^{(j)}(\boldsymbol{\xi}, \boldsymbol{\eta}) &= \chi_{C\bar{\omega}^{(j)}}(\boldsymbol{\eta}) \{ h_N^{(j,O,O)}(\boldsymbol{\eta}, \boldsymbol{\xi}) + \mathcal{D}^{(j,O)}(\boldsymbol{\eta}) \cdot \nabla_{\boldsymbol{\xi}}((2\pi\mu_O)^{-1} \log |\boldsymbol{\xi}|^{-1}) \} \\ &\quad + \chi_{\omega^{(j)}}(\boldsymbol{\eta}) \{ h_N^{(j,I,O)}(\boldsymbol{\eta}, \boldsymbol{\xi}) + \mathcal{D}^{(j,I)}(\boldsymbol{\eta}) \cdot \nabla_{\boldsymbol{\xi}}((2\pi\mu_O)^{-1} \log |\boldsymbol{\xi}|^{-1}) \} \, . \end{aligned}$$

Let $|\boldsymbol{\xi}| > 2$ and write the problem for $r_N^{(j)}$ with respect to $\boldsymbol{\eta}$ as

$$\mu_O \Delta_{\boldsymbol{\eta}} r_N^{(j)}(\boldsymbol{\xi}, \boldsymbol{\eta}) = 0 , \quad \boldsymbol{\eta} \in C\bar{\omega}^{(j)} ,$$

$$\mu_{I_j} \Delta_{\boldsymbol{\eta}} r_N^{(j)}(\boldsymbol{\xi}, \boldsymbol{\eta}) = 0 , \quad \boldsymbol{\eta} \in \omega^{(j)} ,$$

$$\begin{aligned}\mu_{I_j} \frac{\partial r_N^{(j)}}{\partial n_{\boldsymbol{\eta}}}(\boldsymbol{\xi}, \boldsymbol{\eta})\big|_{\boldsymbol{\eta} \in \partial\omega^{(j)-}} &- \mu_O \frac{\partial r_N^{(j)}}{\partial n_{\boldsymbol{\eta}}}(\boldsymbol{\xi}, \boldsymbol{\eta})\big|_{\boldsymbol{\eta} \in \partial\omega^{(j)+}} \\ &= \frac{(\mu_{I_j} - \mu_O)}{\mu_O} \Big\{ \frac{\partial}{\partial n_{\boldsymbol{\xi}}}((2\pi)^{-1} \log|\boldsymbol{\xi} - \boldsymbol{\eta}|) - \frac{\partial}{\partial n_{\boldsymbol{\xi}}}((2\pi)^{-1} \log|\boldsymbol{\xi}|) \Big\} ,\end{aligned} \tag{6.16}$$

$$r_N^{(j)}(\boldsymbol{\xi}, \boldsymbol{\eta})\big|_{\boldsymbol{\eta} \in \partial\omega^{(j)+}} = r_N^{(j)}(\boldsymbol{\xi}, \boldsymbol{\eta})\big|_{\boldsymbol{\eta} \in \partial\omega^{(j)-}} ,$$

$$r_N^{(j)}(\boldsymbol{\xi}, \boldsymbol{\eta}) \to 0, \quad \text{as} \quad |\boldsymbol{\eta}| \to \infty .$$

We have that the right-hand side of (6.16) is self-balanced and

$$\begin{aligned}&\left| \frac{(\mu_{I_j} - \mu_O)}{\mu_O} \left\{ \frac{\partial}{\partial n_{\boldsymbol{\xi}}}((2\pi)^{-1} \log|\boldsymbol{\xi} - \boldsymbol{\eta}|) - \frac{\partial}{\partial n_{\boldsymbol{\xi}}}((2\pi)^{-1} \log|\boldsymbol{\xi}|) \right\} \right| \\ &\quad \le \text{const } |\boldsymbol{\eta}||\boldsymbol{\xi}|^{-2} \le \text{const } |\boldsymbol{\xi}|^{-2} ,\end{aligned}$$

where $\boldsymbol{\eta} \in \partial\omega^{(j)}$, $|\boldsymbol{\eta}| \le 1$, and $|\boldsymbol{\xi}| > 2$. Now an application of Lemma 4.1 leads to the estimate for $r_N^{(j)}$

$$|r_N^{(j)}(\boldsymbol{\xi}, \boldsymbol{\eta})| \le \text{const } |\boldsymbol{\xi}|^{-2}(|\boldsymbol{\eta}| + 1)^{-1} ,$$

for $|\boldsymbol{\xi}| > 2$, $\boldsymbol{\eta} \in C\bar{\omega}^{(j)} \cup \omega^{(j)}$.

(iv) *The asymptotics of* $h_N^{(j,O,I)}$. We once again refer to (3.8), for the relation

$$h_N^{(j,O,I)}(\boldsymbol{\xi}, \boldsymbol{\eta}) = h_N^{(j,I,O)}(\boldsymbol{\eta}, \boldsymbol{\xi}) - \frac{1}{2\pi} \left\{ \frac{1}{\mu_{I_j}} - \frac{1}{\mu_O} \right\} \log|\boldsymbol{\xi} - \boldsymbol{\eta}| \quad \text{for } \boldsymbol{\xi} \in C\bar{\omega}^{(j)}, \boldsymbol{\eta} \in \omega^{(j)} .$$

For $|\boldsymbol{\xi}| > 2$, $\boldsymbol{\eta} \in \omega^{(j)}$, this can be rewritten as

$$h_N^{(j,O,I)}(\boldsymbol{\xi}, \boldsymbol{\eta}) = h_N^{(j,I,O)}(\boldsymbol{\eta}, \boldsymbol{\xi}) - \frac{1}{2\pi} \left\{ \frac{1}{\mu_{I_j}} - \frac{1}{\mu_O} \right\} \{\log|\boldsymbol{\xi}| - \boldsymbol{\eta} \cdot \nabla_{\boldsymbol{\xi}}(\log|\boldsymbol{\xi}|)\} + \left(\frac{1}{|\boldsymbol{\xi}|^2} \right) .$$

By combining this with (6.2) we obtain (6.3). □

The proof of Lemma 6.1 is then completed by applying (3.7), (6.1), and (6.3).

7. Uniform approximation of N_ε in a domain with multiple inclusions

The aim of the current section is to present the uniform asymptotic approximation for N_ε in a domain with several inclusions.

THEOREM 7.1. *The approximation of Green's function for the transmission problem of antiplane shear in* $\bigcup_l \omega_\varepsilon^{(l)} \cup \Omega_\varepsilon \subset \mathbb{R}^2$ *is given by*

$$N_\varepsilon(\mathbf{x},\mathbf{y}) = N^{(\Omega)}(\mathbf{x},\mathbf{y}) + \sum_{j=1}^{M} \mathcal{N}^{(j)}(\boldsymbol{\xi}_j,\boldsymbol{\eta}_j) + M(2\pi\mu_O)^{-1}\log(\varepsilon^{-1}|\mathbf{x}-\mathbf{y}|)$$
$$+\varepsilon\sum_{j=1}^{M}\{\mathcal{D}^{(j)}(\boldsymbol{\xi}_j)\cdot\nabla_{\mathbf{x}}R^{(\Omega)}(\mathbf{O}^{(j)},\mathbf{y}) + \mathcal{D}^{(j)}(\boldsymbol{\eta}_j)\cdot\nabla_{\mathbf{y}}R^{(\Omega)}(\mathbf{x},\mathbf{O}^{(j)})\} + O(\varepsilon^2) \tag{7.1}$$

uniformly for $\mathbf{x},\mathbf{y} \in \bigcup_l \omega_\varepsilon^{(l)} \cup \Omega_\varepsilon$.

PROOF. *Formal asymptotic algorithm.* First, we give a plausible argument concerning the representation of N_ε. We propose the function N_ε to be given in the form

$$N_\varepsilon(\mathbf{x},\mathbf{y}) = -\frac{1}{2\pi}\left(\frac{\chi_{\Omega_\varepsilon}(\mathbf{y})}{\mu_O} + \sum_{l=1}^{M}\frac{\chi_{\omega_\varepsilon^{(l)}}(\mathbf{y})}{\mu_{I_l}}\right)\log|\mathbf{x}-\mathbf{y}| - R_\varepsilon(\mathbf{x},\mathbf{y}), \tag{7.2}$$

where for $\mathbf{y} \in \bigcup_{l=1}^{M}\omega_\varepsilon^{(l)} \cup \Omega_\varepsilon$, R_ε is a solution of

$$\mu_O\Delta_{\mathbf{x}}R_\varepsilon(\mathbf{x},\mathbf{y}) = 0, \quad \mathbf{x}\in\Omega_\varepsilon, \tag{7.3}$$

$$\mu_{I_i}\Delta_{\mathbf{x}}R_\varepsilon(\mathbf{x},\mathbf{y}) = 0, \quad \mathbf{x}\in\omega_\varepsilon^{(i)}, i=1,\dots,M, \tag{7.4}$$

$$\mu_O\frac{\partial R_\varepsilon}{\partial n_{\mathbf{x}}}(\mathbf{x},\mathbf{y}) = -\frac{\mu_O}{2\pi}\left(\frac{\chi_{\Omega_\varepsilon}(\mathbf{y})}{\mu_O} + \sum_{l=1}^{M}\frac{\chi_{\omega_\varepsilon^{(l)}}(\mathbf{y})}{\mu_{I_l}}\right)\frac{\partial(\log|\mathbf{x}-\mathbf{y}|)}{\partial n_{\mathbf{x}}} + \frac{1}{|\partial\Omega|}, \quad \mathbf{x}\in\partial\Omega, \tag{7.5}$$

and for $i=1,\dots,M$, R_ε satisfies the transmission conditions

$$\mu_{I_i}\frac{\partial R_\varepsilon}{\partial n_{\mathbf{x}}}(\mathbf{x},\mathbf{y})\big|_{\mathbf{x}\in\partial\omega_\varepsilon^{(i)}-} - \mu_O\frac{\partial R_\varepsilon}{\partial n_{\mathbf{x}}}(\mathbf{x},\mathbf{y})\big|_{\mathbf{x}\in\partial\omega_\varepsilon^{(i)}+}$$
$$= -\frac{\mu_{I_i}-\mu_O}{2\pi}\left(\frac{\chi_{\Omega_\varepsilon}(\mathbf{y})}{\mu_O} + \sum_{l=1}^{M}\frac{\chi_{\omega_\varepsilon^{(l)}}(\mathbf{y})}{\mu_{I_l}}\right)\frac{\partial(\log|\mathbf{x}-\mathbf{y}|)}{\partial n_{\mathbf{x}}}, \tag{7.6}$$

$$R_\varepsilon(\mathbf{x},\mathbf{y})\big|_{\mathbf{x}\in\partial\omega_\varepsilon^{(i)}+} = R_\varepsilon(\mathbf{x},\mathbf{y})\big|_{\mathbf{x}\in\partial\omega_\varepsilon^{(i)}-}. \tag{7.7}$$

In the above problem we have that R_ε is subject to the orthogonality condition

$$\int_{\partial\Omega} R_\varepsilon(\mathbf{x},\mathbf{y})dS_{\mathbf{x}} = -\frac{1}{2\pi}\left(\frac{\chi_{\Omega_\varepsilon}(\mathbf{y})}{\mu_O} + \sum_{l=1}^{M}\frac{\chi_{\omega_\varepsilon^{(l)}}(\mathbf{y})}{\mu_{I_l}}\right)\int_{\partial\Omega}\log|\mathbf{x}-\mathbf{y}|\,dS_{\mathbf{x}}. \tag{7.8}$$

First-order approximation for R_ε. If we allow $\mathbf{y}$ to be located inside one of the M inclusions, then we assume $\mathbf{y}\in\omega_\varepsilon^{(m)}$, where m is fixed, $1\le m\le M$. We first rewrite the boundary conditions for R_ε.

The boundary condition for R_ε when $\mathbf{x} \in \partial\Omega$, $\mathbf{y} \in \Omega_\varepsilon \cup \omega_\varepsilon^{(m)}$ is equivalent to

$$\mu_O \frac{\partial R_\varepsilon}{\partial n_{\mathbf{x}}}(\mathbf{x},\mathbf{y}) = \chi_{\Omega_\varepsilon}(\mathbf{y}) \left\{ -\frac{1}{2\pi} \frac{\partial(\log|\mathbf{x}-\mathbf{y}|)}{\partial n_{\mathbf{x}}} + \frac{1}{|\partial\Omega|} \right\}$$
$$+\chi_{\omega_\varepsilon^{(m)}}(\mathbf{y}) \left\{ -\frac{\mu_O}{2\pi} \frac{\partial}{\partial n_{\mathbf{x}}} \left[\left\{ \frac{1}{\mu_{I_m}} - \frac{1}{\mu_O} \right\} \log|\mathbf{x}-\mathbf{y}| + \frac{1}{\mu_O} \log|\mathbf{x}-\mathbf{y}| \right] + \frac{1}{|\partial\Omega|} \right\} .$$

Using scaled variables, we can also rewrite the transmission conditions (7.6) on $\partial\omega_\varepsilon^{(q)}$, $q \neq m$, as

$$\mu_{I_q} \frac{\partial R_\varepsilon}{\partial n_{\mathbf{x}}}(\mathbf{x},\mathbf{y})\big|_{\mathbf{x}\in\partial\omega_\varepsilon^{(q)-}} - \mu_O \frac{\partial R_\varepsilon}{\partial n_{\mathbf{x}}}(\mathbf{x},\mathbf{y})\big|_{\mathbf{x}\in\partial\omega_\varepsilon^{(q)+}}$$
$$= -\chi_{\Omega_\varepsilon}(\mathbf{y}) \frac{\mu_{I_q}-\mu_O}{2\pi\mu_O} \frac{\partial(\log|\boldsymbol{\xi}_q - \boldsymbol{\eta}_q|)}{\partial n_{\mathbf{x}}}$$
$$+\chi_{\omega_\varepsilon^{(m)}}(\mathbf{y}) \Big\{ -\frac{\mu_{I_q}-\mu_O}{2\pi} \frac{\partial}{\partial n_{\mathbf{x}}} \left[\left(\frac{1}{\mu_{I_m}} - \frac{1}{\mu_O} \right) \log|\mathbf{x}-\mathbf{y}| + \frac{1}{\mu_O} \log|\boldsymbol{\xi}_q - \boldsymbol{\eta}_q| \right] \Big\} ,$$

where $\mathbf{y} \in \Omega_\varepsilon \cup \omega_\varepsilon^{(m)}$. Finally, the transmission condition (7.6) on the mth inclusion becomes

$$\mu_{I_m} \frac{\partial R_\varepsilon}{\partial n_{\mathbf{x}}}(\mathbf{x},\mathbf{y})\big|_{\mathbf{x}\in\partial\omega_\varepsilon^{(m)-}} - \mu_O \frac{\partial R_\varepsilon}{\partial n_{\mathbf{x}}}(\mathbf{x},\mathbf{y})\big|_{\mathbf{x}\in\partial\omega_\varepsilon^{(m)+}}$$
$$= -\frac{\mu_{I_m}-\mu_O}{2\pi} \left(\frac{\chi_{\Omega_\varepsilon}(\mathbf{y})}{\mu_O} + \frac{\chi_{\omega_\varepsilon^{(m)}}(\mathbf{y})}{\mu_{I_m}} \right) \frac{\partial(\log|\boldsymbol{\xi}_m - \boldsymbol{\eta}_m|)}{\partial n_{\mathbf{x}}} ,$$

for $\mathbf{y} \in \Omega_\varepsilon \cup \omega_\varepsilon^{(m)}$. We therefore have the representation

$$R_\varepsilon(\mathbf{x},\mathbf{y}) = \chi_{\Omega_\varepsilon}(\mathbf{y}) \Big\{ R^{(\Omega)}(\mathbf{x},\mathbf{y}) + \sum_{k=1}^{M} h_N^{(k,O)}(\boldsymbol{\xi}_k, \boldsymbol{\eta}_k) \Big\}$$
$$+\chi_{\omega_\varepsilon^{(m)}}(\mathbf{y}) \Big\{ R^{(\Omega)}(\mathbf{x},\mathbf{y}) + h_N^{(m,I)}(\boldsymbol{\xi}_m, \boldsymbol{\eta}_m) \tag{7.9}$$
$$+ \sum_{\substack{k\neq m \\ 1\leq k\leq M}} h_N^{(k,O)}(\boldsymbol{\xi}_k, \boldsymbol{\eta}_k) \Big\} + Z_\varepsilon(\mathbf{x},\mathbf{y}) .$$

The construction of boundary layer terms. The function Z_ε, given in (7.9), is harmonic for $\mathbf{x} \in \bigcup_{l=1}^{M} \omega_\varepsilon^{(l)} \cup \Omega_\varepsilon$, $\mathbf{y} \in \Omega_\varepsilon \cup \omega_\varepsilon^{(m)}$ and is continuous across the boundaries of the small inclusions. The normal derivative of Z_ε on the exterior boundary is

$$\mu_O \frac{\partial Z_\varepsilon}{\partial n_{\mathbf{x}}}(\mathbf{x},\mathbf{y}) = -\mu_O \chi_{\Omega_\varepsilon}(\mathbf{y}) \sum_{k=1}^{M} \frac{\partial h_N^{(k,O)}}{\partial n_{\mathbf{x}}}(\boldsymbol{\xi}_k, \boldsymbol{\eta}_k)$$
$$-\mu_O \chi_{\omega_\varepsilon^{(m)}}(\mathbf{y}) \frac{\partial}{\partial n_{\mathbf{x}}} \Big[\frac{1}{2\pi} \left(\frac{1}{\mu_{I_m}} - \frac{1}{\mu_O} \right) \log|\mathbf{x}-\mathbf{y}|$$
$$+h_N^{(m,I)}(\boldsymbol{\xi}_m, \boldsymbol{\eta}_m) + \sum_{\substack{k\neq m \\ 1\leq k\leq M}} h_N^{(k,O)}(\boldsymbol{\xi}_k, \boldsymbol{\eta}_k) \Big] ,$$

for $\mathbf{x} \in \partial\Omega$, and the traction transmission conditions for Z_ε on the small inclusions are

$$\mu_{I_m} \frac{\partial Z_\varepsilon}{\partial n_{\mathbf{x}}}(\mathbf{x},\mathbf{y})\big|_{\mathbf{x}\in\partial\omega_\varepsilon^{(m)-}} - \mu_O \frac{\partial Z_\varepsilon}{\partial n_{\mathbf{x}}}(\mathbf{x},\mathbf{y})\big|_{\mathbf{x}\in\partial\omega_\varepsilon^{(m)+}}$$
$$= -(\mu_{I_m} - \mu_O)(\chi_{\Omega_\varepsilon}(\mathbf{y}) + \chi_{\omega_\varepsilon^{(m)}}(\mathbf{y}))\frac{\partial}{\partial n_{\mathbf{x}}}\Big\{R^{(\Omega)}(\mathbf{x},\mathbf{y}) + \sum_{\substack{k\neq m\\ 1\leq k\leq M}} h_N^{(k,O)}(\boldsymbol{\xi}_k,\boldsymbol{\eta}_k)\Big\}$$

and

$$\mu_{I_q} \frac{\partial Z_\varepsilon}{\partial n_{\mathbf{x}}}(\mathbf{x},\mathbf{y})\big|_{\mathbf{x}\in\partial\omega_\varepsilon^{(q)-}} - \mu_O \frac{\partial Z_\varepsilon}{\partial n_{\mathbf{x}}}(\mathbf{x},\mathbf{y})\big|_{\mathbf{x}\in\partial\omega_\varepsilon^{(q)+}}$$
$$= -(\mu_{I_q} - \mu_O)\Big\{\chi_{\Omega_\varepsilon}(\mathbf{y})\frac{\partial}{\partial n_{\mathbf{x}}}\Big[R^{(\Omega)}(\mathbf{x},\mathbf{y}) + \sum_{\substack{k\neq q\\ 1\leq k\leq M}} h_N^{(k,O)}(\boldsymbol{\xi}_k,\boldsymbol{\eta}_k)\Big]$$
$$+\chi_{\omega_\varepsilon^{(m)}}(\mathbf{y})\frac{\partial}{\partial n_{\mathbf{x}}}\Big[\frac{1}{2\pi}\Big(\frac{1}{\mu_{I_m}} - \frac{1}{\mu_O}\Big)\log|\mathbf{x}-\mathbf{y}| + R^{(\Omega)}(\mathbf{x},\mathbf{y})$$
$$+h_N^{(m,I)}(\boldsymbol{\xi}_m,\boldsymbol{\eta}_m) + \sum_{\substack{k\neq m\\ k\neq q\\ 1\leq k\leq M}} h_N^{(k,O)}(\boldsymbol{\xi}_k,\boldsymbol{\eta}_k)\Big]\Big\},$$

where $1 \leq q \leq M$, $q \neq m$, and $\mathbf{y} \in \Omega_\varepsilon \cup \omega_\varepsilon^{(m)}$. By expanding the first-order derivatives of the logarithmic term near $\chi_{\omega_\varepsilon^{(m)}}(\mathbf{y})$, about $\mathbf{y} = \mathbf{O}^{(m)}$ up to $O(\varepsilon^2)$, in the exterior boundary condition and the traction condition on $\partial\omega_\varepsilon^{(q)}$, $q \neq m$, we obtain

$$\mu_O \frac{\partial Z_\varepsilon}{\partial n_{\mathbf{x}}}(\mathbf{x},\mathbf{y}) = -\mu_O\chi_{\Omega_\varepsilon}(\mathbf{y})\sum_{k=1}^{M}\frac{\partial h_N^{(k,O)}}{\partial n_{\mathbf{x}}}(\boldsymbol{\xi}_k,\boldsymbol{\eta}_k)$$
$$-\mu_O\chi_{\omega_\varepsilon^{(m)}}(\mathbf{y})\Big\{\frac{\partial}{\partial n_{\mathbf{x}}}\Big[\frac{1}{2\pi}\Big(\frac{1}{\mu_{I_m}} - \frac{1}{\mu_O}\Big)\Big(\log|\mathbf{x}-\mathbf{O}^{(m)}|$$
$$-\frac{(\mathbf{y}-\mathbf{O}^{(m)})\cdot(\mathbf{x}-\mathbf{O}^{(m)})}{|\mathbf{x}-\mathbf{O}^{(m)}|^2}\Big) + h_N^{(m,I)}(\boldsymbol{\xi}_m,\boldsymbol{\eta}_m) + \sum_{\substack{k\neq m\\ 1\leq k\leq M}} h_N^{(k,O)}(\boldsymbol{\xi}_k,\boldsymbol{\eta}_k)\Big] + O(\varepsilon^2)\Big\},$$

for $\mathbf{x} \in \partial\Omega$, and

$$\mu_{I_q} \frac{\partial Z_\varepsilon}{\partial n_{\mathbf{x}}}(\mathbf{x},\mathbf{y})\big|_{\mathbf{x}\in\partial\omega_\varepsilon^{(q)-}} - \mu_O \frac{\partial Z_\varepsilon}{\partial n_{\mathbf{x}}}(\mathbf{x},\mathbf{y})\big|_{\mathbf{x}\in\partial\omega_\varepsilon^{(q)+}}$$
$$= -(\mu_{I_q} - \mu_O)\Big[\chi_{\Omega_\varepsilon}(\mathbf{y})\frac{\partial}{\partial n_{\mathbf{x}}}\Big\{R^{(\Omega)}(\mathbf{x},\mathbf{y}) + \sum_{\substack{k\neq q\\ 1\leq k\leq M}} h_N^{(k,O)}(\boldsymbol{\xi}_k,\boldsymbol{\eta}_k)\Big\}$$
$$+\chi_{\omega_\varepsilon^{(m)}}(\mathbf{y})\Big\{\frac{\partial}{\partial n_{\mathbf{x}}}\Big[\frac{1}{2\pi}\Big(\frac{1}{\mu_{I_m}} - \frac{1}{\mu_O}\Big)\Big(\log|\mathbf{x}-\mathbf{O}^{(m)}| - \frac{(\mathbf{y}-\mathbf{O}^{(m)})\cdot(\mathbf{x}-\mathbf{O}^{(m)})}{|\mathbf{x}-\mathbf{O}^{(m)}|^2}\Big)$$
$$+R^{(\Omega)}(\mathbf{x},\mathbf{y}) + h_N^{(m,I)}(\boldsymbol{\xi}_m,\boldsymbol{\eta}_m) + \sum_{\substack{k\neq m\\ k\neq q\\ 1\leq k\leq M}} h_N^{(k,O)}(\boldsymbol{\xi}_k,\boldsymbol{\eta}_k)\Big] + O(\varepsilon^2)\Big\}\Big],$$

where $\mathbf{y} \in \Omega_\varepsilon \cup \omega_\varepsilon^{(m)}$. Next we apply Lemma 6.1, in order to rewrite the above boundary condition on $\partial\Omega$ for Z_ε as

$$\mu_O \frac{\partial Z_\varepsilon}{\partial n_{\mathbf{x}}}(\mathbf{x},\mathbf{y}) = -\mu_O \chi_{\Omega_\varepsilon}(\mathbf{y}) \sum_{k=1}^{M} \frac{\partial}{\partial n_{\mathbf{x}}}\Big\{ \frac{\varepsilon \mathcal{D}^{(k,O)}(\boldsymbol{\eta}_k)}{2\pi\mu_O} \cdot \frac{\mathbf{x}-\mathbf{O}^{(k)}}{|\mathbf{x}-\mathbf{O}^{(k)}|^2}$$
$$+O\Big(\sum_{k=1}^{M} \varepsilon^3 (|\mathbf{y}-\mathbf{O}^{(k)}|+\varepsilon)^{-1}\Big)\Big\} + \chi_{\omega_\varepsilon^{(m)}}(\mathbf{y})\Big\{ -\mu_O \frac{\partial}{\partial n_{\mathbf{x}}}\Big[\frac{\varepsilon \mathcal{D}^{(m,I)}(\boldsymbol{\eta}_m)}{2\pi\mu_O} \cdot \frac{\mathbf{x}-\mathbf{O}^{(m)}}{|\mathbf{x}-\mathbf{O}^{(m)}|^2}$$
$$+(\mu_{I_m}^{-1} - \mu_O^{-1})\log\varepsilon + \sum_{\substack{k\neq m\\ 1\le k\le M}} \frac{\varepsilon \mathcal{D}^{(k,O)}(\boldsymbol{\eta}_k)}{2\pi\mu_O} \cdot \frac{\mathbf{x}-\mathbf{O}^{(k)}}{|\mathbf{x}-\mathbf{O}^{(k)}|^2}\Big] + O(\varepsilon^2)\Big\},$$

for $\mathbf{x} \in \partial\Omega$, $\mathbf{y} \in \Omega_\varepsilon \cup \omega_\varepsilon^{(m)}$. The same lemma in combination with the Taylor expansion about $\mathbf{x} = \mathbf{O}^{(m)}$ of the first-order derivatives for the function $R^{(\Omega)}$ leads to

$$\mu_{I_m} \frac{\partial Z_\varepsilon}{\partial n_{\mathbf{x}}}(\mathbf{x},\mathbf{y})\big|_{\mathbf{x}\in\partial\omega_\varepsilon^{(m)-}} - \mu_O \frac{\partial Z_\varepsilon}{\partial n_{\mathbf{x}}}(\mathbf{x},\mathbf{y})\big|_{\mathbf{x}\in\partial\omega_\varepsilon^{(m)+}}$$
$$= -(\chi_{\Omega_\varepsilon}(\mathbf{y}) + \chi_{\omega_\varepsilon^{(m)}}(\mathbf{y}))(\mu_{I_m} - \mu_O)\boldsymbol{n}^{(m)} \cdot \nabla_{\mathbf{x}} R^{(\Omega)}(\mathbf{O}^{(m)},\mathbf{y}) + O(\varepsilon)$$
$$\mathbf{y} \in \Omega_\varepsilon \cup \omega_\varepsilon^{(m)},$$

where $\boldsymbol{n}^{(m)}$ is the unit outward normal to the inclusion $\omega_\varepsilon^{(m)}$. Similarly, on the qth inclusion, $1 \le q \le M$, $q \neq m$, we derive

$$\mu_{I_q} \frac{\partial Z_\varepsilon}{\partial n_{\mathbf{x}}}(\mathbf{x},\mathbf{y})\big|_{\mathbf{x}\in\partial\omega_\varepsilon^{(q)-}} - \mu_O \frac{\partial Z_\varepsilon}{\partial n_{\mathbf{x}}}(\mathbf{x},\mathbf{y})\big|_{\mathbf{x}\in\partial\omega_\varepsilon^{(q)+}}$$
$$= -(\chi_{\Omega_\varepsilon}(\mathbf{y}) + \chi_{\omega_\varepsilon^{(m)}}(\mathbf{y}))(\mu_{I_q} - \mu_O)\boldsymbol{n}^{(q)} \cdot \nabla_{\mathbf{x}} R^{(\Omega)}(\mathbf{O}^{(q)},\mathbf{y}) + O(\varepsilon)$$
$$\mathbf{y} \in \Omega_\varepsilon \cup \omega_\varepsilon^{(m)}.$$

Then we approximate Z_ε by
(7.10)
$$Z_\varepsilon(\mathbf{x},\mathbf{y}) = -(\chi_{\Omega_\varepsilon}(\mathbf{y}) + \chi_{\omega_\varepsilon^{(m)}}(\mathbf{y})) \sum_{\substack{k\neq m\\ 1\le k\le M}} \varepsilon\{\mathcal{D}^{(k,O)}(\boldsymbol{\eta}_k) \cdot \nabla_{\mathbf{y}} R^{(\Omega)}(\mathbf{x},\mathbf{O}^{(k)})$$
$$+\mathcal{D}^{(k)}(\boldsymbol{\xi}_k) \cdot \nabla_{\mathbf{x}} R^{(\Omega)}(\mathbf{O}^{(k)},\mathbf{y})\} - \varepsilon\{\chi_{\Omega_\varepsilon}(\mathbf{y})\mathcal{D}^{(m,O)}(\boldsymbol{\eta}_m) \cdot \nabla_{\mathbf{y}} R^{(\Omega)}(\mathbf{x},\mathbf{O}^{(m)})$$
$$+(\chi_{\Omega_\varepsilon}(\mathbf{y}) + \chi_{\omega_\varepsilon^{(m)}}(\mathbf{y}))\mathcal{D}^{(m)}(\boldsymbol{\xi}_m) \cdot \nabla_{\mathbf{x}} R^{(\Omega)}(\mathbf{O}^{(m)},\mathbf{y})\}$$
$$-\chi_{\omega_\varepsilon^{(m)}}(\mathbf{y})\{\varepsilon\mathcal{D}^{(m,I)}(\boldsymbol{\eta}_m) \cdot \nabla_{\mathbf{y}} R^{(\Omega)}(\mathbf{x},\mathbf{O}^{(m)}) + (\mu_{I_m}^{-1} - \mu_O^{-1})\log\varepsilon\} - r_\varepsilon^{(m)}(\mathbf{x},\mathbf{y}).$$

Combined formula for N_ε. The substitution of (7.9), (7.10) into (7.2), for $\mathbf{y} \in \Omega_\varepsilon \cup \omega_\varepsilon^{(m)}$, yields
(7.11)

$$N_\varepsilon(\mathbf{x},\mathbf{y}) = \chi_{\Omega_\varepsilon}(\mathbf{y}) \Big\{ -(2\pi\mu_O)^{-1} \log|\mathbf{x}-\mathbf{y}| - R^{(\Omega)}(\mathbf{x},\mathbf{y}) - \sum_{k=1}^{M} h_N^{(k,O)}(\boldsymbol{\xi}_k, \boldsymbol{\eta}_k)$$

$$+\varepsilon \sum_{k=1}^{M} \{\mathcal{D}^{(k)}(\boldsymbol{\xi}_k) \cdot \nabla_{\mathbf{x}} R^{(\Omega)}(\mathbf{O}^{(k)}, \mathbf{y}) + \mathcal{D}^{(k,O)}(\boldsymbol{\eta}_k) \cdot \nabla_{\mathbf{y}} R^{(\Omega)}(\mathbf{x}, \mathbf{O}^{(k)})\} \Big\}$$

$$+\chi_{\omega_\varepsilon^{(m)}}(\mathbf{y}) \Big\{ -(2\pi\mu_{I_m})^{-1} \log|\mathbf{x}-\mathbf{y}| - R^{(\Omega)}(\mathbf{x},\mathbf{y}) - h_N^{(m,I)}(\boldsymbol{\xi}_m, \boldsymbol{\eta}_m)$$

$$-\sum_{\substack{k\neq m\\ 1\le k\le M}} h_N^{(k,O)}(\boldsymbol{\xi}_k, \boldsymbol{\eta}_k) + (\mu_{I_m}^{-1} - \mu_O^{-1})(2\pi)^{-1} \log\varepsilon + \varepsilon\, \mathcal{D}^{(m)}(\boldsymbol{\xi}_m) \cdot \nabla_{\mathbf{x}} R^{(\Omega)}(\mathbf{O}^{(m)}, \mathbf{y})$$

$$+\varepsilon \mathcal{D}^{(m,I)}(\boldsymbol{\eta}_m) \cdot \nabla_{\mathbf{y}} R^{(\Omega)}(\mathbf{x}, \mathbf{O}^{(m)}) + \varepsilon \sum_{\substack{k\neq m\\ 1\le k\le M}} \Big[\mathcal{D}^{(k)}(\boldsymbol{\xi}_k) \cdot \nabla_{\mathbf{x}} R^{(\Omega)}(\mathbf{O}^{(k)}, \mathbf{y})$$

$$+\mathcal{D}^{(k,O)}(\boldsymbol{\eta}_k) \cdot \nabla_{\mathbf{y}} R^{(\Omega)}(\mathbf{x}, \mathbf{O}^{(k)}) \Big] \Big\} + r_\varepsilon^{(m)}(\mathbf{x},\mathbf{y}) \,.$$

Then, by the definitions of $N^{(\Omega)}$, $\mathcal{N}^{(m,O)}$, and $\mathcal{N}^{(m,I)}$, this is equivalent to

$$N_\varepsilon(\mathbf{x},\mathbf{y}) = \chi_{\Omega_\varepsilon}(\mathbf{y}) \Big\{ N^{(\Omega)}(\mathbf{x},\mathbf{y}) + \sum_{k=1}^{M} \mathcal{N}^{(k,O)}(\boldsymbol{\xi}_k, \boldsymbol{\eta}_k) + M(2\pi\mu_O)^{-1} \log(\varepsilon^{-1}|\mathbf{x}-\mathbf{y}|)$$

$$+\varepsilon \sum_{k=1}^{M} \{\mathcal{D}^{(k)}(\boldsymbol{\xi}_k) \cdot \nabla_{\mathbf{x}} R^{(\Omega)}(\mathbf{O}^{(k)}, \mathbf{y}) + \mathcal{D}^{(k,O)}(\boldsymbol{\eta}_k) \cdot \nabla_{\mathbf{y}} R^{(\Omega)}(\mathbf{x}, \mathbf{O}^{(k)})\} \Big\}$$

$$+\sum_{j=1}^{M} \chi_{\omega_\varepsilon^{(j)}}(\mathbf{y}) \Big\{ N^{(\Omega)}(\mathbf{x},\mathbf{y}) + \mathcal{N}^{(j,I)}(\boldsymbol{\xi}_j, \boldsymbol{\eta}_j) + \sum_{\substack{k\neq j\\ 1\le k\le M}} \mathcal{N}^{(k,O)}(\boldsymbol{\xi}_k, \boldsymbol{\eta}_k)$$

$$+M(2\pi\mu_O)^{-1} \log(\varepsilon^{-1}|\mathbf{x}-\mathbf{y}|) + \varepsilon \sum_{k=1}^{M} \mathcal{D}^{(k)}(\boldsymbol{\xi}_k) \cdot \nabla_{\mathbf{x}} R^{(\Omega)}(\mathbf{O}^{(k)}, \mathbf{y})$$

$$+\varepsilon \mathcal{D}^{(j,I)}(\boldsymbol{\eta}_j) \cdot \nabla_{\mathbf{y}} R^{(\Omega)}(\mathbf{x}, \mathbf{O}^{(j)}) + \varepsilon \sum_{\substack{k\neq j\\ 1\le k\le M}} \mathcal{D}^{(k,O)}(\boldsymbol{\eta}_k) \cdot \nabla_{\mathbf{y}} R^{(\Omega)}(\mathbf{x}, \mathbf{O}^{(k)}) \Big\} + r_\varepsilon(\mathbf{x},\mathbf{y}) \,.$$

Remainder estimates for the approximation of N_ε. We represent r_ε as

$$r_\varepsilon(\mathbf{x},\mathbf{y}) = \chi_{\Omega_\varepsilon}(\mathbf{y}) \mathfrak{M}_\varepsilon(\mathbf{x},\mathbf{y}) + \sum_{j=1}^{M} \chi_{\omega_\varepsilon^{(j)}}(\mathbf{y}) \mathfrak{h}_\varepsilon^{(j)}(\mathbf{x},\mathbf{y}) \,, \tag{7.12}$$

where $\mathfrak{M}_\varepsilon$ and $\mathfrak{h}_\varepsilon^{(j)}$, $j = 1, \dots, M$, are defined below. In what follows, we estimate $\mathfrak{M}_\varepsilon$ and $\mathfrak{h}_\varepsilon^{(j)}$, $j = 1, \dots, M$, in order to estimate r_ε.

Remainder estimate for the function $\mathfrak{M}_\varepsilon(\mathbf{x}, \mathbf{y})$. First, let $\mathbf{y} \in \Omega_\varepsilon$. According to (7.11), the function $\mathfrak{M}_\varepsilon$ is a solution of

$$\mu_O \Delta_{\mathbf{x}} \mathfrak{M}_\varepsilon(\mathbf{x}, \mathbf{y}) = 0, \quad \mathbf{x} \in \Omega_\varepsilon ,$$

$$\mu_{I_i} \Delta_{\mathbf{x}} \mathfrak{M}_\varepsilon(\mathbf{x}, \mathbf{y}) = 0, \quad \mathbf{x} \in \omega_\varepsilon^{(i)}, \ i = 1, \dots, M ,$$

and on the exterior contour is subject to

(7.13)

$$\mu_O \frac{\partial \mathfrak{M}_\varepsilon}{\partial n_{\mathbf{x}}}(\mathbf{x}, \mathbf{y}) = -\varepsilon \mu_O \sum_{j=1}^{M} \frac{\partial \mathcal{D}^{(j,O)}}{\partial n_{\mathbf{x}}}(\boldsymbol{\xi}_j) \cdot \nabla_{\mathbf{x}} R^{(\Omega)}(\mathbf{O}^{(j)}, \mathbf{y})$$
$$+\mu_O \sum_{j=1}^{M} \left\{ \frac{\partial h_N^{(j,O)}}{\partial n_{\mathbf{x}}}(\boldsymbol{\xi}_j, \boldsymbol{\eta}_j) - \varepsilon \mathcal{D}^{(j,O)}(\boldsymbol{\eta}_j) \cdot \frac{\partial}{\partial n_{\mathbf{x}}} \nabla_{\mathbf{y}} R^{(\Omega)}(\mathbf{x}, \mathbf{O}^{(j)}) \right\}, \qquad \mathbf{x} \in \partial\Omega ,$$

with transmission conditions on $\partial\omega_\varepsilon^{(i)}$ of the form

(7.14)

$$\mu_{I_i} \frac{\partial \mathfrak{M}_\varepsilon}{\partial n_{\mathbf{x}}}(\mathbf{x}, \mathbf{y})\big|_{\mathbf{x}\in\partial\omega_\varepsilon^{(i)-}} - \mu_O \frac{\partial \mathfrak{M}_\varepsilon}{\partial n_{\mathbf{x}}}(\mathbf{x}, \mathbf{y})\big|_{\mathbf{x}\in\partial\omega_\varepsilon^{(i)+}}$$
$$= (\mu_{I_i} - \mu_O) \frac{\partial R^{(\Omega)}}{\partial n_{\mathbf{x}}}(\mathbf{x}, \mathbf{y}) + (\mu_{I_i} - \mu_O) \sum_{\substack{j \neq i \\ 1 \le j \le M}} \frac{\partial h_N^{(j,O)}}{\partial n_{\mathbf{x}}}(\boldsymbol{\xi}_j, \boldsymbol{\eta}_j)$$
$$-\varepsilon \sum_{j=1}^{M} \left\{ \mu_{I_i} \frac{\partial \mathcal{D}^{(j)}}{\partial n_{\mathbf{x}}}(\boldsymbol{\xi}_j)\big|_{\mathbf{x}\in\partial\omega_\varepsilon^{(i)-}} - \mu_O \frac{\partial \mathcal{D}^{(j)}}{\partial n_{\mathbf{x}}}(\boldsymbol{\xi}_j)\big|_{\mathbf{x}\in\partial\omega_\varepsilon^{(i)+}} \right\} \cdot \nabla_{\mathbf{x}} R^{(\Omega)}(\mathbf{O}^{(j)}, \mathbf{y})$$
$$-(\mu_{I_i} - \mu_O)\varepsilon \sum_{j=1}^{M} \mathcal{D}^{(j,O)}(\boldsymbol{\eta}_j) \cdot \nabla_{\mathbf{y}} \frac{\partial R^{(\Omega)}}{\partial n_{\mathbf{x}}}(\mathbf{x}, \mathbf{O}^{(j)}) ,$$

$$\mathfrak{M}_\varepsilon(\mathbf{x}, \mathbf{y})\big|_{\mathbf{x}\in\partial\omega_\varepsilon^{(i)-}} = \mathfrak{M}_\varepsilon(\mathbf{x}, \mathbf{y})\big|_{\mathbf{x}\in\partial\omega_\varepsilon^{(i)+}} ,$$

for $1 \le i \le M$, where $\mathbf{y} \in \Omega_\varepsilon$ and

$$\int_{\partial\Omega} \mathfrak{M}_\varepsilon(\mathbf{x}, \mathbf{y}) dS_{\mathbf{x}} = 0 . \tag{7.15}$$

Before estimating the boundary conditions we observe that

$$\int_{\partial\Omega} \mu_O \frac{\partial \mathfrak{M}_\varepsilon}{\partial n_{\mathbf{x}}}(\mathbf{x}, \mathbf{y})\, dS_{\mathbf{x}} = 0 , \tag{7.16}$$

$$\int_{\partial\omega_\varepsilon^{(i)}} \left(\mu_{I_i} \frac{\partial \mathfrak{M}_\varepsilon}{\partial n_{\mathbf{x}}}(\mathbf{x}, \mathbf{y})\big|_{\mathbf{x}\in\partial\omega_\varepsilon^{(i)-}} - \mu_O \frac{\partial \mathfrak{M}_\varepsilon}{\partial n_{\mathbf{x}}}(\mathbf{x}, \mathbf{y})\big|_{\mathbf{x}\in\partial\omega_\varepsilon^{(i)+}} \right) dS_{\mathbf{x}} = 0 , \tag{7.17}$$

for $i = 1, \dots, M$.

Estimation of the right-hand side of (7.13) *on* $\partial\Omega$. Since $\mathbf{x} \in \partial\Omega$, $|\mathbf{x} - \mathbf{O}^{(j)}| \ge 1$ for $j = 1, \dots, M$, and by Lemma 4.2, the estimate

$$\varepsilon \mu_O \Big| \sum_{j=1}^{M} \frac{\partial \mathcal{D}^{(j,O)}}{\partial n_{\mathbf{x}}}(\boldsymbol{\xi}_j) \cdot \nabla_{\mathbf{x}} R^{(\Omega)}(\mathbf{O}^{(j)}, \mathbf{y}) \Big| \le \text{const}\ \varepsilon^2$$

holds for $\mathbf{x} \in \partial\Omega$, $\mathbf{y} \in \Omega_\varepsilon$. Using Lemma 6.1(a), one has for $\mathbf{x} \in \partial\Omega$,

$$\mu_O \Big| \frac{\partial h_N^{(j,O)}}{\partial n_{\mathbf{x}}}(\boldsymbol{\xi}_j, \boldsymbol{\eta}_j) - \varepsilon \mathcal{D}^{(j,O)}(\boldsymbol{\eta}_j) \cdot \frac{\partial}{\partial n_{\mathbf{x}}} \nabla_{\mathbf{y}} R^{(\Omega)}(\mathbf{x}, \mathbf{O}^{(j)}) \Big|$$
$$= \mu_O \Big| \frac{\partial h_N^{(j,O)}}{\partial n_{\mathbf{x}}}(\boldsymbol{\xi}_j, \boldsymbol{\eta}_j) - \frac{\varepsilon \mathcal{D}^{(j,O)}(\boldsymbol{\eta}_j)}{2\pi \mu_O} \cdot \frac{\partial}{\partial n_{\mathbf{x}}} \Big(\frac{\mathbf{x} - \mathbf{O}^{(j)}}{|\mathbf{x} - \mathbf{O}^{(j)}|^2} \Big) \Big|$$
$$\leq \text{const } \varepsilon^3 |\mathbf{x} - \mathbf{O}^{(j)}|^{-2} (|\mathbf{y} - \mathbf{O}^{(j)}| + \varepsilon)^{-1} \leq \text{const } \varepsilon^3 (|\mathbf{y} - \mathbf{O}^{(j)}| + \varepsilon)^{-1}$$

where we have also made use of the boundary condition (3.1) for $R^{(\Omega)}$. The previous two inequalities then give

$$\mu_O \Big| \frac{\partial \mathfrak{M}_\varepsilon}{\partial n_{\mathbf{x}}}(\mathbf{x}, \mathbf{y}) \Big| \leq \text{const } \varepsilon^2 , \quad \mathbf{x} \in \partial\Omega, \ \mathbf{y} \in \Omega_\varepsilon . \tag{7.18}$$

Estimate for the right-hand side of (7.14) *on* $\partial\omega_\varepsilon^{(i)}$, $i = 1, \dots, M$. The regular part $R^{(\Omega)}$ of $N^{(\Omega)}$ is smooth for $\mathbf{x}, \mathbf{y} \in \Omega$, and we can expand the first-order derivatives of this function about the center of the small inclusion $\omega_\varepsilon^{(i)}$ to obtain

$$\Big| (\mu_{I_i} - \mu_O) \frac{\partial R^{(\Omega)}}{\partial n_{\mathbf{x}}}(\mathbf{x}, \mathbf{y}) - \varepsilon \Big\{ \mu_{I_i} \frac{\partial \mathcal{D}^{(i,I)}}{\partial n_{\mathbf{x}}}(\boldsymbol{\xi}_i) \big|_{\mathbf{x} \in \partial\omega_\varepsilon^{(i)}-} -$$
$$- \mu_O \frac{\partial \mathcal{D}^{(i,O)}}{\partial n_{\mathbf{x}}}(\boldsymbol{\xi}_i) \big|_{\mathbf{x} \in \partial\omega_\varepsilon^{(i)}+} \Big\} \cdot \nabla_{\mathbf{x}} R^{(\Omega)}(\mathbf{O}^{(i)}, \mathbf{y}) \Big|$$
$$= \Big| (\mu_{I_i} - \mu_O) \boldsymbol{n}^{(i)} \cdot (\nabla_{\mathbf{x}} R^{(\Omega)}(\mathbf{x}, \mathbf{y}) - \nabla_{\mathbf{x}} R^{(\Omega)}(\mathbf{O}^{(i)}, \mathbf{y})) \Big|$$
$$\leq \text{const } \varepsilon , \quad \mathbf{x} \in \partial\omega_\varepsilon^{(i)}, \ \mathbf{y} \in \Omega_\varepsilon . \tag{7.19}$$

With the use of Lemma 4.2, we have

$$\varepsilon \Big| \sum_{\substack{j \neq i \\ 1 \leq j \leq M}} \Big\{ \mu_{I_i} \frac{\partial \mathcal{D}^{(j,O)}}{\partial n_{\mathbf{x}}}(\boldsymbol{\xi}_j) \big|_{\mathbf{x} \in \partial\omega_\varepsilon^{(i)}-} - \mu_O \frac{\partial \mathcal{D}^{(j,O)}}{\partial n_{\mathbf{x}}}(\boldsymbol{\xi}_j) \big|_{\mathbf{x} \in \partial\omega_\varepsilon^{(i)}+} \Big\} \cdot \nabla_{\mathbf{x}} R^{(\Omega)}(\mathbf{O}^{(j)}, \mathbf{y}) \Big|$$
$$\leq \text{const } \varepsilon^2, \quad \text{for } \mathbf{x} \in \partial\omega_\varepsilon^{(i)}, \ \mathbf{y} \in \Omega_\varepsilon , \tag{7.20}$$

and the same lemma also gives

$$\Big| \varepsilon \sum_{j=1}^{M} \mathcal{D}^{(j,O)}(\boldsymbol{\eta}_j) \cdot \frac{\partial \nabla_{\mathbf{y}} R^{(\Omega)}}{\partial n_{\mathbf{x}}}(\mathbf{x}, \mathbf{O}^{(j)}) \Big|$$
$$\leq \text{const} \sum_{j=1}^{M} \varepsilon^2 (|\mathbf{y} - \mathbf{O}^{(j)}| + \varepsilon)^{-1} , \quad \text{for } \mathbf{x} \in \partial\omega_\varepsilon^{(i)}, \mathbf{y} \in \Omega_\varepsilon . \tag{7.21}$$

Due to Lemma 6.1(a),

$$\Big| \sum_{\substack{j \neq i \\ 1 \leq j \leq M}} \frac{\partial h_N^{(j,O)}}{\partial n_{\mathbf{x}}}(\boldsymbol{\xi}, \boldsymbol{\eta}) \Big| \leq \text{const} \sum_{\substack{j \neq i \\ 1 \leq j \leq M}} \varepsilon^2 (|\mathbf{y} - \mathbf{O}^{(j)}| + \varepsilon)^{-1}, \quad \text{for } \mathbf{x} \in \partial\omega_\varepsilon^{(i)}, \ \mathbf{y} \in \Omega_\varepsilon .$$

Therefore, this estimate with (7.19), (7.20), and (7.21) leads to

$$\Big| \mu_{I_i} \frac{\partial \mathfrak{M}_\varepsilon}{\partial n_{\mathbf{x}}}(\mathbf{x}, \mathbf{y}) \big|_{\mathbf{x} \in \partial\omega_\varepsilon^{(i)}-} - \mu_O \frac{\partial \mathfrak{M}_\varepsilon}{\partial n_{\mathbf{x}}}(\mathbf{x}, \mathbf{y}) \big|_{\mathbf{x} \in \partial\omega_\varepsilon^{(i)}+} \Big|$$
$$\leq \text{const } \varepsilon, \quad \mathbf{x} \in \partial\omega_\varepsilon^{(i)}, \ i = 1, \dots, M, \ \mathbf{y} \in \Omega_\varepsilon .$$

Then, by Lemma 5.1 and the preceding estimate with (7.15)–(7.18), we obtain

$$|\mathfrak{M}_\varepsilon(\mathbf{x},\mathbf{y})| \le \text{const}\ \varepsilon^2, \quad \mathbf{x} \in \bigcup_{l=1}^{M} \omega_\varepsilon^{(l)} \cup \Omega_\varepsilon,\ \mathbf{y} \in \Omega_\varepsilon\ . \tag{7.22}$$

Remainder estimate for $\mathfrak{h}_\varepsilon^{(j)}$, $j = 1, \dots, M$. Let $\mathbf{y} \in \omega_\varepsilon^{(j)}$, where j is fixed, $1 \le j \le M$. Then by (7.1), the remainder term $\mathfrak{h}_\varepsilon^{(j)}$ solves

$$\mu_O \Delta_\mathbf{x} \mathfrak{h}_\varepsilon^{(j)}(\mathbf{x},\mathbf{y}) = 0, \quad \mathbf{x} \in \Omega_\varepsilon\ ,$$

$$\mu_{I_i} \Delta_\mathbf{x} \mathfrak{h}_\varepsilon^{(j)}(\mathbf{x},\mathbf{y}) = 0, \quad \mathbf{x} \in \omega_\varepsilon^{(i)},\ i = 1, \dots, M\ ,$$

and satisfies on the exterior boundary

(7.23)

$$\mu_O \frac{\partial \mathfrak{h}_\varepsilon^{(j)}}{\partial n_\mathbf{x}}(\mathbf{x},\mathbf{y}) = -\mu_O \Bigg\{ \frac{\partial \mathcal{N}^{(j,I)}}{\partial n_\mathbf{x}}(\boldsymbol{\xi}_j, \boldsymbol{\eta}_j) + \frac{\partial}{\partial n_\mathbf{x}}((2\pi\mu_O)^{-1} \log(\varepsilon^{-1}|\mathbf{x}-\mathbf{y}|))$$

$$- \sum_{\substack{k \ne m \\ 1 \le k \le M}} \frac{\partial h_N^{(k,O)}}{\partial n_\mathbf{x}}(\boldsymbol{\xi}_k, \boldsymbol{\eta}_k) + \varepsilon \frac{\partial \mathcal{D}^{(j,O)}}{\partial n_\mathbf{x}}(\boldsymbol{\xi}_j) \cdot \nabla_\mathbf{x} R^{(\Omega)}(\mathbf{O}^{(j)}, \mathbf{y})$$

$$+ \varepsilon \mathcal{D}^{(j,I)}(\boldsymbol{\eta}_j) \cdot \frac{\partial}{\partial n_\mathbf{x}} \nabla_\mathbf{y} R^{(\Omega)}(\mathbf{x}, \mathbf{O}^{(j)}) + \varepsilon \sum_{\substack{k \ne j \\ 1 \le k \le M}} \Bigg[\frac{\partial \mathcal{D}^{(k,O)}}{\partial n_\mathbf{x}}(\boldsymbol{\xi}_k) \cdot \nabla_\mathbf{x} R^{(\Omega)}(\mathbf{O}^{(k)}, \mathbf{y})$$

$$+ \mathcal{D}^{(k,O)}(\boldsymbol{\eta}_k) \cdot \frac{\partial}{\partial n_\mathbf{x}} \nabla_\mathbf{y} R^{(\Omega)}(\mathbf{x}, \mathbf{O}^{(j)}) \Bigg] \Bigg\}, \quad \mathbf{x} \in \partial\Omega\ .$$

Also, on the boundary of the ith inclusion, $\mathfrak{h}_\varepsilon^{(j)}$ is subject to

(7.24)

$$\mu_{I_i} \frac{\partial \mathfrak{h}_\varepsilon^{(j)}}{\partial n_\mathbf{x}}(\mathbf{x},\mathbf{y})\big|_{\mathbf{x} \in \partial\omega_\varepsilon^{(i)-}} - \mu_O \frac{\partial \mathfrak{h}_\varepsilon^{(j)}}{\partial n_\mathbf{x}}(\mathbf{x},\mathbf{y})\big|_{\mathbf{x} \in \partial\omega_\varepsilon^{(i)+}}$$

$$= (\mu_{I_i} - \mu_O) \Bigg[\frac{\partial R^{(\Omega)}}{\partial n_\mathbf{x}}(\mathbf{x},\mathbf{y}) - \varepsilon \mathcal{D}^{(j,I)}(\boldsymbol{\eta}_j) \cdot \frac{\partial}{\partial n_\mathbf{x}} \nabla_\mathbf{y} R^{(\Omega)}(\mathbf{x}, \mathbf{O}^{(j)})$$

$$- \varepsilon \sum_{\substack{k \ne j \\ 1 \le k \le M}} \mathcal{D}^{(k,O)}(\boldsymbol{\eta}_k) \cdot \frac{\partial}{\partial n_\mathbf{x}} \nabla_\mathbf{y} R^{(\Omega)}(\mathbf{x}, \mathbf{O}^{(k)}) + \sum_{\substack{k \ne j \\ 1 \le k \le M}} (1 - \delta_{ik}) \frac{\partial h_N^{(k,O)}}{\partial n_\mathbf{x}}(\boldsymbol{\xi}_k, \boldsymbol{\eta}_k)$$

$$- (1 - \delta_{ij})(2\pi\mu_O)^{-1} \frac{\partial}{\partial n_\mathbf{x}}(\log(\varepsilon^{-1}|\mathbf{x}-\mathbf{y}|)) - (1 - \delta_{ij}) \frac{\partial \mathcal{N}^{(j,I)}}{\partial n_\mathbf{x}}(\boldsymbol{\xi}_j, \boldsymbol{\eta}_j) \Bigg]$$

$$- \varepsilon \Bigg\{ \mu_{I_i} \frac{\partial \mathcal{D}^{(j)}}{\partial n_\mathbf{x}}(\boldsymbol{\xi}_j)\big|_{\mathbf{x} \in \partial\omega_\varepsilon^{(i)-}} - \mu_O \frac{\partial \mathcal{D}^{(j)}}{\partial n_\mathbf{x}}(\boldsymbol{\xi}_j)\big|_{\mathbf{x} \in \partial\omega_\varepsilon^{(i)+}} \Bigg\} \cdot \nabla_\mathbf{x} R^{(\Omega)}(\mathbf{O}^{(j)}, \mathbf{y})$$

$$- \varepsilon \sum_{\substack{k \ne j \\ 1 \le k \le M}} \Bigg\{ \mu_{I_i} \frac{\partial \mathcal{D}^{(k)}}{\partial n_\mathbf{x}}(\boldsymbol{\xi}_k)\big|_{\mathbf{x} \in \partial\omega_\varepsilon^{(i)-}} - \mu_O \frac{\partial \mathcal{D}^{(k)}}{\partial n_\mathbf{x}}(\boldsymbol{\xi}_k)\big|_{\mathbf{x} \in \partial\omega_\varepsilon^{(i)+}} \Bigg\} \cdot \nabla_\mathbf{x} R^{(\Omega)}(\mathbf{O}^{(k)}, \mathbf{y})\ ,$$

$$\mathfrak{h}_\varepsilon^{(j)}(\mathbf{x},\mathbf{y})\big|_{\mathbf{x} \in \partial\omega_\varepsilon^{(i)-}} = \mathfrak{h}_\varepsilon^{(j)}(\mathbf{x},\mathbf{y})\big|_{\mathbf{x} \in \partial\omega_\varepsilon^{(i)+}}\ ,$$

for $i = 1, \dots, M$, where $\mathbf{y} \in \omega_\varepsilon^{(j)}$ and

$$\int_{\partial\Omega} \mathfrak{h}_\varepsilon^{(j)}(\mathbf{x},\mathbf{y}) dS_\mathbf{x} = 0\ . \tag{7.25}$$

From this problem we can see

$$(7.26)\qquad \int_{\partial\Omega} \mu_O \frac{\partial \mathfrak{h}_\varepsilon^{(j)}}{\partial n_{\mathbf{x}}}(\mathbf{x},\mathbf{y})\, dS_{\mathbf{x}} = 0\,,$$

$$(7.27)\qquad \int_{\partial\omega_\varepsilon^{(j)}} \left(\mu_{I_j} \frac{\partial \mathfrak{h}_\varepsilon^{(j)}}{\partial n_{\mathbf{x}}}(\mathbf{x},\mathbf{y})\big|_{\mathbf{x}\in\partial\omega_\varepsilon^{(j)-}} - \mu_O \frac{\partial \mathfrak{h}_\varepsilon^{(j)}}{\partial n_{\mathbf{x}}}(\mathbf{x},\mathbf{y})\big|_{\mathbf{x}\in\partial\omega_\varepsilon^{(j)+}} \right) dS_{\mathbf{x}} = 0\,,$$

for $i = 1,\dots,M$.

Before estimating the discrepancies in the boundary conditions, we note for $\mathbf{y}\in\omega_\varepsilon^{(j)}$, by Lemma 4.2

$$(7.28)\qquad \Big| \sum_{\substack{k\neq j\\ 1\le k\le M}} \mathcal{D}^{(k,O)}(\boldsymbol{\eta}_k)\cdot \frac{\partial}{\partial n_{\mathbf{x}}} \nabla_{\mathbf{y}} R^{(\Omega)}(\mathbf{x},\mathbf{O}^{(k)})\Big| \le \text{const}\ \varepsilon \qquad \text{holds for } \mathbf{x}\in\partial\Omega_\varepsilon\,,$$

whereas the same lemma in conjunction with Lemma 6.1(a) leads to

$$(7.29)\qquad \Big| \frac{\partial h_N^{(k,O)}}{\partial n_{\mathbf{x}}}(\boldsymbol{\xi}_k,\boldsymbol{\eta}_k)\Big| \le \text{const}\ \varepsilon^2\,, \qquad \text{for } \mathbf{x}\in \bigcup_{\substack{i\neq k\\ 1\le i\le M}} \partial\omega_\varepsilon^{(i)}\cup\partial\Omega\,.$$

Estimate of the right-hand side of (7.23) *on* $\partial\Omega$. The definition of $\mathcal{N}^{(j,I)}$ in Section 3 and the boundary condition (3.1) for $R^{(\Omega)}$ give us

$$\begin{aligned}
&\mu_O\Big| \frac{\partial}{\partial n_{\mathbf{x}}}\Big\{ \mathcal{N}^{(j,I)}(\boldsymbol{\xi}_j,\boldsymbol{\eta}_j) + (2\pi\mu_O)^{-1}\log(\varepsilon^{-1}|\mathbf{x}-\mathbf{y}|) + \varepsilon\mathcal{D}^{(j,I)}(\boldsymbol{\eta}_j)\cdot\nabla_{\mathbf{y}} R^{(\Omega)}(\mathbf{x},\mathbf{O}^{(j)})\Big\}\Big| \\
&\quad = \mu_O\Big| \frac{\partial}{\partial n_{\mathbf{x}}}\{ -(2\pi)^{-1}(\mu_{I_j}^{-1}-\mu_O^{-1})\log|\mathbf{x}-\mathbf{y}| - h_N^{(j,I)}(\boldsymbol{\xi}_j,\boldsymbol{\eta}_j) \\
&\qquad + \varepsilon(2\pi\mu_O)^{-1}\mathcal{D}^{(j,I)}(\boldsymbol{\eta}_j)\cdot(\mathbf{x}-\mathbf{O}^{(j)})|\mathbf{x}-\mathbf{O}^{(j)}|^{-2}\}\Big| \qquad \text{for } \mathbf{x}\in\partial\Omega,\ \mathbf{y}\in\omega_\varepsilon^{(j)}\,.
\end{aligned}$$

Using the asymptotics of $h_N^{(j,I)}$ at infinity contained in Lemma 6.1(b), we obtain

$$(7.30)\qquad \begin{aligned}
&\mu_O\Big| \frac{\partial}{\partial n_{\mathbf{x}}}\Big\{ \mathcal{N}^{(j,I)}(\boldsymbol{\xi}_j,\boldsymbol{\eta}_j) + (2\pi\mu_O)^{-1}\log(\varepsilon^{-1}|\mathbf{x}-\mathbf{y}|) + \varepsilon\mathcal{D}^{(j,I)}(\boldsymbol{\eta}_j)\cdot\nabla_{\mathbf{y}} R^{(\Omega)}(\mathbf{x},\mathbf{O}^{(j)})\Big\}\Big| \\
&\qquad \le \text{const}\ \varepsilon^2\,, \qquad \text{for } \mathbf{x}\in\partial\Omega,\ \mathbf{y}\in\omega_\varepsilon^{(j)}\,.
\end{aligned}$$

Then from Lemma 4.2 we have

$$\begin{aligned}
&\varepsilon\mu_O\Big| \frac{\partial}{\partial n_{\mathbf{x}}}\Big\{ \mathcal{D}^{(j,O)}(\boldsymbol{\xi}_j)\cdot\nabla_{\mathbf{x}} R^{(\Omega)}(\mathbf{O}^{(j)},\mathbf{y}) + \sum_{\substack{k\neq j\\ 1\le k\le M}} \mathcal{D}^{(k,O)}(\boldsymbol{\xi}_k)\cdot\nabla_{\mathbf{x}} R^{(\Omega)}(\mathbf{O}^{(k)},\mathbf{y})\Big\}\Big| \\
&\qquad \le \text{const}\ \varepsilon^2, \quad \mathbf{x}\in\partial\Omega,\ \mathbf{y}\in\omega_\varepsilon^{(j)}\,.
\end{aligned}$$

Therefore, this along with (7.28), (7.29), and (7.30) yields

$$(7.31)\qquad \mu_O\Big| \frac{\partial \mathfrak{h}_\varepsilon^{(j)}}{\partial n_{\mathbf{x}}}(\mathbf{x},\mathbf{y})\Big| \le \text{const}\ \varepsilon^2\,, \quad \mathbf{x}\in\partial\Omega,\ \mathbf{y}\in\omega_\varepsilon^{(j)}\,.$$

Estimate for the right-hand side of (7.24) *on* $\partial\omega_\varepsilon^{(i)}$, $i = 1,\dots,M$. Consider first the situation when $i = j$. Since $R^{(\Omega)}$ is smooth for $\mathbf{x},\mathbf{y}\in\Omega$, we can take the

Taylor expansion of its first-order derivatives about $\mathbf{x} = \mathbf{O}^{(j)}$ to obtain

$$
\begin{aligned}
&\Big|(\mu_{I_j} - \mu_O)\frac{\partial R^{(\Omega)}}{\partial n_{\mathbf{x}}}(\mathbf{x}, \mathbf{y}) - \varepsilon\Big\{\mu_{I_j}\frac{\partial \mathcal{D}^{(j)}}{\partial n_{\mathbf{x}}}(\boldsymbol{\xi}_j)\big|_{\mathbf{x}\in\partial\omega_\varepsilon^{(j)}-} \\
&\qquad\qquad - \mu_O\frac{\partial \mathcal{D}^{(j)}}{\partial n_{\mathbf{x}}}(\boldsymbol{\xi}_j)\big|_{\mathbf{x}\in\partial\omega^{(j)}+}\Big\}\cdot\nabla R^{(\Omega)}(\mathbf{O}^{(j)}, \mathbf{y})\Big| \\
&= |\mu_{I_j} - \mu_O||\boldsymbol{n}^{(j)}\cdot\nabla_{\mathbf{x}}R^{(\Omega)}(\mathbf{x}, \mathbf{y}) - \boldsymbol{n}^{(j)}\cdot\nabla_{\mathbf{x}}R^{(\Omega)}(\mathbf{O}^{(j)}, \mathbf{y})\Big| \\
&\le \text{const}\ \varepsilon, \quad \mathbf{x}\in\partial\omega_\varepsilon^{(j)},\ \mathbf{y}\in\omega_\varepsilon^{(j)}\ ,
\end{aligned}
\tag{7.32}
$$

where the boundary condition (3.9) for the dipole fields of the inclusion $\partial\omega_\varepsilon^{(i)}$ was also used. Then, owing to Lemma 4.2 for $i = j$, one has

$$
\begin{aligned}
&\Big|\varepsilon(\mu_{I_j} - \mu_O)\mathcal{D}^{(j,I)}(\boldsymbol{\eta}_j)\cdot\frac{\partial}{\partial n_{\mathbf{x}}}\nabla_{\mathbf{y}}R^{(\Omega)}(\mathbf{x}, \mathbf{O}^{(j)}) \\
&\quad -\varepsilon\sum_{\substack{k\neq j\\ 1\le k\le M}}\Big\{\mu_{I_j}\frac{\partial\mathcal{D}^{(k,O)}}{\partial n_{\mathbf{x}}}(\boldsymbol{\xi}_k)\big|_{\mathbf{x}\in\partial\omega_\varepsilon^{(j)}-} - \mu_O\frac{\partial\mathcal{D}^{(k,O)}}{\partial n_{\mathbf{x}}}(\boldsymbol{\xi}_k)\big|_{\mathbf{x}\in\partial\omega^{(j)}+}\Big\}\cdot\nabla_{\mathbf{x}}R^{(\Omega)}(\mathbf{O}^{(k)}, \mathbf{y})\Big| \\
&\le \text{const}\ \varepsilon, \quad \mathbf{x}\in\partial\omega_\varepsilon^{(j)},\ \mathbf{y}\in\omega_\varepsilon^{(j)}\ .
\end{aligned}
$$

This and (7.28), (7.29), (7.32) lead to the estimate

$$
\Big|\mu_{I_j}\frac{\partial\mathfrak{h}_\varepsilon^{(j)}}{\partial n_{\mathbf{x}}}(\mathbf{x}, \mathbf{y})\big|_{\mathbf{x}\in\partial\omega_\varepsilon^{(j)}-} - \mu_O\frac{\partial\mathfrak{h}_\varepsilon^{(j)}}{\partial n_{\mathbf{x}}}(\mathbf{x}, \mathbf{y})\big|_{\mathbf{x}\in\partial\omega_\varepsilon^{(j)}+}\Big| \le \text{const}\ \varepsilon, \quad \mathbf{y}\in\omega_\varepsilon^{(j)}\ .
\tag{7.33}
$$

It remains to consider the case $i \neq j$. In this situation, we require Lemma 6.1(b) to obtain

$$
\begin{aligned}
&\Big|\frac{\partial\mathcal{N}^{(j,I)}}{\partial n_{\mathbf{x}}}(\boldsymbol{\xi}_j, \boldsymbol{\eta}_j) + (2\pi\mu_O)^{-1}\frac{\partial}{\partial n_{\mathbf{x}}}(\log(\varepsilon^{-1}|\mathbf{x}-\mathbf{y}|)) \\
&\qquad\qquad + \varepsilon\mathcal{D}^{(j,I)}(\boldsymbol{\eta}_j)\cdot\frac{\partial}{\partial n_{\mathbf{x}}}\nabla_{\mathbf{y}}R^{(\Omega)}(\mathbf{x}, \mathbf{O}^{(j)})\Big| \\
&= \varepsilon\Big|\mathcal{D}^{(j,I)}(\boldsymbol{\eta}_j)\cdot\Big\{\frac{\partial}{\partial n_{\mathbf{x}}}\nabla_{\mathbf{x}}((2\pi\mu_O)^{-1}\log|\mathbf{x}-\mathbf{O}^{(j)}|^{-1}) - \frac{\partial}{\partial n_{\mathbf{x}}}\nabla_{\mathbf{y}}R^{(\Omega)}(\mathbf{x}, \mathbf{O}^{(j)})\Big\}\Big| \\
&\le \text{const}\ \varepsilon\ , \quad \mathbf{x}\in\partial\omega_\varepsilon^{(i)},\ \mathbf{y}\in\omega_\varepsilon^{(j)},\ i\neq j\ .
\end{aligned}
\tag{7.34}
$$

The boundary conditions for the dipole fields (3.9) give

$$
\begin{aligned}
&\Big|(\mu_{I_i} - \mu_O)\frac{\partial R^{(\Omega)}}{\partial n_{\mathbf{x}}}(\mathbf{x}, \mathbf{y}) \\
&\quad - \varepsilon\Big\{\mu_{I_i}\frac{\partial\mathcal{D}^{(i)}}{\partial n_{\mathbf{x}}}(\boldsymbol{\xi}_i)\big|_{\mathbf{x}\in\partial\omega_\varepsilon^{(i)}-} - \mu_O\frac{\partial\mathcal{D}^{(i)}}{\partial n_{\mathbf{x}}}(\boldsymbol{\xi}_i)\big|_{\mathbf{x}\in\partial\omega_\varepsilon^{(i)}+}\Big\}\cdot\nabla_{\mathbf{x}}R^{(\Omega)}(\mathbf{O}^{(i)}, \mathbf{y})\Big| \\
&\le |\mu_{I_i} - \mu_O||\boldsymbol{n}^{(i)}\cdot(\nabla_{\mathbf{x}}R^{(\Omega)}(\mathbf{x}, \mathbf{y}) - \nabla_{\mathbf{x}}R^{(\Omega)}(\mathbf{O}^{(i)}, \mathbf{y}))| \\
&\le \text{const}\ \varepsilon, \quad \mathbf{x}\in\partial\omega_\varepsilon^{(i)},\ \mathbf{y}\in\omega_\varepsilon^{(j)},\ i\neq j\ .
\end{aligned}
\tag{7.35}
$$

Then Lemma 4.2 allows one to deduce

$$\varepsilon\Big|\Big\{\mu_{I_i}\frac{\partial\mathcal{D}^{(j,O)}}{\partial n_{\mathbf{x}}}(\boldsymbol{\xi}_j)\big|_{\mathbf{x}\in\partial\omega_\varepsilon^{(i)-}}-\mu_O\frac{\partial\mathcal{D}^{(j,O)}}{\partial n_{\mathbf{x}}}(\boldsymbol{\xi}_j)\big|_{\mathbf{x}\in\partial\omega_\varepsilon^{(i)+}}\Big\}\cdot\nabla_{\mathbf{y}}R^{(\Omega)}(\mathbf{O}^{(j)},\mathbf{y})$$
$$+\sum_{\substack{k\neq j\\ k\neq i\\ 1\le k\le M}}\Big\{\mu_{I_i}\frac{\partial\mathcal{D}^{(k,O)}}{\partial n_{\mathbf{x}}}(\boldsymbol{\xi}_k)\big|_{\mathbf{x}\in\partial\omega_\varepsilon^{(i)-}}-\mu_O\frac{\partial\mathcal{D}^{(k,O)}}{\partial n_{\mathbf{x}}}(\boldsymbol{\xi}_k)\big|_{\mathbf{x}\in\partial\omega_\varepsilon^{(i)+}}\Big\}\cdot\nabla_{\mathbf{x}}R^{(\Omega)}(\mathbf{O}^{(k)},\mathbf{y})\Big|$$
$$\le \text{const }\varepsilon^2,\quad \mathbf{x}\in\partial\omega_\varepsilon^{(i)},\ \mathbf{y}\in\omega_\varepsilon^{(j)},\ i\neq j\,.$$

Thus, from the preceding inequality and (7.28), (7.29), (7.34), (7.35), we have

$$\Big|\mu_{I_i}\frac{\partial\mathfrak{h}_\varepsilon^{(j)}}{\partial n_{\mathbf{x}}}(\mathbf{x},\mathbf{y})\big|_{\mathbf{x}\in\partial\omega_\varepsilon^{(i)-}}-\mu_O\frac{\partial\mathfrak{h}_\varepsilon^{(j)}}{\partial n_{\mathbf{x}}}(\mathbf{x},\mathbf{y})\big|_{\mathbf{x}\in\partial\omega_\varepsilon^{(i)+}}\Big|\le\text{const }\varepsilon,\quad \mathbf{y}\in\omega_\varepsilon^{(j)},\ i\neq j\,.$$

We can conclude from this estimate, the conditions (7.25)–(7.27), inequalities (7.31), (7.33), and Lemma 5.1 that

$$\Big|\mathfrak{h}_\varepsilon^{(j)}(\mathbf{x},\mathbf{y})\Big|\le\text{const }\varepsilon^2,\quad \mathbf{x}\in\bigcup_{l=1}^{M}\omega_\varepsilon^{(l)}\cup\Omega_\varepsilon,\ \mathbf{y}\in\omega_\varepsilon^{(j)},\ j=1,\dots,M\,.$$

Finally, the above estimate for $\mathfrak{h}_\varepsilon^{(j)}$, $j=1,\dots,M$, and (7.22), combined with (7.12) complete the proof. □

8. Numerical simulations

In the current section, we implement the asymptotic formulae derived in Section 7 for Green's function N_ε in numerical simulations. The numerical computations are carried out for the regular part R_ε of Green's function in $\bigcup_l\omega_\varepsilon^{(l)}\cup\Omega_\varepsilon$ for the transmission problem. In other words, let this regular part be given by the formula

$$R_\varepsilon(\mathbf{x},\mathbf{y})=-\frac{1}{2\pi}\left(\frac{\chi_{\Omega_\varepsilon}(\mathbf{y})}{\mu_O}+\sum_{l=1}^{M}\frac{\chi_{\omega_\varepsilon^{(l)}}(\mathbf{y})}{\mu_{I_l}}\right)\log|\mathbf{x}-\mathbf{y}|-N_\varepsilon(\mathbf{x},\mathbf{y})\,. \tag{8.1}$$

Then in accordance with the boundary value problem for N_ε given in Section 2, the function R_ε, which we choose to consider for our numerical schemes, is a solution of the problem (7.3)–(7.8).

8.1. Asymptotic formulae for R_ε. From formula (8.1) and Theorem 7.1, we can immediately state the asymptotic formulae for the regular part R_ε that will be used in the examples below. When $\mathbf{y}\in\Omega_\varepsilon$, we have from (7.1) that R_ε admits the asymptotic representation

(8.2)
$$R_\varepsilon(\mathbf{x},\mathbf{y})=R^{(\Omega)}(\mathbf{x},\mathbf{y})+\sum_{j=1}^{M}h_N^{(j,O)}(\boldsymbol{\xi}_j,\boldsymbol{\eta}_j)$$
$$-\varepsilon\sum_{j=1}^{M}\{\mathcal{D}^{(j)}(\boldsymbol{\xi}_j)\cdot\nabla_{\mathbf{x}}R^{(\Omega)}(\mathbf{O}^{(j)},\mathbf{y})+\mathcal{D}^{(j,O)}(\boldsymbol{\eta}_j)\cdot\nabla_{\mathbf{y}}R^{(\Omega)}(\mathbf{x},\mathbf{O}^{(j)})\}+O(\varepsilon^2)\,.$$

When $\mathbf{y} \in \omega_\varepsilon^{(j)}$ where j is fixed, $j = 1, \dots, M$, R_ε has the form

(8.3)

$$R_\varepsilon(\mathbf{x}, \mathbf{y}) = R^{(\Omega)}(\mathbf{x}, \mathbf{y}) + h_N^{(j,I)}(\boldsymbol{\xi}_j, \boldsymbol{\eta}_j) + \sum_{\substack{k \neq j \\ 1 \le k \le M}} h_N^{(k,O)}(\boldsymbol{\xi}_k, \boldsymbol{\eta}_k)$$

$$+(2\pi)^{-1}(\mu_O^{-1} - \mu_{I_j}^{-1}) \log \varepsilon - \varepsilon \mathcal{D}^{(j,I)}(\boldsymbol{\eta}_j) \cdot \nabla_{\mathbf{y}} R^{(\Omega)}(\mathbf{x}, \mathbf{O}^{(j)})$$

$$-\varepsilon \sum_{\substack{k \neq j \\ 1 \le k \le M}} \mathcal{D}^{(k,O)}(\boldsymbol{\eta}_k) \cdot \nabla_{\mathbf{y}} R^{(\Omega)}(\mathbf{x}, \mathbf{O}^{(k)}) - \varepsilon \sum_{p=1}^{M} \mathcal{D}^{(p)}(\boldsymbol{\xi}_p) \cdot \nabla_{\mathbf{x}} R^{(\Omega)}(\mathbf{O}^{(p)}, \mathbf{y}) + O(\varepsilon^2)\,.$$

Before proceeding with examples where these formulae will be implemented, we first discuss the numerical settings.

8.2. Numerical settings: Description of the geometry and physical parameters. Let Ω be a disk of radius 150m with center at the origin and occupied by a material with shear modulus $\mu_O = 5.6 \times 10^{10}\text{Nm}^{-2}$, which is that of cast iron. We set the number of inclusions $M = 6$ and assume that the $\omega_\varepsilon^{(j)}$, $j = 1, \dots, 6$, are circular. We summarize the data corresponding to the inclusions in Table 1.

TABLE 1. Data for the inclusions $\omega_\varepsilon^{(j)}$, $j = 1, \dots, 6$.

Inclusion	Center	Radius (m)	Shear Modulus ($\times 10^{10}\text{Nm}^{-2}$)	Material
$\omega_\varepsilon^{(1)}$	(−90m, 40m)	27	2.6316	Aluminum
$\omega_\varepsilon^{(2)}$	(−50m,−50m)	24	4.0741	Copper
$\omega_\varepsilon^{(3)}$	(−30m, 10m)	9	7.7519	Iron
$\omega_\varepsilon^{(4)}$	(20m, 70m)	19.5	7.5188	High Strength Alloy Steel
$\omega_\varepsilon^{(5)}$	(50m, 0m)	22.5	8.0078	Steel AISI 4348
$\omega_\varepsilon^{(6)}$	(70m, −80m)	15	9.0496	Nimonic Alloy 90

8.3. Model solutions used in the numerical simulations.

Neumann's function for the disk Ω. In our examples, we need the Neumann function $N^{(\Omega)}$ for a disk of radius R ($R = 150$m for our demonstrations), which is given by

$$N^{(\Omega)}(\mathbf{x}, \mathbf{y}) = -\frac{1}{2\pi\mu_O} \log|\mathbf{x} - \mathbf{y}| - \frac{1}{2\pi\mu_O} \log\left(\left|\mathbf{x} - \frac{R^2}{|\mathbf{y}|^2}\mathbf{y}\right| |\mathbf{y}|\right)\,,$$

and the regular part $R^{(\Omega)}$ of this function is defined by the formula

$$R^{(\Omega)}(\mathbf{x}, \mathbf{y}) = (2\pi\mu_O)^{-1} \log|\mathbf{x} - \mathbf{y}| - N^{(\Omega)}(\mathbf{x}, \mathbf{y})\,.$$

The regular part of Green's function for the transmission problem in a plane with an inclusion at the origin. Now we state the form of the regular part $h_N^{(j)}$ of the Green's function for the transmission problem in the infinite plane with a circular inclusion at the origin of radius a_j. The solution is constructed using the involution procedure which is discussed in [**4**]. The representation of this function is dependent on the position of the point $\boldsymbol{\eta}_j$. When $\boldsymbol{\eta}_j \in C\bar{\omega}^{(j)}$,

$$h^{(j,O)}(\boldsymbol{\xi}_j, \boldsymbol{\eta}_j) = -\frac{1}{2\pi\mu_O} \frac{\mu_{I_j} - \mu_O}{\mu_{I_j} + \mu_O} \log\left(\frac{1}{|\boldsymbol{\xi}_j|}\left|\boldsymbol{\xi}_j - \frac{a_j^2}{|\boldsymbol{\eta}_j|^2}\boldsymbol{\eta}_j\right|\right) \quad \text{for} \quad \boldsymbol{\xi}_j \in C\bar{\omega}^{(j)}\,,$$

and

$$h^{(j,O)}(\boldsymbol{\xi}_j, \boldsymbol{\eta}_j) = \frac{1}{2\pi\mu_O}\frac{\mu_{I_j} - \mu_O}{\mu_{I_j} + \mu_O}(\log|\boldsymbol{\xi}_j - \boldsymbol{\eta}_j|^{-1} + \log|\boldsymbol{\eta}_j|)\,, \quad \boldsymbol{\xi}_j \in \omega^{(j)}\,.$$

For the case $\boldsymbol{\eta}_j \in \omega^{(j)}$, $\boldsymbol{\xi}_j \in C\bar{\omega}^{(j)}$

$$h^{(j,I)}(\boldsymbol{\xi}_j, \boldsymbol{\eta}_j) = \frac{1}{2\pi}\frac{\mu_{I_j} - \mu_O}{\mu_{I_j} + \mu_O}\left(\frac{1}{\mu_{I_j}}\log|\boldsymbol{\xi}_j - \boldsymbol{\eta}_j| + \frac{1}{\mu_O}\log|\boldsymbol{\xi}_j|\right)\,,$$

and for $\boldsymbol{\eta}_j, \boldsymbol{\xi}_j \in \omega^{(j)}$ we have

$$h^{(j,I)}(\boldsymbol{\xi}_j, \boldsymbol{\eta}_j) = \frac{1}{2\pi}\frac{\mu_{I_j} - \mu_O}{\mu_{I_j} + \mu_O}\left(\frac{1}{\mu_{I_j}}\log\left(\frac{|\boldsymbol{\eta}_j|}{a_j}\left|\boldsymbol{\xi}_j - \frac{a_j^2}{|\boldsymbol{\eta}_j|^2}\boldsymbol{\eta}_j\right|\right) + \frac{1}{\mu_O}\log a_j\right)\,.$$

The dipole fields for the circular inclusion in the infinite plane. Here we give the vector function $\mathcal{D}^{(j)}$ whose components are the dipole fields for the circular inclusion of radius a_j in the infinite plane and

$$\mathcal{D}^{(j)}(\boldsymbol{\xi}_j) = \chi_{C\bar{\omega}^{(j)}}(\boldsymbol{\xi}_j)\mathcal{D}^{(j,O)}(\boldsymbol{\xi}_j) + \chi_{\omega^{(j)}}(\boldsymbol{\xi}_j)\mathcal{D}^{(j,I)}(\boldsymbol{\xi}_j)\,.$$

We have

$$\mathcal{D}^{(j,O)}(\boldsymbol{\xi}_j) = \frac{\mu_{I_j} - \mu_O}{\mu_{I_j} + \mu_O}\frac{a_j^2\,\boldsymbol{\xi}_j}{|\boldsymbol{\xi}_j|^2}\,, \qquad \text{for } \boldsymbol{\xi}_j \in C\bar{\omega}^{(j)}\,,$$

and

$$\mathcal{D}^{(j,I)}(\boldsymbol{\xi}_j) = \frac{\mu_{I_j} - \mu_O}{\mu_{I_j} + \mu_O}\boldsymbol{\xi}_j\,, \qquad \text{for } \boldsymbol{\xi}_j \in \omega^{(j)}\,.$$

8.4. Example 1.

The case of the force applied outside the inclusions. For our first example, we look at the case when $\mathbf{y} \in \Omega_\varepsilon$. We therefore base our computations on the asymptotic formula (8.2) when comparing with those of COMSOL. The coordinates of the point force are given as $\mathbf{y} = (-10\text{m}, -80\text{m})$. We plot the modulus of the gradient of the regular part R_ε in Figure 2(a) according to the analytical formulae (8.2). Figure 2(b) is the same quantity computed using the method of finite elements in COMSOL. Both figures are very similar; the maximum absolute error between these computations is 7.666×10^{-16} occurring on the exterior boundary.

8.5. Example 2.

The case of the force positioned inside an inclusion. In the second example, we aim to compare the computations produced by formula (8.3) with those generated by COMSOL. Now the point force is assumed to be situated at $\mathbf{y} = (60\text{m}, 0\text{m})$, in the inclusion $\omega_\varepsilon^{(5)}$, containing the steel AISI 4340. Figure 3(a), gives the surface plot for the modulus of the gradient of the regular part, provided by formula (8.3). The numerical solution given in COMSOL is shown in Figure 3(b). The maximum absolute error here is 7.98×10^{-16}, which occurs on the boundary of the inclusion $\omega_\varepsilon^{(1)}$. We conclude that the asymptotic formulae and numerical computations are in good agreement with each other.

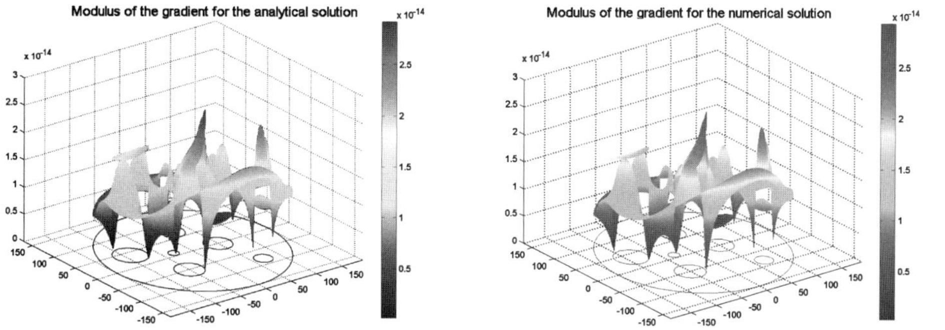

(a) (b)

FIGURE 2. (a) Computations based on asymptotic formula (8.2) and (b) numerical solution for the absolute value of the gradient of the regular part of Green's function for the transmission problem. Here $\mathbf{y} = (-10\text{m}, -80\text{m})$ and the mesh contains 44,784 elements. The plots are practically indistinguishable.

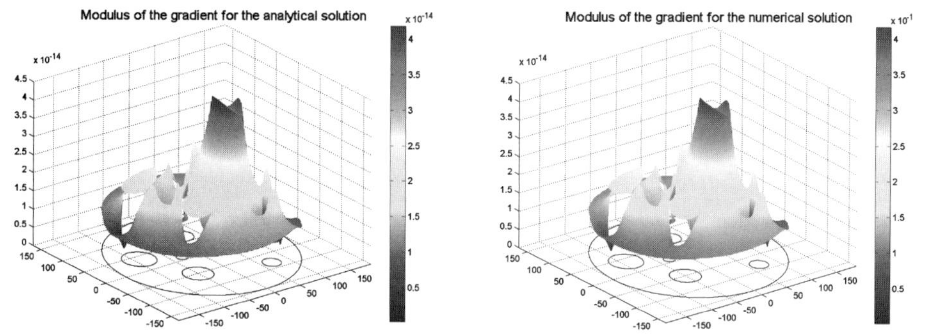

(a) (b)

FIGURE 3. (a) Computations based on asymptotic formula (8.3) and (b) numerical solution for the absolute value of the gradient of the regular part of Green's function for the transmission problem. Here $\mathbf{y} = (60\text{m}, 0\text{m})$ and lies in the steel AISI 4340 inclusion. The mesh contains 44,784 elements. There is once again a good similarity between the surface plots.

References

[1] C. Carstensen, E. P. Stephan, Adaptive boundary-element methods for transmission problems, *J. Austral. Math. Soc. B* **38**, 336–367 (1997).

[2] M. Costabel, E. Stephan, A direct integral equation method for transmission problems, *Journal of Mathematical Analysis and Applications* **106**, 367–413 (1985).

[3] J. Hadamard, Sur le problème d'analyse relatif à l'équilibre des plaques élastiques encastrées. *Mémoire couronné en 1907 par l'Académie des Sciences* **33**, no. 4, 515–629.

[4] T. Honein, G. Herrmann, On bonded inclusions with circular or straight boundaries in plane elastostatics, *J. Appl. Mech. Trans. ASME* **57**, 850–856.

[5] V. A. Kondratiev, O. A. Oleinik, On the behavior at infinity of solutions of elliptic systems with a finite energy integral, *Archive for Rational Mechanics and Analysis* **99**, no. 1, 75-89 (1987).

[6] V. Maz′ya, A. Movchan, Uniform asymptotic formulae for Green's functions in singularly perturbed domains, *Journal of Computational and Applied Mathematics* **208**, no. 1, 194–206 (2007).
[7] V. Maz′ya, A. Movchan, Uniform asymptotics of Green's kernels for mixed and Neumann problems in domains with small holes and inclusions, *Sobolev Spaces in Mathematics III*, Springer New-York, vol. 10, 277–316 (2009).
[8] V. Maz′ya, A. Movchan, Uniform asymptotic formulae for Green's kernels in regularly and singularly perturbed domains, *C. R. Acad. Sci. Paris.* Ser. I **343**, 185–190 (2006).
[9] V. Maz′ya, A. Movchan, M. Nieves, Uniform asymptotic formulae for Green's tensors in elastic singularly perturbed domains, *Asymptotic Analysis* **52**, nos. 3/4, 173–206 (2007).
[10] V. Maz′ya, A. Movchan, M. Nieves, Uniform asymptotic formulae for Green's tensors in elastic singularly perturbed domains with multiple inclusions, *Rendiconti della accademis nazionale delle scienze detta dei XL, Memorie di matematica e applicazioni*, Serie V, Vol. XXX, Parte I, 103–158 (2006).

Department of Mathematical Sciences, University of Liverpool, Liverpool L69 3BX, United Kingdom and Department of Mathematics, Linköping University, SE-581 83 Linköping, Sweden.
E-mail address: vlmaz@mai.liu.se

Department of Mathematical Sciences, University of Liverpool, Liverpool L69 3BX, United Kingdom
E-mail address: abm@liv.ac.uk

Department of Mathematical Sciences, University of Liverpool, Liverpool L69 3BX, United Kingdom
E-mail address: mf0u1036@liv.ac.uk

Amer. Math. Soc. Transl.
Volume **231**, 2010

Lower Bound on the Density of States for Periodic Schrödinger Operators

Sergey Morozov, Leonid Parnovski, and Irina Pchelintseva

ABSTRACT. We consider Schrödinger operators $-\Delta + V$ in $\mathbb{R}^d$ $(d \geqslant 2)$ with smooth periodic potentials V and obtain a uniform lower bound on the (non-integrated) density of states for large values of the spectral parameter. It turns out that as the spectral parameter increases, the density of states approaches the one for $-\Delta$.

1. Introduction

Let $H = -\Delta + V$ be a Schrödinger operator in $L_2(\mathbb{R}^d)$ with a smooth periodic potential V. We will assume throughout that $d \geqslant 2$. The *integrated density of states (IDS)* for H is defined as

$$N(\lambda) := \lim_{L\to\infty} L^{-d} N(\lambda; H_D^{(L)}), \quad \lambda \in \mathbb{R}. \tag{1.1}$$

Here $H_D^{(L)}$ is the restriction of H to the cube $[0, L]^d$ with the Dirichlet boundary conditions, and $N(\lambda; \cdot)$ is the counting function of the discrete spectrum below λ. The existence of the limit in (1.1) is well known; see, e.g., [**RS78, Shu79**]. For $H_0 := -\Delta$ the IDS can be easily computed explicitly (e.g., using the representation (2.6) below):

$$N_0(\lambda) = \begin{cases} (2\pi)^{-d} d^{-1} \omega_d \lambda^{d/2}, & \lambda > 0, \\ 0, & \lambda \leqslant 0. \end{cases} \tag{1.2}$$

Here $\omega_d = 2\pi^{d/2}/\Gamma(d/2)$ is the surface area of the unit sphere S^{d-1} in $\mathbb{R}^d$.

The asymptotic behavior of the function (1.1) for large values of the spectral parameter was recently studied in a number of publications; see [**Skr87**], [**Kar97**], [**HM98**], [**Kar00**], [**Sob05**], [**PS09**], and references therein.

Our article concerns the high-energy behavior of the Radon–Nikodym derivative of IDS

$$g := \mathrm{d}N/\,\mathrm{d}\lambda,$$

2010 *Mathematics Subject Classification.* Primary 35P20, 35P15, 35J10.

Key words and phrases. Density of states, periodic Schrödinger operator.

This paper is a result of observations and suggestions made by Yu. Karpeshina; we are very grateful to her for sharing them with us and allowing us to use them.

The authors were supported by the EPSRC grant EP/F029721/1.

which is called the *density of states (DOS)* (see [**RS78**]). Our main result is that for big values of λ

$$g(\lambda) \geqslant g_0(\lambda)\big(1 - o(1)\big), \tag{1.3}$$

where

$$g_0(\lambda) = \mathrm{d}N_0(\lambda)/\,\mathrm{d}\lambda = (2\pi)^{-d}\omega_d\lambda^{(d-2)/2}/2.$$

We remark that (1.3) should be understood in the sense of measures; in particular, we do not claim that $g(\lambda)$ is everywhere differentiable.

It has been proved in [**Par08**] that the spectrum of H contains a semiaxis $[\lambda_0, +\infty)$; this statement is known as the Bethe-Sommerfeld conjecture (see the references in [**Par08**] for the history of this problem). This result has an obvious reformulation in terms of IDS: each point $\lambda \geqslant \lambda_0$ is a point of growth of N. It was also proved in [**Par08**] that for each $n \in \mathbb{N}$ and $\varepsilon = \lambda^{-n}$ we have

$$N(\lambda + \varepsilon) - N(\lambda) \ll \varepsilon\lambda^{(d-2)/2}. \tag{1.4}$$

Later, when the second author discussed the results and methods of [**Par08**] with Yu. Karpeshina, she suggested that using the technique from that paper, one should be able to prove the opposite bound

$$N(\lambda + \varepsilon) - N(\lambda) \gg \varepsilon\lambda^{(d-2)/2} \tag{1.5}$$

when λ is sufficiently large, not just with $\varepsilon = \lambda^{-n}$ (when the proof is relatively straightforward, given [**Par08**]), but also uniformly over all $\varepsilon \in (0, 1]$. In our paper we prove that for big λ

$$N(\lambda + \varepsilon) - N(\lambda) \geqslant \frac{\omega_d}{2(2\pi)^d}\varepsilon\lambda^{(d-2)/2}\big(1 - o(1)\big). \tag{1.6}$$

Note that (1.6) implies the claimed bound (1.3).

The main result of the paper is

THEOREM 1.1. *For sufficiently big λ and any $\varepsilon > 0$ the integrated density of states of H satisfies* (1.6).

The proof of Theorem 1.1 is heavily based on the technique of [**Par08**] and uses various statements proved therein. In order to minimize the size of the paper, we will try to quote as many results as we can from [**Par08**], possibly with some minor modifications where necessary.

The article is organized as follows. In Section 2 we introduce the necessary notation and quote the results of [**Par08**] which we need for the proof of Theorem 1.1. Sections 3 and 4 contain some auxiliary results, and the proof is finished in Section 5.

2. Preliminaries

We study the Schrödinger operator

$$H = -\Delta + V(\mathbf{x}), \quad \mathbf{x} \in \mathbb{R}^d, \tag{2.1}$$

with the potential V being infinitely smooth and periodic with the lattice of periods Λ. We denote the lattice dual to Λ by $\Lambda^\dagger$; fundamental cells of these lattices are denoted by Ω and $\Omega^\dagger$, respectively. We introduce

$$Q := \sup\big\{|\boldsymbol{\xi}|\,\big|\,\boldsymbol{\xi} \in \Omega^\dagger\big\}. \tag{2.2}$$

Let

$$\mathbf{D} := -i\nabla, \quad \mathbf{D}(\mathbf{k}) := \mathbf{D} + \mathbf{k}. \tag{2.3}$$

The Floquet-Bloch decomposition allows us to represent our operator (2.1) as a direct integral (see, e.g., [**RS78**]):

$$H = \int_{\Omega^\dagger} \bigoplus H(\mathbf{k})\, \mathrm{d}\mathbf{k}, \tag{2.4}$$

where

$$H(\mathbf{k}) = \mathbf{D}(\mathbf{k})^2 + V(\mathbf{x}) \tag{2.5}$$

is the family of "fiber" operators acting in $L_2(\Omega)$. The domain $\mathfrak{D}$ of each $H(\mathbf{k})$ is the set of periodic functions from $H^2(\Omega)$. The spectrum of H is the union over $\mathbf{k} \in \Omega^\dagger$ of the spectra of the operators $H(\mathbf{k})$. Let $\{\lambda_j(\mathbf{k})\}$, $j \in \mathbb{N}$, be the set of eigenvalues of $H(\mathbf{k})$ (counting with multiplicities). Then the integrated density of states (1.1) admits the following representation:

$$N(\lambda) := (2\pi)^{-d} \int_{\Omega^\dagger} \#\{j : \lambda_j(\mathbf{k}) < \lambda\}\, \mathrm{d}\mathbf{k}; \tag{2.6}$$

see, e.g., [**RS78**].

We denote by $|\cdot|_\circ$ the surface area Lebesgue measure on the unit sphere S^{d-1} in $\mathbb{R}^d$ and put $\omega_d := |S^{d-1}|_\circ = 2\pi^{d/2}/\Gamma(d/2)$. By $\mathrm{vol}(\cdot)$ we denote the Lebesgue measure in $\mathbb{R}^d$. We write $B(R)$ for the ball of radius R centered at the origin. The identity matrix is denoted by $\mathbf{I}$. By $\lambda = \rho^2$ we denote a point on the spectral axis. We also denote by v the L_∞-norm of the potential V, and put $J := [\lambda - 20v, \lambda + 20v]$.

Any vector $\boldsymbol{\xi} \in \mathbb{R}^d$ can be uniquely decomposed as $\boldsymbol{\xi} = \mathbf{n} + \mathbf{k}$ with $\mathbf{n} \in \Lambda^\dagger$ and $\mathbf{k} \in \Omega^\dagger$. We call $\mathbf{n} =: [\boldsymbol{\xi}]$ the "integer part" of $\boldsymbol{\xi}$ and $\mathbf{k} =: \{\boldsymbol{\xi}\}$ the "fractional part" of $\boldsymbol{\xi}$. For $\boldsymbol{\xi} \in \mathbb{R}^d \setminus \{\mathbf{0}\}$ we define $r = r(\boldsymbol{\xi}) := |\boldsymbol{\xi}|$ and $\boldsymbol{\xi}' := \boldsymbol{\xi}/|\boldsymbol{\xi}|$. For any $h \in L_2(\Omega)$ we introduce its Fourier coefficients

$$h_{\mathbf{n}} := (\mathrm{vol}\,\Omega)^{-1/2} \int_\Omega h(\mathbf{x}) \exp\big(-i\langle \mathbf{n}, \mathbf{x}\rangle\big)\, \mathrm{d}\mathbf{x}, \quad \mathbf{n} \in \Lambda^\dagger. \tag{2.7}$$

Given two positive functions f and g, we say that $f \gg g$, or $g \ll f$, or $g = O(f)$, if the ratio g/f is bounded. We say $f \asymp g$ if $f \gg g$ and $f \ll g$. Whenever we use O, o, $\gg$, $\ll$, or $\asymp$ notation, the constants involved can depend on d and norms of the potential in various Sobolev spaces H^s; the same is also the case when we use the expression "sufficiently large".

Let

$$\mathcal{A} := \Big\{\boldsymbol{\xi} \in \mathbb{R}^d,\ \big||\boldsymbol{\xi}|^2 - \lambda\big| \leqslant 40v\Big\}. \tag{2.8}$$

Notice that the definition of $\mathcal{A}$ obviously implies that if $\boldsymbol{\xi} \in \mathcal{A}$, then $\big||\boldsymbol{\xi}| - \rho\big| \ll \rho^{-1}$. We put

$$R = R(\rho) := \rho^{1/(36d^2(d+2))} \tag{2.9}$$

(so that the condition stated after equation (5.15) in [**Par08**] is satisfied). For $j \in \mathbb{N}$ let

$$\Theta'_j := \Lambda^\dagger \cap B(jR) \setminus \{\mathbf{0}\}.$$

Let $M := 5d^2 + 7d$. We introduce the set

$$\mathcal{B} := \Big\{\boldsymbol{\xi} \in \mathcal{A} \Big| |\langle \boldsymbol{\xi}, \boldsymbol{\eta}'\rangle| > \rho^{1/2}, \text{ for all } \boldsymbol{\eta} \in \Theta'_{6M}\Big\}. \tag{2.10}$$

In other words, $\mathcal{B}$ consists of all points $\boldsymbol{\xi} \in \mathcal{A}$ the projections of which to the directions of all vectors $\boldsymbol{\eta} \in \Theta'_{6M}$ have lengths larger than $\rho^{1/2}$. We also denote $\mathcal{D} := \mathcal{A} \setminus \mathcal{B}$.

In the rest of the section we quote some results from [**Par08**] which will be used in this paper. Our approach is slightly different from that of [**Par08**]. In particular, we consider the lattice of periods Λ, not equal to $(2\pi\mathbb{Z})^d$. We also use a different form of the Floquet-Bloch decomposition (so that the operators on fibers (2.5) are defined on the same domain). This leads to several straightforward changes in the formulation of the results from [**Par08**]. These changes are:

(1) The lattices $(2\pi\mathbb{Z})^d$ and $\mathbb{Z}^d$ are replaced by Λ and $\Lambda^\dagger$, respectively. The "integer" and "fractional" parts are now defined with respect to $\Lambda^\dagger$ (see above).
(2) The matrices $\mathbf{F}$ and $\mathbf{G}$ are replaced by the unit matrix $\mathbf{I}$ throughout.
(3) The Fourier transform is now defined by (2.7), and the exponentials $e_{\mathbf{m}}$ introduced at the beginning of Section 5 in [**Par08**] are redefined as

$$e_{\mathbf{m}}(\mathbf{x}) := (\operatorname{vol}\Omega)^{-1/2}\, \mathrm{e}^{i\langle \mathbf{m}, \mathbf{x}\rangle}, \quad \mathbf{m} \in \Lambda^\dagger.$$

(4) The operators $H(\mathbf{k})$ are now given by (2.5) on the common domain $\mathfrak{D}$.

The main result we will need follows from Corollary 7.15 of [**Par08**]:

PROPOSITION 2.1. *There exist mappings $f, g : \mathcal{A} \to \mathbb{R}$ which satisfy the following properties:*

(i) *$f(\boldsymbol{\xi})$ is an eigenvalue of $H(\mathbf{k})$ with $\{\boldsymbol{\xi}\} = \mathbf{k}$; $\big|f(\boldsymbol{\xi}) - |\boldsymbol{\xi}|^2\big| \leqslant 2v$. The mapping f is an injection (if we count all eigenvalues with multiplicities) and all eigenvalues of $H(\mathbf{k})$ inside J are in the image of f.*

(ii) *If $\boldsymbol{\xi} \in \mathcal{A}$, then $\big|f(\boldsymbol{\xi}) - g(\boldsymbol{\xi})\big| \leqslant \rho^{-d-3}$.*

(iii) *For any $\boldsymbol{\xi} \in \mathcal{B}$*

$$\begin{aligned} g(\boldsymbol{\xi}) &= |\boldsymbol{\xi}|^2 \\ &+ \sum_{j=1}^{2M} \sum_{\boldsymbol{\eta}_1, \dots, \boldsymbol{\eta}_j \in \Theta'_M} \sum_{2 \leqslant n_1 + \dots + n_j \leqslant 2M} C_{n_1, \dots, n_j} \langle \boldsymbol{\xi}, \boldsymbol{\eta}_1 \rangle^{-n_1} \dots \langle \boldsymbol{\xi}, \boldsymbol{\eta}_j \rangle^{-n_j}. \end{aligned} \tag{2.11}$$

REMARK 2.2. Formula (2.11) implies that

$$\partial g/\partial r(\boldsymbol{\xi}) \asymp \rho, \quad \text{for any} \quad \boldsymbol{\xi} \in \mathcal{B}. \tag{2.12}$$

For each positive $\delta \leqslant v$ we denote by $\mathcal{A}(\delta)$, $\mathcal{B}(\delta)$, and $\mathcal{D}(\delta)$ the intersections of $g^{-1}\big([\rho^2 - \delta, \rho^2 + \delta]\big)$ with $\mathcal{A}$, $\mathcal{B}$, and $\mathcal{D}$, respectively.

It is proved in Lemma 8.1 of [**Par08**] that

$$\operatorname{vol}\big(\mathcal{D}(\delta)\big) \ll \rho^{d-7/3}\delta. \tag{2.13}$$

The following statement (Corollary 8.5 of [**Par08**]) gives a sufficient condition for the continuity of f:

LEMMA 2.3. *There exists a constant C_1 with the following properties. Let*

$$I := \big\{\boldsymbol{\xi}(t) : t \in [t_{\min}, t_{\max}]\big\} \subset \mathcal{B}(v)$$

be a straight interval of length $L < \rho^{-1}\delta$. Suppose that there is a point $t_0 \in [t_{\min}, t_{\max}]$ with the property that for each nonzero $\mathbf{n} \in \Lambda^\dagger$, $g\big(\boldsymbol{\xi}(t_0) + \mathbf{n}\big)$ is either *outside the interval*

$$\Big[g\big(\boldsymbol{\xi}(t_0)\big) - C_1\rho^{-d-3} - C_1\rho L,\ g\big(\boldsymbol{\xi}(t_0)\big) + C_1\rho^{-d-3} + C_1\rho L\Big]$$

or is not defined. Then $f\big(\boldsymbol{\xi}(t)\big)$ is a continuous function of t.

By inspection of the proof of Lemma 8.3 of [**Par08**] we obtain

LEMMA 2.4. *For large enough ρ and $\delta < \rho^{-1}$ the following estimates hold uniformly over $\mathbf{a} \in \Lambda^\dagger \setminus \{\mathbf{0}\}$: if $d \geqslant 3$,*

$$\operatorname{vol}\Big(\mathcal{B}(\delta) \cap \big(\mathcal{B}(\delta) + \mathbf{a}\big)\Big) \ll (\delta^2\rho^{d-3} + \delta\rho^{-d}); \tag{2.14}$$

if $d = 2$,

$$\operatorname{vol}\Big(\mathcal{B}(\delta) \cap \big(\mathcal{B}(\delta) + \mathbf{a}\big)\Big) \begin{cases} \ll \delta^{3/2}, & |\mathbf{a}| \leqslant 2\rho - 1, \\ \ll \delta^{3/2} + \delta\rho^{-2}, & \big||\mathbf{a}| - 2\rho\big| < 1, \\ = 0, & |\mathbf{a}| \geqslant 2\rho + 1. \end{cases} \tag{2.15}$$

3. Prevalence of regular directions

In this section we prove that for most directions $\boldsymbol{\xi}'$ the image of the function f of Proposition 2.1 is an isolated eigenvalue of $H\big(\{\boldsymbol{\xi}\}\big)$ continuously depending on $|\xi|$ if it belongs to a neighborhood of ρ^2.

LEMMA 3.1. *For ρ big enough and*

$$0 < \delta \leqslant \rho^{-d-3}$$

there exists a set $\mathcal{F} = \mathcal{F}(\rho)$ on the unit sphere S^{d-1} in $\mathbb{R}^d$ with

$$|\mathcal{F}|_\circ \geqslant \omega_d\big(1 - o(1)\big) \tag{3.1}$$

such that $f(\boldsymbol{\xi})$ is a simple eigenvalue of $H\big(\{\boldsymbol{\xi}\}\big)$ continuously depending on $r := |\boldsymbol{\xi}|$ for every $\boldsymbol{\xi} = (r, \boldsymbol{\xi}') \in f^{-1}\big([\rho^2 - \delta, \rho^2 + \delta)\big)$ with $\boldsymbol{\xi}' := \boldsymbol{\xi}/|\boldsymbol{\xi}| \in \mathcal{F}$.

PROOF. It is enough to consider $\delta := \rho^{-d-3}$. For each $\boldsymbol{\xi}' \in S^{d-1}$ let

$$I_{\boldsymbol{\xi}'}(\delta) := \{r\boldsymbol{\xi}', r > 0\} \cap \mathcal{B}(\delta). \tag{3.2}$$

Let $\mathcal{F}_1 := \big\{\boldsymbol{\xi}' \in S^{d-1} \big| I_{\boldsymbol{\xi}'}(\delta) \neq \emptyset,\ \overline{I_{\boldsymbol{\xi}'}(\delta)} \cap \mathcal{D}(\delta) = \emptyset\big\}$.

For any $\boldsymbol{\eta} \in \Theta'_{6M}$ the area of the set of points $\boldsymbol{\xi}' \in S^{d-1}$ satisfying

$$\big|\langle r\boldsymbol{\xi}', \boldsymbol{\eta}'\rangle\big| \leqslant \rho^{1/2}$$

is evidently $O(\rho^{-1/2})$ if $r \geqslant \rho/2$ (the latter is true for all $r\boldsymbol{\xi}' \in \mathcal{A}$). Since the number of elements in Θ'_{6M} is $O(R^d)$, by (2.9) and (2.10) we have

$$|S^{d-1} \setminus \mathcal{F}_1|_\circ = o(1). \tag{3.3}$$

By definition $\mathcal{B}(\delta) = \mathcal{B} \cap g^{-1}\big([\rho^2 - \delta, \rho^2 + \delta]\big)$; hence (2.12) implies that for big ρ the length $l_{\boldsymbol{\xi}'}(\delta)$ of $I_{\boldsymbol{\xi}'}(\delta)$ satisfies

$$l_{\boldsymbol{\xi}'}(\delta) \asymp \delta\rho^{-1}, \quad \boldsymbol{\xi}' \in \mathcal{F}_1. \tag{3.4}$$

Let

$$\mathcal{F} := \big\{\boldsymbol{\xi}' \in \mathcal{F}_1 \,\big|\, f \text{ is continuous on } I_{\boldsymbol{\xi}'}(\delta)\big\}$$

and

$$\mathcal{E}(\delta) := \big\{\boldsymbol{\xi} \in \mathcal{B}(\delta) \big| \boldsymbol{\xi}' \in \mathcal{F}_1 \setminus \mathcal{F}\big\}.$$

Lemma 2.3 tells us that for each point $\boldsymbol{\xi} \in \mathcal{E}(\delta)$ there is a nonzero vector $\mathbf{n} \in \Lambda^\dagger$ such that

$$|g(\boldsymbol{\xi}+\mathbf{n}) - g(\boldsymbol{\xi})| \leqslant C_1\big(\rho^{-d-3} + \rho l_{\boldsymbol{\xi}'}(\delta)\big) \ll (\rho^{-d-3}+\delta). \tag{3.5}$$

Since $|g(\boldsymbol{\xi}) - \rho^2| \leqslant \delta$, this implies

$$|g(\boldsymbol{\xi}+\mathbf{n}) - \rho^2| \leqslant C_2(\rho^{-d-3}+\delta) =: \delta_1 \ll \rho^{-d-3} = \delta,$$

and thus $\boldsymbol{\xi}+\mathbf{n} \in \mathcal{A}(\delta_1)$; notice that $C_2 > 1$ and so $\delta_1 > \delta$. Therefore, each point $\boldsymbol{\xi} \in \mathcal{E}(\delta)$ also belongs to the set $\big(\mathcal{A}(\delta_1) - \mathbf{n}\big)$ for a nonzero $\mathbf{n} \in \Lambda^\dagger$; obviously, $|\mathbf{n}| \ll \rho$. In other words,

$$\mathcal{E}(\delta) \subset \bigcup_{\mathbf{n}\in\Lambda^\dagger\cap B(C\rho),\, \mathbf{n}\neq 0} \big(\mathcal{A}(\delta_1) - \mathbf{n}\big) = \bigcup_{\mathbf{n}\neq 0}\big(\mathcal{B}(\delta_1)-\mathbf{n}\big) \cup \bigcup_{\mathbf{n}\neq 0}\big(\mathcal{D}(\delta_1)-\mathbf{n}\big). \tag{3.6}$$

To proceed further, we need more notation. Denote $\mathcal{D}_0(\delta_1)$ to be the set of all points $\boldsymbol{\nu}$ from $\mathcal{D}(\delta_1)$ for which there is no nonzero $\mathbf{n} \in \Lambda^\dagger$ satisfying $\boldsymbol{\nu} - \mathbf{n} \in \mathcal{B}(\delta)$; $\mathcal{D}_1(\delta_1)$ to be the set of all points $\boldsymbol{\nu}$ from $\mathcal{D}(\delta_1)$ for which there is a unique nonzero $\mathbf{n} \in \Lambda^\dagger$ satisfying $\boldsymbol{\nu} - \mathbf{n} \in \mathcal{B}(\delta)$; and $\mathcal{D}_2(\delta_1)$ to be the rest of the points from $\mathcal{D}(\delta_1)$ (i.e., $\mathcal{D}_2(\delta_1)$ consists of all points $\boldsymbol{\nu}$ from $\mathcal{D}(\delta_1)$ for which there exist at least two different nonzero vectors $\mathbf{n}_1, \mathbf{n}_2 \in \Lambda^\dagger$ satisfying $\boldsymbol{\nu} - \mathbf{n}_j \in \mathcal{B}(\delta)$). Then Lemma 8.7 of [**Par08**] implies that we can rewrite (3.6) as

$$\mathcal{E}(\delta) \subset \bigcup_{\mathbf{n}\neq 0}\big(\mathcal{B}(\delta_1)-\mathbf{n}\big) \cup \bigcup_{\mathbf{n}\neq 0}\big(\mathcal{D}_1(\delta_1)-\mathbf{n}\big). \tag{3.7}$$

From this we conclude that

$$\mathcal{E}(\delta) \subset \bigcup_{\mathbf{n}\neq 0}\Big(\big(\mathcal{B}(\delta_1)-\mathbf{n}\big)\cap\mathcal{B}(\delta)\Big) \cup \bigcup_{\mathbf{n}\neq 0}\Big(\big(\mathcal{D}_1(\delta_1)-\mathbf{n}\big)\cap\mathcal{B}(\delta)\Big), \tag{3.8}$$

since $\mathcal{E}(\delta) \subset \mathcal{B}(\delta)$.

The definition of the set $\mathcal{D}_1(\delta_1)$ and (2.13) imply that

$$\begin{aligned} \mathrm{vol}\Bigg(\bigcup_{\mathbf{n}\neq 0}\Big(\big(\mathcal{D}_1(\delta_1)-\mathbf{n}\big)\cap\mathcal{B}(\delta)\Big)\Bigg) &\leqslant \mathrm{vol}\,\big(\mathcal{D}_1(\delta_1)\big) \\ &\leqslant \mathrm{vol}\,\big(\mathcal{D}(\delta_1)\big) \ll \delta_1\rho^{d-7/3} \ll \delta\rho^{d-7/3}. \end{aligned} \tag{3.9}$$

For $d \geqslant 3$ Lemma 2.4, inequality $\delta < \delta_1$, and the fact that the union in (3.8) consists of no more than $C\rho^d$ terms imply

$$\mathrm{vol}\Bigg(\bigcup_{\mathbf{n}\neq 0}\Big(\big(\mathcal{B}(\delta_1)-\mathbf{n}\big)\cap\mathcal{B}(\delta)\Big)\Bigg) \ll \rho^d(\delta_1^2\rho^{d-3} + \delta_1\rho^{-d}) \ll \delta(\rho^{d-6}+1). \tag{3.10}$$

For $d = 2$ we obtain by Lemma 2.4

$$\begin{aligned}
&\operatorname{vol}\Big(\bigcup_{\mathbf{n}\in\Lambda^\dagger\setminus\{\mathbf{0}\}} \big(\mathcal{B}(\delta) \cap (\mathcal{B}(\delta_1) + \mathbf{n})\big)\Big) \\
&\leqslant \sum_{\substack{\mathbf{n}\in\Lambda^\dagger\setminus\{\mathbf{0}\} \\ |\mathbf{n}|\leqslant 2\rho-1}} \operatorname{vol}\big(\mathcal{B}(\delta) \cap (\mathcal{B}(\delta_1) + \mathbf{n})\big) \\
&\quad + \sum_{\substack{\mathbf{n}\in\Lambda^\dagger\setminus\{\mathbf{0}\} \\ ||\mathbf{n}|-2\rho|<1}} \operatorname{vol}\big(\mathcal{B}(\delta) \cap (\mathcal{B}(\delta_1) + \mathbf{n})\big) \\
&\ll \delta_1^{3/2}\rho^2 + \rho(\delta_1^{3/2} + \delta_1\rho^{-2}) \ll \delta\rho^{-1/2},
\end{aligned} \tag{3.11}$$

where we have used that

$$\#\Big\{\mathbf{n} \in \Lambda^\dagger \Big| \big||\mathbf{n}| - 2\rho\big| < 1\Big\} \ll \rho.$$

Applying (3.9), (3.10), and (3.11) to (3.8), we obtain for all $d \geqslant 2$

$$\operatorname{vol}\mathcal{E}(\delta) \ll \delta\rho^{d-7/3}. \tag{3.12}$$

By definition,

$$\mathcal{E}(\delta) = \bigcup_{\boldsymbol{\xi}'\in\mathcal{F}_1\setminus\mathcal{F}} I_{\boldsymbol{\xi}'}(\delta).$$

Hence by (3.4)

$$|\mathcal{F}_1 \setminus \mathcal{F}|_\circ \ll \delta^{-1}\rho^{2-d}\operatorname{vol}\mathcal{E}(\delta). \tag{3.13}$$

Combining (3.12) and (3.13), we conclude that for big ρ

$$|\mathcal{F}_1 \setminus \mathcal{F}|_\circ = o(1). \tag{3.14}$$

We have

$$|S^{d-1} \setminus \mathcal{F}|_\circ = |S^{d-1} \setminus \mathcal{F}_1|_\circ + |\mathcal{F}_1 \setminus \mathcal{F}|_\circ. \tag{3.15}$$

Substituting (3.3) and (3.14) into (3.15), we obtain (3.1).

Now we notice that for every $\boldsymbol{\xi}' \in \mathcal{F}$ the interval $I_{\boldsymbol{\xi}'}(\delta)$ has the following property: for each point $\boldsymbol{\xi} \in I_{\boldsymbol{\xi}'}(\delta)$ and each nonzero vector $\mathbf{n} \in \Lambda^\dagger$ such that $\boldsymbol{\xi} + \mathbf{n} \in \mathcal{A}$ we have $\big|g(\boldsymbol{\xi} + \mathbf{n}) - g(\boldsymbol{\xi})\big| > 2\rho^{-d-3}$. This implies $f(\boldsymbol{\xi} + \mathbf{n}) - f(\boldsymbol{\xi}) \neq 0$. Therefore, $f(\boldsymbol{\xi})$ is a simple eigenvalue of $H\big(\{\boldsymbol{\xi}\}\big)$ for each $\boldsymbol{\xi} \in I_{\boldsymbol{\xi}'}(\delta)$. The lemma is proved. □

4. Some properties of operators on the fibers

In this section we discuss some properties of operators $H(\mathbf{k})$, $\mathbf{k} \in \Omega^\dagger$. In Lemma 4.1 we prove that the Fourier coefficients of the eigenfunctions of these operators satisfy certain decay estimates if the corresponding eigenvalues are big enough. Using this, we obtain an estimate on the rate of change of such eigenvalues with $\mathbf{k}$ in Lemma 4.2.

For $m \in \mathbb{R}$ let

$$V^{(m)} := \Big(\sum_{\mathbf{n}\in\Lambda^\dagger} |\mathbf{n}|^{2m}|V_{\mathbf{n}}|^2\Big)^{1/2}.$$

Since V is smooth, $V^{(m)}$ is finite for any $m \geqslant 0$. Recall that Q is defined by (2.2).

LEMMA 4.1. *Fix $m \in \mathbb{N}$ and $\varkappa \in (0,1)$. For $\mathbf{k} \in \Omega^\dagger$ let ψ be a normalized eigenfunction of $H(\mathbf{k})$:*

$$H(\mathbf{k})\psi = \zeta\psi \tag{4.1}$$

with the eigenvalue

$$\zeta \geqslant \max\left\{36Q^2\varkappa^{-2}, (1+m\varkappa)^{2/(d-1)}\varkappa^{-2d/(d-1)}\right\}. \tag{4.2}$$

Then there exists $M_m = M_m(d,\Lambda,V) \in \mathbb{R}_+$ such that for all $\mathbf{n} \in \Lambda^\dagger$ with

$$|\mathbf{n}| \geqslant (1+m\varkappa)\sqrt{\zeta} \tag{4.3}$$

the Fourier coefficients of ψ satisfy

$$|\psi_\mathbf{n}| < M_m \varkappa^{-m}|\mathbf{n}|^{-(3m+1)/2}. \tag{4.4}$$

PROOF. We proceed by induction. Suppose that either $m = 1$, or $m > 1$ and the statement is proved for $m-1$. Substituting the Fourier series

$$\psi(\mathbf{x}) = (\operatorname{vol}\Omega)^{-1/2} \sum_{\mathbf{n}\in\Lambda^\dagger} \psi_\mathbf{n} \exp\big(i\langle\mathbf{n},\mathbf{x}\rangle\big), \quad \mathbf{x} \in \Omega,$$

into (4.1) and equating the coefficients at $\exp\big(i\langle\mathbf{n},\mathbf{x}\rangle\big)$ on both sides, we obtain by (2.5)

$$|\mathbf{n}+\mathbf{k}|^2\psi_\mathbf{n} + \sum_{\mathbf{l}\in\Lambda^\dagger} V_{\mathbf{n}-\mathbf{l}}\psi_\mathbf{l} = \zeta\psi_\mathbf{n}. \tag{4.5}$$

Since $|\mathbf{k}| \leqslant Q$, by (4.2) and (4.3) we have

$$2|\mathbf{n}||\mathbf{k}| \leqslant \varkappa|\mathbf{n}|^2/6 + 6\varkappa^{-1}Q^2 \leqslant \varkappa|\mathbf{n}|^2/3. \tag{4.6}$$

For $\varkappa \in (0,1)$, it follows from (4.3) that

$$|\mathbf{n}|^2 - \zeta \geqslant \big(1-(1+\varkappa)^{-2}\big)|\mathbf{n}|^2 = \varkappa(2+\varkappa)(1+\varkappa)^{-2}|\mathbf{n}|^2 \geqslant \varkappa|\mathbf{n}|^2/2. \tag{4.7}$$

Combining (4.6) and (4.7), we obtain

$$|\mathbf{n}+\mathbf{k}|^2 - \zeta \geqslant |\mathbf{n}|^2 - 2|\mathbf{n}||\mathbf{k}| - \zeta \geqslant \varkappa|\mathbf{n}|^2/6,$$

and thus by (4.5)

$$|\psi_\mathbf{n}| < 6\varkappa^{-1}|\mathbf{n}|^{-2}\sum_{\mathbf{l}\in\Lambda^\dagger}|V_{\mathbf{n}-\mathbf{l}}\psi_\mathbf{l}|. \tag{4.8}$$

If $m = 1$, we estimate the sum on the right-hand side by $V^{(0)}$ using the Cauchy–Schwarz inequality (since ψ is normalized) and obtain (4.4) with $M_1 := 6V^{(0)}$.

If $m > 1$, we estimate

$$\sum_{\mathbf{l}\in\Lambda^\dagger:\,|\mathbf{l}-\mathbf{n}|\leqslant|\mathbf{n}|^{1/d}} |V_{\mathbf{n}-\mathbf{l}}\psi_\mathbf{l}| \leqslant \sup_{\mathbf{m}:\,|\mathbf{m}|\geqslant|\mathbf{n}|-|\mathbf{n}|^{1/d}} |\psi_\mathbf{m}| \sum_{\mathbf{l}\in\Lambda^\dagger:\,|\mathbf{l}|\leqslant|\mathbf{n}|^{1/d}} |V_\mathbf{l}|. \tag{4.9}$$

By (4.3), (4.2), and monotonicity of the function $q(t) = t - t^{1/d}$ for $t > 1$ we have

$$|\mathbf{n}| - |\mathbf{n}|^{1/d} \geqslant (1+m\varkappa)\sqrt{\zeta} - \big((1+m\varkappa)\sqrt{\zeta}\big)^{1/d} \geqslant \big(1+(m-1)\varkappa\big)\sqrt{\zeta}.$$

According to the induction hypothesis

$$\sup_{\mathbf{m}:\,|\mathbf{m}|\geqslant|\mathbf{n}|-|\mathbf{n}|^{1/d}} |\psi_\mathbf{m}| \leqslant \varkappa^{1-m}M_{m-1}\big(1-|\mathbf{n}|^{(1-d)/d}\big)^{1-3m/2}|\mathbf{n}|^{1-3m/2}. \tag{4.10}$$

Since $\varkappa \in (0,1)$, from (4.3) and (4.2) we conclude

$$|\mathbf{n}| \geqslant (1+m\varkappa)\sqrt{\zeta} \geqslant (\varkappa^{-1}+m)^{d/(d-1)} > 2^{d/(d-1)};$$

hence

$$(1-|\mathbf{n}|^{(1-d)/d})^{1-3m/2} < 2^{3m/2-1}. \tag{4.11}$$

Let

$$W := \sup_{r>1} r^{-d}\#\{\mathbf{l}\in\Lambda^{\dagger}\,|\,|\mathbf{l}|\leqslant r\}.$$

Clearly, $W<\infty$. By the Cauchy–Schwarz inequality

$$\sum_{\mathbf{l}\in\Lambda^{\dagger}:\,|\mathbf{l}|\leqslant|\mathbf{n}|^{1/d}} |V_{\mathbf{l}}| \leqslant W^{1/2}V^{(0)}|\mathbf{n}|^{1/2}. \tag{4.12}$$

Substituting (4.10), (4.11), and (4.12) into (4.9), we get

$$\sum_{\mathbf{l}\in\Lambda^{\dagger}:\,|\mathbf{l}-\mathbf{n}|\leqslant|\mathbf{n}|^{1/d}} |V_{\mathbf{n}-\mathbf{l}}\psi_{\mathbf{l}}| < 2^{3m/2-1}\varkappa^{1-m}W^{1/2}V^{(0)}M_{m-1}|\mathbf{n}|^{3(1-m)/2}. \tag{4.13}$$

On the other hand, since $\|\psi\|=1$, applying the Cauchy–Schwarz inequality, we obtain

$$\begin{aligned}\sum_{\mathbf{l}\in\Lambda^{\dagger}:\,|\mathbf{l}-\mathbf{n}|>|\mathbf{n}|^{1/d}} |V_{\mathbf{n}-\mathbf{l}}\psi_{\mathbf{l}}| &< |\mathbf{n}|^{3(1-m)/2}\sum_{\mathbf{l}\in\Lambda^{\dagger}:\,|\mathbf{l}|>|\mathbf{n}|^{1/d}} |\mathbf{l}|^{3(m-1)d/2}|V_{\mathbf{l}}||\psi_{\mathbf{n}-\mathbf{l}}| \\ &\leqslant |\mathbf{n}|^{3(1-m)/2}\Big(\sum_{\mathbf{l}\in\Lambda^{\dagger}:\,|\mathbf{l}|>|\mathbf{n}|^{1/d}} |\mathbf{l}|^{3(m-1)d}|V_{\mathbf{l}}|^2\Big)^{1/2} \leqslant V^{(3(m-1)d/2)}|\mathbf{n}|^{3(1-m)/2}.\end{aligned} \tag{4.14}$$

Inserting (4.13) and (4.14) into (4.8), we arrive at (4.4) with

$$M_m := 6(2^{3m/2-1}W^{1/2}V^{(0)}M_{m-1}+V^{(3(m-1)d/2)}).$$

□

LEMMA 4.2. *For any $\eta\in(0,1)$ there exists $\zeta_0>0$ such that if $\zeta(\mathbf{k})\geqslant\zeta_0$ is a simple eigenvalue of $H(\mathbf{k})$ for some $\mathbf{k}\in\Omega^{\dagger}$, then*

$$|\nabla_{\mathbf{k}}\zeta| \leqslant 2(1+\eta)\sqrt{\zeta}. \tag{4.15}$$

PROOF. Let $\psi(\mathbf{k})$ be the eigenfunction corresponding to $\zeta(\mathbf{k})$ with

$$\|\psi(\mathbf{k})\| = 1. \tag{4.16}$$

Then

$$\nabla_{\mathbf{k}}\zeta(\mathbf{k}) = \nabla_{\mathbf{k}}\big(\psi(\mathbf{k}),H(\mathbf{k})\psi(\mathbf{k})\big) = \Big(\psi(\mathbf{k}),\big(\nabla_{\mathbf{k}}H(\mathbf{k})\big)\psi(\mathbf{k})\Big). \tag{4.17}$$

By (2.5) and (2.3),

$$\nabla_{\mathbf{k}}H(\mathbf{k}) = 2\mathbf{D}(\mathbf{k}).$$

Substituting this into (4.17), we obtain

$$|\nabla_{\mathbf{k}}\zeta(\mathbf{k})| \leqslant 2\|\mathbf{D}(\mathbf{k})\psi(\mathbf{k})\| = 2\Big(\sum_{\mathbf{n}\in\Lambda^{\dagger}}|\mathbf{n}+\mathbf{k}|^2|\psi_{\mathbf{n}}(\mathbf{k})|^2\Big)^{1/2}. \tag{4.18}$$

Let

$$m := \big[(d+1)/3\big]+1 \tag{4.19}$$

and

$$\varkappa := \eta/(2m+1). \tag{4.20}$$

We assume that

$$\zeta := \zeta(\mathbf{k}) \geqslant \max\big\{36Q^2\varkappa^{-2},(1+m\varkappa)^{2/(d-1)}\varkappa^{-2d/(d-1)}\big\}. \tag{4.21}$$

Since, by (2.2), $|\mathbf{k}| \leqslant Q$, by (4.16), (4.21), and (4.20) we have

$$\sum_{|\mathbf{n}|<(1+m\varkappa)\sqrt{\zeta}} |\mathbf{n}+\mathbf{k}|^2 |\psi_{\mathbf{n}}(\mathbf{k})|^2 < \big(1+(m+1/6)\varkappa\big)^2 \zeta < (1+\eta/2)^2\zeta. \tag{4.22}$$

For $|\mathbf{n}| \geqslant (1+m\varkappa)\sqrt{\zeta}$ we apply Lemma 4.1 obtaining

$$\sum_{|\mathbf{n}|\geqslant(1+m\varkappa)\sqrt{\zeta}} |\mathbf{n}+\mathbf{k}|^2 |\psi_{\mathbf{n}}(\mathbf{k})|^2 \leqslant M_m^2 \varkappa^{-2m} \sum_{|\mathbf{n}|\geqslant(1+m\varkappa)\sqrt{\zeta}} |\mathbf{n}+\mathbf{k}|^2 |\mathbf{n}|^{-3m-1}. \tag{4.23}$$

By (4.19) the right-hand side of (4.23) is finite and is $O(\zeta^{-1/2})$. Thus, choosing ζ_0 big enough, by (4.18), (4.22), and (4.23) we obtain (4.15). □

5. Proof of Theorem 1.1

We are now ready to finish the proof of the main result. It is enough to prove

THEOREM 5.1. *For any $\alpha \in (0,1)$ there exists $\rho_0 > 0$ big enough such that for all $\rho \geqslant \rho_0$*

$$N(\rho^2+\delta) - N(\rho^2-\delta) \geqslant (1-\alpha)(2\pi)^{-d}\omega_d \delta \rho^{d-2} \tag{5.1}$$

for any

$$0 < \delta \leqslant \rho^{-d-3}. \tag{5.2}$$

Indeed, the original statement of Theorem 1.1 can be obtained by partitioning the interval $[\lambda, \lambda+\varepsilon]$ into subintervals with lengths not exceeding $2\lambda^{(-d-3)/2}$ and adding up estimates (5.1) on these subintervals (with ρ^2 being respective middle points).

PROOF. We first express the growth of IDS in terms of the function f of Proposition 2.1(i) using (2.6):

$$N(\rho^2+\delta) - N(\rho^2-\delta) = (2\pi)^{-d} \operatorname{vol}\big(f^{-1}[\rho^2-\delta, \rho^2+\delta)\big). \tag{5.3}$$

We can write

$$\operatorname{vol}\big(f^{-1}[\rho^2-\delta, \rho^2+\delta)\big) = \int_{S^{d-1}} \int_0^\infty \chi(r, \boldsymbol{\xi}') r^{d-1}\, \mathrm{d}r\, \mathrm{d}\boldsymbol{\xi}', \tag{5.4}$$

where χ is the indicator function of $f^{-1}\big([\rho^2-\delta, \rho^2+\delta)\big)$. To obtain a lower bound we can restrict the integration in (5.4) to $\boldsymbol{\xi}' \in \mathcal{F}$ defined in Lemma 3.1. Then for any $\eta \in (0,1)$ there exists $\rho_0 > 0$ such that for any $\rho \geqslant \rho_0$ we have

$$|\mathcal{F}|_\circ \geqslant (1-\eta)\omega_d, \tag{5.5}$$

and for any $\boldsymbol{\xi}' \in \mathcal{F}$ the support of $\chi(\cdot, \boldsymbol{\xi}')$ contains an interval $[r_1, r_2]$ with

$$(1-\eta)\rho \leqslant r_1 < r_1 + (1-\eta)\rho^{-1}\delta \leqslant r_2. \tag{5.6}$$

Indeed, the first inequality in (5.6) follows from Proposition 2.1(ii), (iii). The last inequality in (5.6) follows from Lemmata 3.1 and 4.2.

Thus for all $\rho \geqslant \rho_0$ by (5.5) and (5.6) we obtain

$$\begin{aligned} \int_{S^{d-1}} \int_0^\infty \chi(r, \boldsymbol{\xi}') r^{d-1}\, \mathrm{d}r\, \mathrm{d}\boldsymbol{\xi}' &\geqslant \int_{\mathcal{F}} (1-\eta)^d \rho^{d-2} \delta\, \mathrm{d}\boldsymbol{\xi}' \\ &\geqslant (1-\eta)^{d+1} \omega_d \rho^{d-2} \delta. \end{aligned} \tag{5.7}$$

Combining (5.3), (5.4), and (5.7) and choosing η small enough, we arrive at (5.1). The theorem is proved. □

References

[HM98] Bernard Helffer and Abderemane Mohamed, *Asymptotic of the density of states for the Schrödinger operator with periodic electric potential*, Duke Math. J. **92** (1998), no. 1, 1–60. MR1609321 (99e:35166)

[Kar97] Yulia E. Karpeshina, *Perturbation theory for the Schrödinger operator with a periodic potential*, Lecture Notes in Mathematics, vol. 1663, Springer-Verlag, Berlin, 1997. MR1472485 (2000i:35002)

[Kar00] ______, *On the density of states for the periodic Schrödinger operator*, Ark. Mat. **38** (2000), no. 1, 111–137. MR1749362 (2001g:47088)

[Par08] Leonid Parnovski, *Bethe-Sommerfeld conjecture*, Ann. Henri Poincaré **9** (2008), no. 3, 457–508. MR2419769

[PS09] Leonid Parnovski and Roman Shterenberg, *Asymptotic expansion of the integrated density of states of a two-dimensional periodic Schrödinger operator*, Invent. Math. **176** (2009), no. 2, 275–323.

[RS78] Michael Reed and Barry Simon, *Methods of modern mathematical physics. IV. Analysis of operators*, Academic Press [Harcourt Brace Jovanovich Publishers], New York, 1978. MR0493421 (58 #12429c)

[Shu79] M. A. Shubin, *Spectral theory and the index of elliptic operators with almost-periodic coefficients*, Uspekhi Mat. Nauk **34** (1979), no. 2(206), 95–135. MR535710 (81f:35090)

[Skr87] M. M. Skriganov, *Geometric and arithmetic methods in the spectral theory of multidimensional periodic operators*, Proc. Steklov Inst. Math. (1987), no. 2(171), vi+121, a translation of Trudy Mat. Inst. Steklov **171** (1985). MR0798454 (87h:47110), MR905202 (88g:47038)

[Sob05] Alexander V. Sobolev, *Integrated density of states for the periodic Schrödinger operator in dimension two*, Ann. Henri Poincaré **6** (2005), no. 1, 31–84. MR2119355 (2006b:35056)

Department of Mathematics, University College London, Gower Street, London WC1E 6BT, United Kingdom

E-mail address: morozov@math.ucl.ac.uk

Department of Mathematics, University College London, Gower Street, London WC1E 6BT, United Kingdom

E-mail address: leonid@math.ucl.ac.uk

Department of Mathematics, University College London, Gower Street, London WC1E 6BT, United Kingdom

E-mail address: i.pchelintseva@ucl.ac.uk

Amer. Math. Soc. Transl.
Volume **231**, 2010

On Some Open Problems in Spectral Theory

Eugene Shargorodsky

ABSTRACT. We discuss open questions related to the following four topics:

(a) approximation of the spectra of selfadjoint operators and spectral pollution,

(b) asymptotics of eigenvalues of Toeplitz matrices with piecewise continuous symbols,

(c) pseudospectra of semigroup generators,

(d) an estimate for the number of negative eigenvalues of the derivative of the periodic Hilbert transform perturbed by a nonpositive summable potential.

This paper is loosely based on the talks given by the author in 2009 at the ICMS workshop "Operator Theory and Its Applications" and at the Meeting 1 of the EPSRC Matrix and Operator Pencil Network.

1. Spectra of selfadjoint operators and spectral pollution

Let $\mathcal{H}$ be a Hilbert space and let $A : \mathcal{H} \to \mathcal{H}$ be a closed densely defined operator. In order to determine the spectrum $\mathrm{Spec}(A)$ of A, one can use the following projection method. Let $(\mathcal{L}_k)_{k\in\mathbb{N}}$ be a sequence of closed linear (finite-dimensional) subspaces of $\mathrm{Dom}(A)$ such that the corresponding orthogonal projections $P_k : \mathcal{H} \to \mathcal{L}_k$ converge strongly to the identity operator I. Let $\Lambda(A)$ be the set of all such sequences of subspaces and let $\mathrm{Spec}(A, \mathcal{L}_k)$ be the spectrum of $P_kA : \mathcal{L}_k \to \mathcal{L}_k$. One might hope that

$$\lim_{k\to\infty} \mathrm{Spec}(A, \mathcal{L}_k) = \mathrm{Spec}(A), \tag{1.1}$$

where "lim" is defined in an appropriate sense. The following well-known example (see, e.g., [**1**]) shows that the above equality may be spectacularly wrong.

EXAMPLE 1.1. Let $\mathcal{H} := \ell_2(\mathbb{Z})$, let A be the right shift operator $Ae_n = e_{n+1}$, $n \in \mathbb{Z}$, and let $\mathcal{L}_k := \mathrm{span}\{e_n\}_{n=-k}^{k}$, where $\{e_n\}_{n=-\infty}^{\infty}$ is the standard basis in $\ell_2(\mathbb{Z})$.

2010 *Mathematics Subject Classification.* Primary 47A10, 47B35, 35P05.

Then $\mathrm{Spec}(A) = \{\zeta \in \mathbb{C} : |\zeta| = 1\}$,

$$\mathrm{Spec}(A, \mathcal{L}_k) = \mathrm{Spec}\begin{pmatrix} 0 & 0 & 0 & \dots & 0 & 0 & 0 \\ 1 & 0 & 0 & \dots & 0 & 0 & 0 \\ 0 & 1 & 0 & \dots & 0 & 0 & 0 \\ \vdots & \vdots & \vdots & \ddots & \vdots & \vdots & \vdots \\ 0 & 0 & 0 & \dots & 0 & 0 & 0 \\ 0 & 0 & 0 & \dots & 1 & 0 & 0 \\ 0 & 0 & 0 & \dots & 0 & 1 & 0 \end{pmatrix} = \{0\},$$

and $\mathrm{Spec}(A, \mathcal{L}_k)$ do not provide much information on $\mathrm{Spec}(A)$.

One may still hope that (1.1) holds for selfadjoint operators. Although the operator in Example 1.1 is normal and many properties of selfadjoint operators are similar to those of normal operators, there may still be a difference when it comes to projection methods. Indeed, if A is normal, $A_k := P_k A : \mathcal{L}_k \to \mathcal{L}_k$ might not be normal, while if A is selfadjoint, then so is A_k. The situation for selfadjoint operators is indeed better than for normal ones, but unfortunately (1.1) still fails to hold. *We will assume in the rest of this section that A is a selfadjoint operator.*

If $A = A^*$ is bounded, then

$$\lim_{k\to\infty} \mathrm{Spec}(A, \mathcal{L}_k) \supseteq \mathrm{Spec}(A) \tag{1.2}$$

(see, e.g., [**1**] or [**17**]). This is not the case for unbounded operators.

THEOREM 1.2 ([**13**]). *Let $A = A^*$ be unbounded above. Then for an arbitrary sequence $(\mathcal{L}_k)_{k\in\mathbb{N}} \in \Lambda(A)$ of finite-dimensional subspaces and arbitrary $\varepsilon_k \searrow 0$, $R_k \nearrow +\infty$, there exists $(\mathcal{L}'_k)_{k\in\mathbb{N}} \in \Lambda(A)$ such that*

$$\|P_k - P'_k\| < \varepsilon_k \quad \text{and} \quad \mathrm{Spec}(A, \mathcal{L}'_k) \subset (R_k, +\infty), \quad \forall k \in \mathbb{N},$$

where $P'_k : \mathcal{H} \to \mathcal{L}'_k$ are the corresponding orthogonal projections. A similar statement holds for operators unbounded below.

The difficulty described by Theorem 1.2 does not usually cause problems in "real life". One can usually avoid it by using common sense: If one needs to approximate the spectrum of a differential operator, then it might not be a good idea to include very rapidly oscillating functions in $\mathcal{L}_k$.

The main reason for the failure of (1.1) for both bounded and unbounded selfadjoint operators is the so-called *spectral pollution* (see, e.g., [**7, 8, 10, 13, 15, 16, 17**]) which is a well-known phenomenon in numerical analysis: spurious "eigenvalues" may appear in the gaps of the essential spectrum of A and as a result $\lim_{k\to\infty} \mathrm{Spec}(A, \mathcal{L}_k)$ may contain points that do not belong to $\mathrm{Spec}(A)$. In order to state a general result on spectral pollution for possibly unbounded operators, it is convenient to extend $\mathrm{Spec}(A)$ by adding to it $-\infty$, $+\infty$, or both, if A is unbounded below, above, or from both sides, respectively. We denote the resulting set by $\widehat{\mathrm{Spec}}(A)$. In other words, $\widehat{\mathrm{Spec}}(A)$ is the closure of $\mathrm{Spec}(A)$ in the two-point compactification $\overline{\mathbb{R}} := \mathbb{R} \cup \{\pm\infty\}$ of $\mathbb{R}$. Let $\mathrm{Spec}_{\mathrm{ess}}(A)$ be the essential spectrum of A, i.e.,

$$\mathrm{Spec}_{\mathrm{ess}}(A) := \mathrm{Spec}(A) \setminus \{\text{isolated eigenvalues of finite multiplicity}\},$$

and let $\widehat{\mathrm{Spec}}_{\mathrm{ess}}(A)$ be the extended essential spectrum of A, i.e.,

$$\widehat{\mathrm{Spec}}_{\mathrm{ess}}(A) := \widehat{\mathrm{Spec}}(A) \setminus \{\text{isolated eigenvalues of finite multiplicity}\}.$$

The following theorem from [**13**] says that spectral pollution may occur at any point in a gap of the (extended) essential spectrum. Similar results have been known for a long time (see, e.g., [**15**]).

THEOREM 1.3. *For any* $\lambda \in \operatorname{conv}\left(\widehat{\operatorname{Spec}}_{\mathrm{ess}}(A)\right) \setminus \widehat{\operatorname{Spec}}_{\mathrm{ess}}(A)$ *there exists an increasing sequence* $(\mathcal{L}_k)_{k\in\mathbb{N}} \in \Lambda(A)$ *such that*

$$\lambda \in \operatorname{Spec}(A, \mathcal{L}_k), \quad \forall k \in \mathbb{N}.$$

A possible way of dealing with spectral pollution in the case of selfadjoint operators is based on the notion of second-order relative spectra which was introduced by E. B. Davies in [**9**]: Suppose $\mathcal{L}$ is a finite-dimensional subspace of $\operatorname{Dom}(A^2)$ and let P be the orthogonal projection onto $\mathcal{L}$. Then the second-order spectrum of A relative to $\mathcal{L}$ is defined as

$$\operatorname{Spec}_2(A, \mathcal{L}) := \{z \in \mathbb{C} : \ P(A - zI)^2 : \mathcal{L} \to \mathcal{L} \ \text{ is not invertible}\}.$$

The requirement $\mathcal{L} \subset \operatorname{Dom}(A^2)$ is often too restrictive, e.g., when A is the operator corresponding to a boundary value problem. The following equivalent definition (see [**13**]) is more convenient in such situations: Let $\mathcal{L}$ be a finite-dimensional subspace of $\operatorname{Dom}(A)$. Then $z \in \mathbb{C}$ is said to belong to $\operatorname{Spec}_2(A, \mathcal{L})$ if there exists $u \in \mathcal{L} \setminus \{0\}$ such that

$$((A - zI)u, (A - \overline{z}I)v) = 0, \quad \forall v \in \mathcal{L}.$$

Although the spectrum of a selfadjoint operator A is a subset of $\mathbb{R}$, the set $\operatorname{Spec}_2(A, \mathcal{L})$ may and usually does contain points from $\mathbb{C}\setminus\mathbb{R}$. The following theorem claims that if a point of $\operatorname{Spec}_2(A, \mathcal{L})$ is close to the real line, then it is close to $\operatorname{Spec}(A)$, i.e., that second-order relative spectra do not pollute.

THEOREM 1.4 ([**13, 17**]). *If* $z \in \operatorname{Spec}_2(A, \mathcal{L})$, *then*

$$\operatorname{Spec}(A) \cap [\operatorname{Re} z - |\operatorname{Im} z|, \operatorname{Re} z + |\operatorname{Im} z|] \neq \emptyset.$$

It is natural to ask whether $\operatorname{Spec}_2(A, \mathcal{L}_k)$, $(\mathcal{L}_k)_{k\in\mathbb{N}} \in \Lambda(A)$ capture the whole spectrum of A.

Question 1. *Is it true that, under suitable conditions on* $(\mathcal{L}_k)_{k\in\mathbb{N}} \in \Lambda(A)$,

$$\lim_{k\to\infty} \operatorname{Spec}_2(A, \mathcal{L}_k) \supseteq \operatorname{Spec}(A)?$$

Boulton gave a partial answer to this question ([**7, 8**]): He proved that

$$\lim_{k\to\infty} \operatorname{Spec}_2(A, \mathcal{L}_k) \supseteq \{\text{isolated eigenvalues of } A \text{ of finite multiplicity}\}$$

holds under very mild restrictions on $(\mathcal{L}_k)_{k\in\mathbb{N}}$, which are automatically satisfied if $A = A^*$ is bounded. This is a surprising result because isolated eigenvalues of finite multiplicity constitute the most unstable part of the spectrum and it is still unknown whether or not $\operatorname{Spec}_2(A, \mathcal{L}_k)$ approximate the essential spectrum. It is worth mentioning that the essential spectrum can often be found analytically and that Boulton's result allows one to approximate the rest of the spectrum.

One can summarize (1.2) and Theorem 1.3 by saying that the standard projection methods "tell the truth, the whole truth", but not necessarily "nothing but the truth". On the other hand, methods utilizing second-order relative spectra "tell the truth", "nothing but the truth" according to Theorem 1.4, and almost "the whole truth" according to Boulton, but it is unknown whether they tell "the whole truth".

2. Toeplitz matrices with piecewise continuous symbols

The following example was studied in [**13**].

Example. Consider $A = aI : L_2([-\pi,\pi]) \to L_2([-\pi,\pi])$, where

$$a(x) = \begin{cases} -\frac{3}{2} + \frac{1}{2}\cos\sqrt{5}\,x\,, & -\pi \le x < y\,, \\ 2 + \cos\sqrt{2}\,x\,, & y \le x < \pi\,, \end{cases}$$

for some $y \in (-\pi, 0]$. The function a and, in particular, the irrational coefficients $\sqrt{5}$ and $\sqrt{2}$ had been chosen to get a "generic" situation and to avoid any possible hidden symmetries.

It is clear that

$$\mathrm{Spec}(A) = \mathrm{Spec}_{\mathrm{ess}}(A) = a\left([-\pi,\pi]\right) = [-2,-1] \cup [1,3]$$

for any $y \in [-\pi/\sqrt{5}, 0]$. (If $y \in (-\pi, -\pi/\sqrt{5})$, then $[-2,-1]$ should be replaced with $[-\frac{3}{2} + \frac{1}{2}\cos\sqrt{5}\,y, -1]$.)

Let $(e_k)_{k\in\mathbb{Z}}$ denote the standard orthonormal basis $e_k(x) := (2\pi)^{-1/2}\exp(ikx)$, $k \in \mathbb{Z}$, in $L_2([-\pi,\pi])$, and let $\mathcal{L}_N := \mathrm{span}\left(\{e_k\}_{k=-n}^{n}\right)$, where $N = 2n+1$, and $\mathcal{L}_N := \mathrm{span}\left(\{e_k\}_{k=-n}^{n-1}\right)$, where $N = 2n$. Then $P_N A : \mathcal{L}_N \to \mathcal{L}_N$ is represented by a finite-dimensional Toeplitz matrix. A detailed discussion of the behavior of the eigenvalues of these matrices is contained in [**13**] and we are not going to reproduce it here. We will only consider the eigenvalues lying in the gap $(-1,1)$.

Let $y = 0$. The eigenvalues of $P_N A : \mathcal{L}_N \to \mathcal{L}_N$ lying in the gap $(-1,1)$ show remarkable stability as $N = 2n+1$ increases. The eigenvalues are also very stable in the case $N = 2n$, but they differ from those corresponding to $N = 2n+1$. In other words, the behavior of the eigenvalues of $P_N A : \mathcal{L}_N \to \mathcal{L}_N$ lying in the gap $(-1,1)$ is "2-periodic" in N.

The following figure shows the result of calculations in the case $y = -\pi/3$.

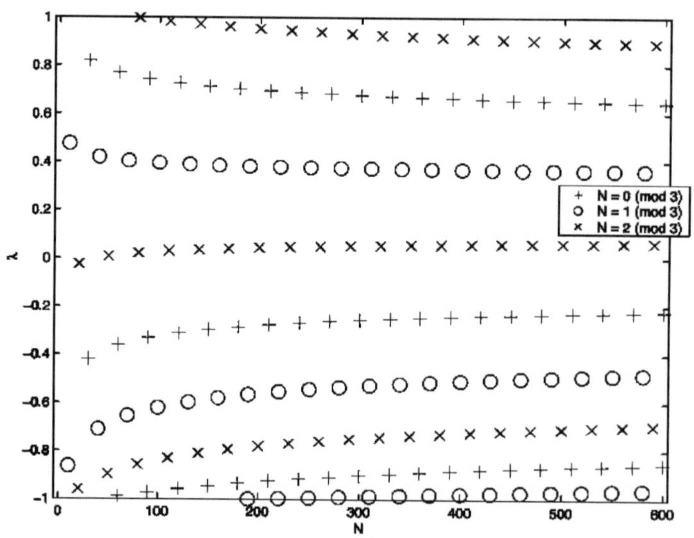

It is very easy to notice that the behavior of the eigenvalues is now "3-periodic" in N. Similar calculations have been performed with $y = -\pi q/p = -\pi/(p/q)$ for the values $p/q = 2$, 4, 5, 6, 7, 21, 21/10, 21/5, 7/3, 8/3, and 5/2. In all these cases the eigenvalues of $P_N A : \mathcal{L}_N \to \mathcal{L}_N$ lying in the gap $(-1,1)$ showed a very clear pattern of "periodic" behavior with the minimal "period"

$$\omega(q/p) := \begin{cases} 2, & \text{if } \ q = 0, \\ p, & \text{if } \ p \text{ and } q \text{ are both odd}, \\ 2p, & \text{if } \ q \neq 0 \text{ and either } p \text{ or } q \text{ is even}, \end{cases}$$

where $p \in \mathbb{N}$ and $q \in \mathbb{Z}$ are co-prime.

The appearance of $\omega(q/p)$ here is probably not very surprising as it is, for any simple fraction p/q, the smallest natural number such that $\omega\times$ (length of $[-\pi, -\pi q/p]$) and $\omega\times$ (length of $[-\pi q/p, \pi]$) are integer multiples of 2π.

The above "periodicity" seems to be related to the periodicity built into the finite-dimensional Toeplitz matrix corresponding to $P_N A : \mathcal{L}_N \to \mathcal{L}_N$. To see the latter clearly, let us consider the following simple example. Let

$$b(x) = \begin{cases} -1, & -\pi \le x < y\,, \\ 1, & y \le x < \pi\,. \end{cases} \tag{2.1}$$

Then the Fourier coefficients of b are

$$\widehat{b}_0 = -\frac{y}{\pi}, \qquad \widehat{b}_n = \frac{i\,(-1)^n}{\pi n}\left(1 - e^{in(\pi - y)}\right), \quad n \in \mathbb{Z}\setminus\{0\}.$$

Now, if $y = -\pi q/p$, then $e^{in(\pi-y)}$ is $\omega(q/p)$-periodic in n.

Question 2. *Do we really have "periodicity" of eigenvalues in the gap? If yes, why?*

In order to make this question precise, one needs to explain what is meant by "periodicity" here, and this is in fact a part of the question. Let

$$\delta(\Lambda, \Sigma) := \sup_{\zeta\in\Lambda}\inf_{\xi\in\Sigma} |\zeta - \xi|, \quad \Lambda, \Sigma \subset \mathbb{C},$$

$$\sigma_N(\Xi) := \operatorname{Spec}(A, \mathcal{L}_N) \cap \Xi, \quad \Xi \subset \mathbb{R}.$$

The word "periodicity" may be taken to mean the following:

$$\delta(\sigma_N(\Xi), \sigma_{N+\omega(q/p)}(\Xi)) = o\Big(\delta(\sigma_N(\Xi), \sigma_{N+m}(\Xi))\Big) \quad \text{as } N \to \infty$$

for any $m = 1, \ldots, \omega(q/p) - 1$ and any compact subset Ξ of the gap in $\operatorname{Spec}(A)$. The term *near periodicity* is used in the recent paper [**14**] to describe a slightly different notion. It is proved there that the eigenvalues of the square of the Toeplitz matrix corresponding to b in (2.1) do indeed behave "periodically" in N. It would be interesting to find out whether the method of [**14**] can be extended to more general symbols. Another promising approach is the one developed in [**4, 5**] where the authors dealt with banded Hermitian Toeplitz matrices, i.e., with the case where the symbol a is a real-valued trigonometric polynomial. Whether this approach can be extended to piecewise continuous symbols remains to be seen.

3. Pseudospectra of semigroup generators

In 1976, J. Globevnik posed the following question (see [**12**]): *Can the norm of the resolvent of an element of a unital Banach algebra take a finite constant value on an open set?* Although he was unable to answer this question, he showed that

(a) the resolvent norm of an element of a unital Banach algebra cannot be constant on an open subset of the unbounded component of the resolvent set,

(b) the resolvent norm of a bounded linear operator on a Banach space cannot be constant on an open set if the underlying space is complex uniformly convex (see **[11]** or **[18, 20]** for the definition).

An easy duality argument shows that one can require in (b) that the dual of the underlying Banach space, rather than the space itself, be complex uniformly convex. Hence the class of spaces to which Globevnik's result applies includes Hilbert spaces and $L^p(S, \Sigma, \mu)$ with $1 \le p \le \infty$, where (S, Σ, μ) is an arbitrary measure space (see **[18]** for details and references).

Unfortunately Globevnik's results went largely unnoticed by the operator theory community and they were partially rediscovered in the 1990s when A. Böttcher highlighted the importance of the above question for the theory of pseudospectra (see **[2, 6]** and **[22]**). However the answer remained unknown until very recently when an example of a bounded linear operator on a Banach space with the resolvent norm that is constant in a neighborhood of zero was constructed in **[18]**. The operator in that example is a weighted bilateral shift on the space $\ell^\infty(\mathbb{Z})$ equipped with the equivalent norm

$$\|x\|_* = \sup_{k \neq 0} |x_k| + |x_0|, \quad x = (x_k)_{k \in \mathbb{Z}}. \tag{3.1}$$

This space is neither reflexive nor complex strictly convex, and this example was modified in **[20]** to produce an example in a separable, reflexive, and strictly convex Banach space.

According to the above, the resolvent norm of a bounded linear operator on a Hilbert space cannot be constant on an open set, and there have been several claims in the literature that the same is true for closed densely defined operators. The following counterexample to those claims was constructed in **[18]**.

Example. Let $B : \ell_2(\mathbb{N}) \to \ell_2(\mathbb{N})$, $B = \operatorname{diag}(B_1, B_2, B_3, \dots)$ with $B_n = \begin{pmatrix} 0 & \alpha_n \\ \beta_n & 0 \end{pmatrix}$, where $2 \le \alpha_n \to \infty$ as $n \to \infty$ and $\beta_n := 1 + 1/\alpha_n$. Then B is a closed operator with the dense domain

$$\operatorname{Dom}(B) = \left\{ x = (x_1, x_2, \dots) \in \ell_2(\mathbb{N}) \,\Big|\, \sum_{n=1}^{\infty} \alpha_n^2 |x_{2n}|^2 < \infty \right\},$$

and

$$\|(B - \lambda I)^{-1}\| = 1, \quad |\lambda| < \frac{1}{2}.$$

Question 3 (J. R. Partington). *Let $\mathcal{H}$ be a Hilbert space and suppose a closed densely defined operator $A : \mathcal{H} \to \mathcal{H}$ is the generator of a semigroup. Can $\|(A - \lambda I)^{-1}\|$ be constant on an open set?*[1]

This question is deliberately stated in a slightly vague manner in order to allow various classes of semigroups, e.g., C_0, analytic, etc.

If A is the generator of a C_0-semigroup, then, according to the Hille-Yosida theorem, $\|(A - \lambda I)^{-1}\| = O(1/\lambda)$ as $\lambda \to +\infty$. Then the following simple result shows that $\|(A - \lambda I)^{-1}\|$ cannot be constant on an open subset Ω of the connected

[1] After this paper was accepted for publication, the author proved that the resolvent norm of the infinitesimal generator of a C_0-semigroup on a complex uniformly convex Banach space cannot be constant on a nonempty open set (E. Shargorodsky, *Pseudospectra of semigroup generators*, Bull. London Math. Soc. (to appear)).

component Ω_0 of the resolvent set of A which contains $[N, +\infty)$ with a sufficiently large $N > 0$.

THEOREM 3.1 ([**18**]). *Let $\Omega \subset \Omega_0 \subset \mathbb{C}$ be open sets and let Ω_0 be connected. Let Z be a Banach space. Suppose $F : \Omega_0 \to Z$ is an analytic vector-valued function,*

$$\|F(\lambda)\| \le M, \quad \forall \lambda \in \Omega,$$

and $\|F(\mu)\| < M$ for some $\mu \in \Omega_0$. Then

$$\|F(\lambda)\| < M, \quad \forall \lambda \in \Omega.$$

So, if the answer to Partington's question is positive, then the resolvent set of A cannot be connected. Note, by the way, that the spectrum of the operator B in the last example is $\left\{\pm\sqrt{\alpha_n + 1}\right\}_{n\in\mathbb{N}}$ and the resolvent set is connected.

4. Negative eigenvalues of a perturbation of the derivative of the periodic Hilbert transform

Our last question is motivated by the study of the Morse index of solutions of Bernoulli free-boundary problems and of Stokes waves in particular (see [**21**]). Let

$$A\left(\sum_{k=-\infty}^{\infty} c_k e^{ikt}\right) := \sum_{k=-\infty}^{\infty} |k| c_k e^{ikt}.$$

This is the derivative of the periodic Hilbert transform from the title of this section. The operator A is a first-order ΨDO on the unit circle $\mathbb{T}$ with the symbol $|\xi|$, i.e., $\sqrt{-\Delta}$ on $\mathbb{T}$.

Question 4. *Is there a function $h : \mathbb{R}_+ \to \mathbb{R}_+$ such that $h(\tau) \to +\infty$ as $\tau \to +\infty$ and*

$$\text{the number of negative eigenvalues of } A - qI \ge h\left(\|q\|_{L^1(\mathbb{T})}\right),$$

$\forall q \ge 0,\ q \in C^\infty(\mathbb{T})$?

This question was stated as an open problem by the author at the LMS Durham Symposium "Operator Theory and Spectral Analysis" in 2005. Although he offered a "decent bottle of wine" for the first correct solution, no one came up with an answer. After repeating the question and the offer in several later talks, the author decided in 2007 to raise the stakes and to upgrade a "decent bottle of wine" to a bottle of Chateau Latour, which is arguably the best wine in the world. Although this much more attractive offer was repeated in several talks since then, the bottle of Chateau Latour is still unclaimed at the time of this writing.

References

1. W. Arveson, *C^*-algebras and numerical linear algebra.* J. Functional Analysis **122** (1994), 330–360.
2. A. Böttcher, *Pseudospectra and singular values of large convolution operators.* J. Integral Equations Appl. **6** (1994), 267–301.
3. A. Böttcher and S. M. Grudsky, *Spectral Properties of Banded Toeplitz Matrices.* SIAM, Philadelphia, 2005.
4. A. Böttcher, S. Grudsky, and E. A. Maksimenko, *Inside the eigenvalues of certain Hermitian Toeplitz band matrices.* J. Computational and Applied Mathematics **233** (2010), 2245–2264.
5. A. Böttcher, S. Grudsky, and E. A. Maksimenko, *On the structure of the eigenvectors of large Hermitian Toeplitz band matrices.* Operator Theory: Adv. and Appl. **210** (2010), 15–36.

6. A. Böttcher, S. Grudsky, and B. Silbermann, *Norms of inverses, spectra, and pseudospectra of large truncated Wiener-Hopf operators and Toeplitz matrices.* New York J. Math. **3** (1997), 1–31.
7. L. Boulton, *Limiting set of second order spectra.* Mathematics of Computation **75** (2006), 1367–1382.
8. L. Boulton, *Non-variational approximation of discrete eigenvalues of self-adjoint operators.* IMA Journal of Numerical Analysis **27** (2007), 102–121.
9. E. B. Davies, *Spectral enclosures and complex resonances for general self-adjoint operators.* LMS J. Comput. Math. **1** (1998), 42–74.
10. E. B. Davies and M. Plum, *Spectral pollution.* IMA J. Numerical Analysis **24** (2004), 417–438.
11. J. Globevnik, *On complex strict and uniform convexity.* Proc. Amer. Math. Soc. **47** (1975), 175–178.
12. J. Globevnik, *Norm-constant analytic functions and equivalent norms.* Ill. J. Math. **20** (1976), 503–506.
13. M. Levitin and E. Shargorodsky, *Spectral pollution and second order relative spectra for self-adjoint operators.* IMA J. Numerical Analysis **24** (2004), 393–416.
14. M. Levitin, A. V. Sobolev, and D. Sobolev, *On the near periodicity of eigenvalues of Toeplitz matrices*, in this volume.
15. A. Pokrzywa, *Method of orthogonal projections and approximation of the spectrum of a bounded operator.* Stud. Math. **65** (1979), 21–29.
16. J. Rappaz, J. Sanchez Hubert, E. Sanchez Palencia, and D. Vassiliev, *On spectral pollution in the finite element approximation of thin elastic "membrane" shells.* Numer. Math. **75** (1997), 473-500.
17. E. Shargorodsky, *Geometry of higher order relative spectra and projection methods.* J. Operator Theory **44** (2000), 43–62.
18. E. Shargorodsky, *On the level sets of the resolvent norm of a linear operator.* Bull. Lond. Math. Soc. **40** (2008), 493–504.
19. E. Shargorodsky, *On the definition of pseudospectra.* Bull. Lond. Math. Soc. **41** (2009), 524–534.
20. E. Shargorodsky and S. Shkarin, *The level sets of the resolvent norm and convexity properties of Banach spaces.* Archiv der Mathematik **93** (2009), 59–66.
21. E. Shargorodsky and J. F. Toland, *Bernoulli free-boundary problems.* Memoirs AMS **196**, 914 (2008), 70 pp.
22. L. N. Trefethen and M. Embree, *Spectra and Pseudospectra: The Behavior of Nonnormal Matrices and Operators.* Princeton University Press, Princeton, NJ, 2005.

Department of Mathematics, King's College London, Strand, London WC2R 2LS, United Kingdom

E-mail address: eugene.shargorodsky@kcl.ac.uk

Titles in This Series

231 **Michael Levitin and Dmitri Vassiliev, Editors,** Operator Theory and Its Applications (In Memory of V. B. Lidskii (1924–2008))

229 **Arina A. Arkhipova and Alexander I. Nazarov, Editors,** Nonlinear Partial Differential Equations and Related Topics

228 **N. N. Uraltseva, Editor,** Proceedings of the St. Petersburg Mathematical Society, Volume XIV

227 Selected Papers on Probability and Statistics

226 **Alexei Alexandrov, Anton Baranov, and Sergey Kislyakov, Editors,** Linear and Complex Analysis (Dedicated to V. P. Havin on the Occasion of His 75th Birthday)

225 **T. Suslina and D. Yafaev, Editors,** Spectral Theory of Differential Operators (M. Sh. Birman 80th Anniversary Collection)

224 **V. M. Buchstaber and I. M. Krichever, Editors,** Geometry, Topology, and Mathematical Physics (S. P. Novikov's Seminar: 2006–2007)

223 Selected Papers on Analysis and Related Topics

222 **N. N. Uraltseva, Editor,** Proceedings of the St. Petersburg Mathematical Society, XIII

221 **Yu. Neretin, M. Olshanetsky, and A. Rosly, Editors,** Seminar on Mathematical Physics, II

220 **M. S. Birman and N. N. Uraltseva, Editors,** Nonlinear Equations and Spectral Theory

219 **N. N. Uraltseva, Editor,** Proceedings of the St. Petersburg Mathematical Society, Volume XII

218 **N. N. Uraltseva, Editor,** Proceedings of the St. Petersburg Mathematical Society, Volume XI

217 **V. Kaimanovich and A. Lodkin, Editors,** Representation Theory, Dynamical Systems, and Asymptotic Combinatorics

216 **M. V. Karasev, Editor,** Quantum Algebras and Poisson Geometry in Mathematical Physics

214 **N. N. Uraltseva, Editor,** Proceedings of the St. Petersburg Mathematical Society, Volume X

213 **Ernest Vinberg, Editor,** Lie Groups and Invariant Theory

212 **V. M. Buchstaber and I. M. Krichever, Editors,** Geometry, Topology, and Mathematical Physics. S. P. Novikov's Seminar: 2002–2003

211 **K. Nomizu, Editor,** Selected Papers on Analysis and Differential Equations

210 **S. G. Gindikin, Editor,** Lie Groups and Symmetric Spaces. In Memory of F. I. Karpelevich

209 **N. N. Uraltseva, Editor,** Proceedings of the St. Petersburg Mathematical Society, Volume IX

208 **M. V. Karasev, Editor,** Asymptotic Methods for Wave and Quantum Problems

207 **Yu. M. Suhov, Editor,** Analytic Methods in Applied Probability

206 **M. A. Shubin and M. S. Agranovich, Editors,** Partial Differential Equations (Mark Viskik's Seminar)

205 **N. N. Uraltseva, Editor,** Proceedings of the St. Petersburg Mathematical Society, Volume VIII

204 **K. Nomizu, Editor,** Selected Papers on Classical Analysis

203 **N. N. Uraltseva, Editor,** Proceedings of the St. Petersburg Mathematical Society, Volume VII

202 **V. Turaev and A. Vershik, Editors,** Topology, Ergodic Theory, Real Algebraic Geometry. Rokhlin's Memorial

201 **Michael Semenov-Tian-Shansky, Editor,** L. D. Faddeev's Seminar on Mathematical Physics

200 **L. Lerman, G. Polotovskiĭ, and L. Shilnikov, Editors,** Methods of Qualitative Theory of Differential Equations and Related Topics

199 **N. N. Uraltseva, Editor,** Proceedings of the St. Petersburg Mathematical Society, Volume VI

198 **R. A. Minlos, Senya Shlosman, and Yu. M. Suhov, Editors,** On Dobrushin's Way. From Probability Theory to Statistical Physics

197 **Vladimir Arnold, Maxim Kontsevich, and Anton Zorich, Editors,** Pseudoperiodic Topology

196 **Ya. Eliashberg, D. Fuchs, T. Ratiu, and A. Weinstein, Editors,** Northern California Symplectic Geometry Seminar

195 **O. V. Belegradek et al.,** Model Theory and Applications

194 **A. Astashkevich and S. Tabachnikov, Editors,** Differential Topology, Infinite-Dimensional Lie Algebras, and Applications (D. B. Fuchs' 60th Anniversary Collection)

193 **N. N. Uraltseva, Editor,** Proceedings of the St. Petersburg Mathematical Society, Volume V

192 **Lev Beklemishev, Mati Pentus, and Nikolai Vereshchagin,** Provability, Complexity, Grammars

191 **A. Yu. Morozov and M. A. Olshanetsky, Editors,** Moscow Seminar in Mathematical Physics

190 **S. Tabachnikov, Editor,** Differential and Symplectic Topology of Knots and Curves

189 **V. Buslaev, M. Solomyak, and D. Yafaev, Editors,** Differential Operators and Spectral Theory (M. Sh. Birman's 70th Anniversary Collection)

188 **O. A. Ladyzhenskaya, Editor,** Proceedings of the St. Petersburg Mathematical Society, Volume IV

187 **M. V. Karasev, Editor,** Coherent Transform, Quantization, and Poisson Geometry

186 **A. Khovanskiĭ, A. Varchenko, and V. Vassiliev, Editors,** Geometry of Differential Equations

185 **B. Feigin and V. Vassiliev, Editors,** Topics in Quantum Groups and Finite-Type Invariants (Mathematics at the Independent University of Moscow)

184 **Peter Kuchment and Vladimir Lin, Editors,** Voronezh Winter Mathematical Schools (Dedicated to Selim Krein)

183 **K. Nomizu, Editor,** Selected Papers on Harmonic Analysis, Groups, and Invariants

182 **V. E. Zakharov, Editor,** Nonlinear Waves and Weak Turbulence

181 **G. I. Olshanski, Editor,** Kirillov's Seminar on Representation Theory

180 **A. Khovanskiĭ, A. Varchenko, and V. Vassiliev, Editors,** Topics in Singularity Theory

179 **V. M. Buchstaber and S. P. Novikov, Editors,** Solitons, Geometry, and Topology: On the Crossroad

178 **V. Kreinovich and G. Mints, Editors,** Problems of Reducing the Exhaustive Search

177 **R. L. Dobrushin, R. A. Minlos, M. A. Shubin, and A. M. Vershik, Editors,** Topics in Statistical and Theoretical Physics (F. A. Berezin Memorial Volume)

176 **E. V. Shikin, Editor,** Some Questions of Differential Geometry in the Large

175 **R. L. Dobrushin, R. A. Minlos, M. A. Shubin, and A. M. Vershik, Editors,** Contemporary Mathematical Physics (F. A. Berezin Memorial Volume)

174 **A. A. Bolibruch, A. S. Merkur′ev, and N. Yu. Netsvetaev, Editors,** Mathematics in St. Petersburg

173 **V. Kharlamov, A. Korchagin, G. Polotovskiĭ, and O. Viro, Editors,** Topology of Real Algebraic Varieties and Related Topics

172 **K. Nomizu, Editor,** Selected Papers on Number Theory and Algebraic Geometry

For a complete list of titles in this series, visit the
AMS Bookstore at **www.ams.org/bookstore/**.